DIFFUSION AND REACTIVITY OF SOLIDS

DIFFUSION AND REACTIVITY OF SOLIDS

JAMES Y. MURDOCH
Editor

Nova Science Publishers, Inc.
New York

For permission to use material from this book please contact us:
Telephone 631-231-7269; Fax 631-231-8175
Web Site: http://www.novapublishers.com

Library of Congress Cataloging-in-Publication Data

Diffusion and reactivity of solids / James Y. Murdoch, editor.
p. cm.
Includes index.
ISBN-13: 978-1-60021-890-3 (hardcover)
ISBN-10: 1-60021-890-3 (hardcover)
1. Kirkendall effect. 2. Reactivity (Chemistry) I. Murdoch, James Y.
QC176.8.D5D535 2008
541'.0421--dc22

2007030058

Published by Nova Science Publishers, Inc. ✢ New York

CONTENTS

Preface		**vii**
Chapter 1	Surface Modification and its Mechanism for Performance Improvements of Cathode Material for Li-Ion Batteries *Zhaoxiang Wang, Na Liu, Jianyong Liu, Ying Bai, Xueping Gao and Liquan Chen*	**1**
Chapter 2	Catalysts Design for Hydrogen Production: Embedded Rhodium Nanoparticles *Paolo Fornasiero, Tiziano Montini and Loredana De Rogatis*	**69**
Chapter 3	AC Measurements of High Ionic Conductivity Due to Oxygen Migrations in Doped Lanthanum Gallates *E. Iguchi, D. I. Savytskii and M. Kurumada*	**115**
Chapter 4	Nanosized Materials as Electrodes for Lithium Ion Batteries *Jesús Santos-Peña, Julián Morales, Enrique Rodríguez-Castellón and Sylvain Franger*	**163**
Chapter 5	Structure and Diffuse Scattering of Superionic Conductor CuI *Takashi Sakuma, Xianglian and Khairul Basar*	**209**
Chapter 6	Oxygen Diffusion in $YBa_2Cu_3O_{7-x}$ and its Potential Applications *Xing Hu, Delin Yang and Jie Hu*	**227**
Index		**243**

PREFACE

This new book is devoted to the physics, chemistry and materials science of diffusion, mass transport, and reactivity of solidsincluding (i) physics and chemistry of defects in solids; (ii) reactions in and on solids, e.g. intercalation, corrosion, oxidation, sintering; (iii) ion transport measurements, mechanisms and theory.

Chapter 1 - The surface of commercial $LiCoO_2$ was coated with a thin layer of amorphous magnesium oxide (MgO). The surface morphology, crystalline structure and electrochemical performances of the modified cathode materials were characterized and compared with that of commercial $LiCoO_2$. It was found that the surface-coated $LiCoO_2$ can provide higher specific capacities than commercial $LiCoO_2$ while its structure and structural stability are not disturbed. Coating the surface of commercial $LiCoO_2$ with a thin layer of amorphous yttrium phosphate (YPO_4) at room temperature can also improve its electrochemical and thermal performances. As the YPO_4-coating was carried out at room temperature, such a surface modification helped us to clarify some important and basic questions.

In order to understand the improvement mechanism of surface coating and study the compatibility of the cathode material with the electrolyte, $LiCoO_2$, commercial and nano-sized, bare and Al_2O_3-coated, were soaked in commercial electrolyte and its solvent. Strong spontaneous reactions were observed between $LiCoO_2$ and the solvent. Significant impacts of the spontaneous reactions on the structure and electrochemical performance of the electrode material were evaluated with X-ray diffraction (XRD) and electrochemical cycling, respectively. Based on these results, some previously proposed improvement mechanisms are challenged: surface coating cannot prevent the dissolution of Li and Co ions from $LiCoO_2$.

The variation of the electronic structures of commercial and MgO-modified $LiCoO_2$ charged to various potentials was studied by X-ray photoelectron spectroscopy. It was found that surface coating suppresses the interaction between $LiCoO_2$ and the electrolyte at the uncharged state and alleviates the electrolyte decomposition at charged states by hindering the formation of oxygen with strong oxidizing power.

In accordance to the authors above understanding to the essence of surface coating and by the revelation of others' reports, the authors proposed that the interaction between the coating material and the electrolyte, rather than the physical separation of the coating layer, helps to improve the electrochemical and thermal performances of commercial $LiCoO_2$. Contrary to the traditional beliefs, addition of nano-$Al2O_3$ in commercial electrolyte remarkably increases the acidity of the latter. Based on extended and comprehensive analysis, a solid super-acid model was proposed. The performance improvement is attributed to the formation of solid superacids such as AlF_3/Al_2O_3 and Li_3AlF_6/Al_2O_3 in the Al_2O_3-added cathode and electrolyte. This model disagrees with previous improvement mechanisms and predicts that some other nano-compounds can also be used as additives for improving the performances of $LiCoO_2$ cathode materials.

Chapter 2 - In a sustainable energy and mobility development, hydrogen will become very important as it is considered one of the key energy carriers in terms of energy source, as fuel for transportation and intermediate in the conversion of renewable energy sources. In addition, hydrogen is also of relevance as a clean fuel for fuel cells. Catalytic technologies will play a major role in the transformation towards hydrogen economy. Here, examples of the development of new catalysts for hydrogen production from fossil fuels (methane) and from bio-masses (ethanol / water solution) are presented.

In particular, the partial oxidation of methane over Rh-based catalysts is discussed as an attractive process for the production of syngas. Despite its high cost, rhodium is widely investigated since it shows high yields, good selectivity and good resistance towards the deactivating effects of coke deposition. Nevertheless, the extreme working conditions encountered in syngas production, such as the high temperature and high space velocity, combined with the necessity of long lifetime for commercialisation of such catalysts, require the development of new catalytic materials with superior thermal stability than those currently available. It is showed that the controlled synthesis of Rh nanoparticles embedded in porous oxides results in catalysts which exhibit high hydrogen yield for partial oxidation of methane. Moreover, the process of encapsulation of the Rh nanoparticles during the synthesis stage largely prevents Rh sintering. Furthermore, the undesirable incorporation of Rh into the Al2O3 lattice, during high temperature oxidation treatments, can also be minimised. Consistently, under the working conditions employed, the embedded Rh nanoparticles present high thermal stability. Small and slow deactivation is observed due to coke formation and sintering of the support. The adoption of appropriate strategies, such as nature and texture modulation of the support or inclusion of extra components in the catalyst formulation, can be employed to minimise these drawbacks. In situ regeneration treatments have proved to strongly extend the embedded catalyst life. The use of preformed metal nanoparticles, protected by a porous layer of nanocomposite oxides, is a successful strategy also for ethanol steam reforming. Notably, the rather low Rh loading, presently used, opens perspectives for technological transfer to industrial applications.

Chapter 3 - Oxygen ionic conduction in oxides results from the self diffusion of O^{2-} ions, the elementary process of which constitutes the migration of an O^{2-} ion from a lattice site to the next vacant lattice site across a saddle point in a diffusion path. The ac experimental method provides important knowledge about the dynamics of O^{2-} migrations in ceramic oxides because there are two effective techniques in the ac method, i.e., impedance analysis and measurements of the dielectric properties. In the impedance analysis, intra-grain conduction can be distinguished from inter-grain conduction and the parameter that represents the degree of the distribution of the relaxation times involved in O^{2-} migrations is provided directly. In the measurements of the dielectric properties, the relaxation processes due to O^{2-} migrations in different zones in a ceramic oxide can be recognized separately and the energy required for O^{2-} migration in each relaxation process is obtained directly. If both the impedance analysis and measurements of the dielectric relaxation processes are conducted together, ionic transport properties in oxides can be elucidated more directly with high-precision. The ionic conductivity of doped lanthanum gallates is very high as compared to that of most other oxides. In order to investigate the reasons for the high ionic conductivity of doped lanthanum gallates, ac measurements have been carried out using a single crystal of $La_{0.95}Sr_{0.05}Ga_{0.9}Mg_{0.1}O_{3-\delta}$ grown by the Czochralski method along with the dc measurements. This single crystal comprises a twin structure that consists of domains and the domain walls. These experiments have succeeded in revealing the dynamics of O^{2-} migrations in the domains and along the domain walls. These two different types of O^{2-} migrations constitute a parallel circuit of two independent *R-C* combinations. This parallel circuit corresponds to the conventional equivalent circuit of the twin structure modeled in the ac treatments. As a consequence of the parallel circuit, the resultant resistivity in the twin structure is considerably low. In order to examine whether this speculation holds in polycrystalline doped lanthanum gallates also, similar experiments have been carried out with $La_{1-x}Sr_xGa_{1.1-x}Zr_{x-0.1}O_{3-\delta}$ ceramics (x = 0.2-0.5). Subsequently, it is observed that the ionic conductive behaviors of these ceramics can be explained in terms of the twin structures in the bulks when the value of x is small.

Chapter 4 - In this work the authors show some results on the research of nanosized materials with potential applications in lithium ion batteries. The study is focussed on positive electrodes such as olivine $LiFePO_4$ and α-$LiFeO_2$ as well as negative electrodes based on iron containing spinels. For the positive electrodes, the nanosized nature was found to enhance the efficiency of the lithium extraction/insertion reaction, due to a reduced path length for the transport of electrons and lithium ions. Moreover cycling properties were improved in the nanomaterials due to the combination of faster reaction kinetics and increased electrolyte-electrode interface. Capacities as high as 140 mAh/g were observed for $LiFePO_4$ when is modified by adding of conductive systems such as copper or carbon. α-$LiFeO_2$ nanobelts showed better electrochemical properties than other lithium ferrite polymorphs. For the spinels, capacities as high as 1400 mAh/g were found. However, the nanometric character induces the formation of a solid electrolyte interface that decreases the reversibility of the reaction with lithium. The three systems are examples of the applicability of nanodesign in the search for new electrodes for rechargeable batteries.

Chapter 5 - The structure and diffuse scattering of CuI that has high ionic conductivity at high temperature have been studied by X-ray diffraction, anomalous X-ray scattering and neutron diffraction methods. The expression of the diffuse scattering intensity including the correlation effects among the thermal displacements of atoms was shown and applied to the analysis of diffuse scattering of γ-, β- and α-CuI. The calculated energy dependence of the intensities of Bragg lines based on the ordered arrangement of Cu atoms could explain the characteristics of the observed scattering intensities of γ-CuI by anomalous X-ray scattering measurement. The model which includes the ordered arrangements of Cu atoms could explain the observed neutron diffuse scattering intensities of γ-CuI at 8 and 290 K. From the structural model with trigonal system the intensities of X-ray and neutron diffuse scattering was estimated based on the disordered arrangement of Cu atoms in β-CuI. Numerical calculations of the diffuse background of α-CuI have been made based on the short range order of copper atoms and the correlation effects among the thermal displacements of atoms. The cubic system of the space group $Fm3m$ with the disordered arrangement of copper atoms could explain the diffuse scattering of α-CuI. The low-energy excitation in CuI by neutron inelastic scattering method was discussed. The temperature dependence of the damping factor of the excitation would be related to that of the ionic conductivity.

Chapter 6 - Oxygen diffusion properties of high temperature superconductor material $YBa_2Cu_3O_{7-x}$ (YBCO) was studied by thermogravimetry (TG), oxygen static adsorption, oxygen permeability and resistance measurement. The non-isothermal TG experiment in air shows that the mass of sample exhibits periodic variation with temperature increase and decrease. The isothermal kinetic TG experiment indicates that the oxygen in-diffusion is faster than out-diffusion. The TG experiments with different heating rates indicates that between 500°~800°C the oxygen desorption activation energy has some relations with the oxygen stoichiometry of the material. The activation energy increases obviously with temperature in the range of 500°~650°C, from 184kJ/mol to 290kJ/mol. But the energy increases smoothly from 293kJ/mol to 315kJ/mol when temperature changing from 650°~800°C. The influences of oxygen partial pressure and temperature on saturated oxygen adsorption of the material were also evaluated by the static oxygen adsorption experiments. The application of YBCO membrane in the process of partial oxidation of methane (POM) to syngas was also investigated. Methane conversion, CO and H_2 selectivity can reach almost 100%, 95%, and 86% respectively at 900°C. However, the stability of YBCO in reducing atmosphere is questionable because of the reduction of copper from the YBCO membrane.

In: Diffusion and Reactivity of Solids
Editor: James Y. Murdoch, pp. 1-67

ISBN: 978-1-60021-890-3

Chapter 1

SURFACE MODIFICATION AND ITS MECHANISM FOR PERFORMANCE IMPROVEMENTS OF CATHODE MATERIAL FOR LI-ION BATTERIES

Zhaoxiang Wang,[1]* Na Liu,[1] Jianyong Liu,[1] Ying Bai,[1] Xueping Gao[2] and Liquan Chen[1]

[1]Laboratory for Solid State Ionics, Institute of Physics, Chinese Academy of Sciences, Beijing 100080, China

[2]Institute of New Energy Materials Chemistry, Nankai University, Tianjin 300071, China

Abstract

The surface of commercial $LiCoO_2$ was coated with a thin layer of amorphous magnesium oxide (MgO). The surface morphology, crystalline structure and electrochemical performances of the modified cathode materials were characterized and compared with that of commercial $LiCoO_2$. It was found that the surface-coated $LiCoO_2$ can provide higher specific capacities than commercial $LiCoO_2$ while its structure and structural stability are not disturbed. Coating the surface of commercial $LiCoO_2$ with a thin layer of amorphous yttrium phosphate (YPO_4) at room temperature can also improve its electrochemical and thermal performances. As the YPO_4-coating was carried out at room temperature, such a surface modification helped us to clarify some important and basic questions.

In order to understand the improvement mechanism of surface coating and study the compatibility of the cathode material with the electrolyte, $LiCoO_2$, commercial and nano-sized, bare and Al_2O_3-coated, were soaked in commercial electrolyte and its solvent. Strong spontaneous reactions were observed between $LiCoO_2$ and the solvent. Significant impacts of the spontaneous reactions on the structure and electrochemical performance of the electrode material were evaluated with X-ray diffraction (XRD) and electrochemical cycling, respectively. Based on these results, some previously proposed improvement mechanisms are challenged: surface coating cannot prevent the dissolution of Li and Co ions from $LiCoO_2$.

The variation of the electronic structures of commercial and MgO-modified $LiCoO_2$ charged to various potentials was studied by X-ray photoelectron spectroscopy. It was found that surface coating suppresses the interaction between $LiCoO_2$ and the electrolyte at the uncharged state and alleviates the electrolyte decomposition at charged states by hindering the formation of oxygen with strong oxidizing power.

In accordance to our above understanding to the essence of surface coating and by the revelation of others' reports, we proposed that the interaction between the coating material and the electrolyte, rather than the physical separation of the coating layer, helps to improve the electrochemical and thermal performances of commercial $LiCoO_2$. Contrary to the traditional beliefs, addition of nano-Al_2O_3 in commercial electrolyte remarkably increases the acidity of the latter. Based on extended and comprehensive analysis, a solid super-acid model was proposed. The performance improvement is attributed to the formation of solid superacids such as AlF_3/Al_2O_3 and Li_3AlF_6/Al_2O_3 in the Al_2O_3-added cathode and electrolyte. This model disagrees with previous improvement mechanisms and predicts that some other nano-compounds can also be used as additives for improving the performances of $LiCoO_2$ cathode materials.

1. Introduction

Lithium ion batteries are undergoing a period of intense commercialization due to their intrinsically superior energy density over other rechargeable battery technologies such as nickel metal hydride. With decades of study, the research on the cathode materials for lithium ion batteries has been focused on hexagonal $LiCoO_2$, spinel $LiMn_2O_4$ and olivine $LiFePO_4$ and their derivatives though there are some other materials. Of these materials, $LiCoO_2$ is considered the most stable among the family of α-$NaFeO_2$ structure materials and is the only cathode material that has been commercialized in large scale.

$LiCoO_2$ cathode materials are typically cycled between the fully-lithiated discharge state $LiCoO_2$ (*ca.* 3.0V *vs* Li) and a roughly half-delithiated charge state Li_xCoO_2 (x = 0.5-0.6; 4.2V *vs* Li) yielding a useable specific capacity below 150 mAh/g. More Li^+ ions can be extracted from $Li_{0.5}CoO_2$ by raising the charge cutoff potential. However over-delithiation is often found to result in significant deterioration of the stability of the material due to a monoclinic to hexagonal (M→H) phase transition.[1] Simultaneous to the research and development of other cathode materials, cobalt is chemically substituted with some transition metal ions to suppress the phase transition as well as to lower the cost of the material of $LiCoO_2$. The investigated dopants include Fe, [2,3] Ti, V, Mn, Ni[4] or binary transition metal ions such as Mn-Ni.[5] Based on the calculation of Ceder,[6] some elements electrochemically inactive in the redox process have also been used as dopants, such as Al,[7-13] Mg [14-16] and B.[10] These substitutions are actually stabilizing the structure of the material at the expense of its specific capacity. For example, the specific capacity of $LiNi_{1-x}Mg_xO_2$ faded from 200 mAh/g at x =0 to 90 mAh/g at x = 0.2 when it was cycled between 3.1 and 4.4V.[17] These results indicate that partial substitution of Co with M in the $LiCo_{1-y}M_yO_2$ (M = metal) system may not improve the electrochemical performance of $LiCoO_2$. Suppression of the phase transition in the bulk alone cannot improve the cycling stability significantly. There must be some other driving forces as the origin of the capacity fading.

There has been evidence[18] that the performance degradation of $LiCoO_2$ is related to the dissolution of its Co^{4+} ions in the electrolyte solution. Aurbach *et al*[19] reported that the electrochemical behaviors of Li_xMO_y (M = Ni, Mn) cathode materials were strongly dependent on their surface chemistry. Clearly, coating $LiCoO_2$ material can modify the properties of its surface exposed to the electrolyte solution and change its cycling performance. Therefore, an alternate approach to improve the electrochemical performance is to change the surface properties of the material by coating its particles with some metal oxides to avoid the undesired reactions on the surface and protect the bulk. This must have been the basic consideration of surface coating. Kweon *et al.*[20] first reported that the electrochemical cycling performance of $LiCoO_2$ at high voltage (> 4.2 V) could be fairly enhanced by Al_2O_3 coating. Later, MgO,[21] ZrO_2,[22] TiO_2,[22] SnO_2[23] CeO_2,[24] ZnO,[25] P_2O_5[26] and SiO_2[27] were used as coating materials to modify the surface chemistry of $LiCoO_2$. To date, surface modification has been extended to $LiMnO_2$, $LiMn_2O_4$, $LiNi_xCo_{1-x}O_2$ with SnO_2, MgO, $LiCoO_2$ $AlPO_4$ and diamond-like carbon (DLC).[23, 28-34,35] However, a common feature of these studies is that the available capacity of the coated material becomes lower than that of the commercial material because most of these coating materials are electrochemically inactive and electrically insulating.

2. Electrochemical Evaluation and Structural Characterization of Surface-Modified $LiCoO_2$

In this section, we improved the performances of commercial $LiCoO_2$ by coating its surface with amorphous magnesium oxide (MgO) and yttrium orthophosphate (YPO_4). Yttrium orthophosphate (YPO_4) has a tetragonal symmetry ($a = b = 6.822$ Å and $c = 6.018$ Å) and belongs to space group I_1/amd. Chains parallel to the *c* axis of corner-sharing structural units built of a (YO_8) dodecahedron and a (PO_4) tetrahedron are linked together by an edge. These chains are further linked together by edge sharing. These features insure the structural stability of YPO_4. In addition, different from other previous coating materials, YPO_4 can deposit on $LiCoO_2$ by a simple replacement reaction between $Y(NO_3)_3{\cdot}6H_2O$ and $Na_3PO_4{\cdot}12H_2O$ without any subsequent processing (*e.g.* annealing). This helps to clarify some controversies about the mechanism of performance improvements of surface-modified $LiCoO_2$ as will be seen in the following sections. Therefore, YPO_4 is used as the coating material in this section.

2.1. Experimental

The $LiCoO_2$ powder (Cellseeds™, C-5, average particle size: 5-6 μm; surface area: 0.40-0.70 m^2/g) was a commercial product of Nippon Chemical Industrial. On coating the $LiCoO_2$ particles with MgO, 6 grams of $LiCoO_2$ powder was added into 300ml 0.1M H_2SO_4. The purpose of slightly corroding $LiCoO_2$ with dilute H_2SO_4 was to produce more active sites for the subsequent coating process. The mixture was mechanically stirred for 10 min before it was filtered and rinsed three times with distilled water and dried at 100°C. After that, 5g $LiCoO_2$ was mixed with 0.125g NaOH in 400ml distilled water and heated and stirred at 50°C for 24 hours. During this process, 0.6g $MgCl_2{\cdot}6H_2O$ (98%) dissolved in water was gradually added into the mixture. Additional NaOH was added in the co-precipitation process to help the formation of $Mg(OH)_2$ and avoid the severe corrosion to $LiCoO_2$ due to the hydrolyzing of $MgCl_2$ (without NaOH, the filtered solution will become purple and Co_2O_3 can be detected in the product, whether or not $LiCoO_2$ was rinsed with dilute H_2SO_4). The mixture was then rinsed and filtered another three times with distilled water. In this way, $LiCoO_2$ particles were coated with $Mg(OH)_2$. $Mg(OH)_2$ was dehydrated by heating the coated material at 600°C for 2 hours in air and hence MgO-coated $LiCoO_2$ was obtained.

On coating $LiCoO_2$ with YPO_4, $Y(NO_3)_3 \cdot 6H_2O$ (99.99%) and commercial $LiCoO_2$ were mixed in distilled water. Then the aqueous solution of $Na_3PO_4 \cdot 12H_2O$ (98%) was gradually added into the mixture. YPO_4 was formed and deposited onto $LiCoO_2$ particles at room temperature ($Y(NO_3)_3 + Na_3PO_4 \longrightarrow YPO_4\downarrow + 3NaNO_3$.). After repeated rinsing and filtering, YPO_4-coated $LiCoO_2$ was obtained. The content of the expected (nominal) YPO_4 on $LiCoO_2$ varies from 1% to 20% that of $LiCoO_2$.

The surface-coated $LiCoO_2$ was mixed with carbon black (CB) and a polymer binder (poly (vinylidene fluoride), PVdF) in 1-methyl-2-pyrrolidone (NMP) at a weight ratio of $LiCoO_2$:CB:PVdF = 85:10:5 to form a slurry. The slurry was uniformly cast on an Al foil by doctor's blade technique. Such prepared electrode sheets were cut into circles and stored in an 80°C vacuum oven for more than 24 hours for later use.

Test cells were assembled in Ar-filled glove box (MBraun) with (surface-coated) $LiCoO_2$ as the working electrode, fresh lithium foil as the counter electrode, 1mol/L $LiPF_6$ in EC/DMC (1:1 v/v) as the electrolyte (EC for ethylene carbonate and DMC for dimethyl carbonate) and Celguard™ 2400 polypropylene as the separator. The cell was left aged for at least five hours before galvanostatically cycled between 2.5V and various charge cutoff voltages on LAND battery tester (Wuhan, China). CH Electrochemical Workstation was used for the cyclic voltammetry (CV) test at a scanning rate of 0.04mV/s between 2.5V and 4.7V.

Samples before and after electrochemical cycling were characterized with scanning electron microscope (SEM, JSM-6301F), X-ray diffractometer (XRD; M18A-HF, MacScience) with Cu Kα radiation. Differential scanning calorimetry (DSC) analysis was carried out on NETSCH STA 449C in air by sealing the charged cathode sheet in an Al crucible in dry Ar and heated from 25°C to 500°C at a rate of 5°C/min. All the operations on the moisture-sensitive samples were carried out in dry argon atmosphere and the samples were kept in dry Ar before the instruments were ready for use.

2.2. Results and Discussion

Figure 1 shows the morphologies of commercial $LiCoO_2$ before and after surface coating. The particle surface of commercial $LiCoO_2$ is very smooth and seems rather "clean". After coating, the particle surface is covered with a layer of uniformly distributed MgO beads. The average size of the MgO beads is about 50 nm. Inductively coupled plasma (ICP) analysis indicated that the Li/Co atomic ratio in the final product was not affected with the slight corrosion with dilute H_2SO_4 in the above washing procedure and that the actual content of MgO *vs* $LiCoO_2$ in the sample was 1.5mol%.

Different from the MgO-coated $LiCoO_2$, the surface of the YPO_4-coated $LiCoO_2$ is smooth and clean in a very large field of view, similar to that of the commercial $LiCoO_2$. No YPO_4 fractures were observed in the sample. However, ICP analysis shows that the actual content of YPO_4 is 3.64wt% (but this sample is still called 5wt%YPO_4-coated $LiCoO_2$ in the following discussion). This means that the coating was homogeneous on the $LiCoO_2$ particles. The actual YPO_4 contents in the other samples were not analyzed.

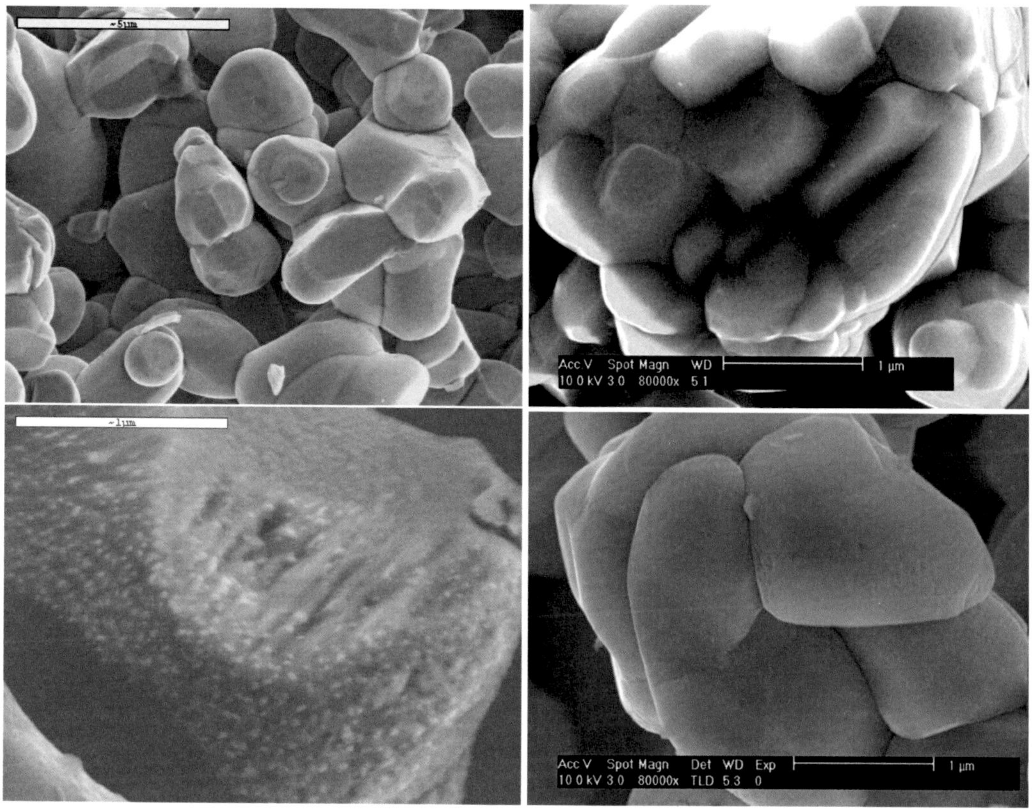

Figure 1. SEM imaging of commercial $LiCoO_2$ (a and b), $LiCoO_2$ coated with 1.5 mol % MgO (c) and with 5 wt % YPO_4 (d).

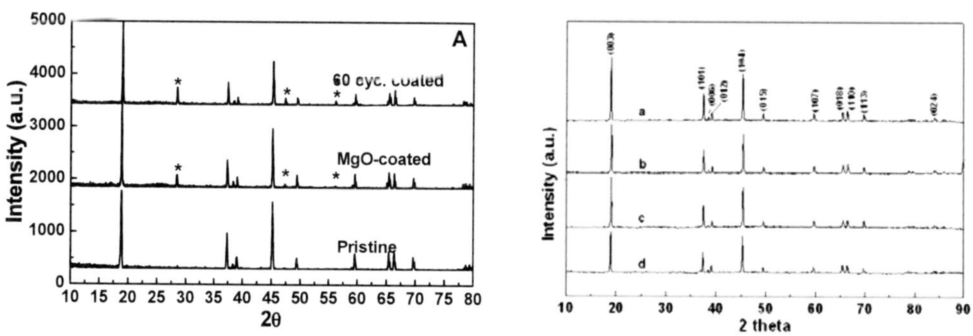

Figure 2. XRD patterns of (A, left) (a) commercial $LiCoO_2$, (b) MgO-coated $LiCoO_2$, and (c) MgO-coated $LiCoO_2$ after 60 cycles (at discharge state). (* for diffractions of Si reference) and (B, right) the XRD patterns of (a) commercial, (b) 5%, (c) 10% and (d) 20%YPO_4-coated $LiCoO_2$

Figure 2 compares the XRD patterns of commercial $LiCoO_2$ and surface-coated $LiCoO_2$. All the diffraction peaks are indexed to hexagonal $LiCoO_2$. Calculations indicate that the a and c values of the MgO-coated $LiCoO_2$ (a = 2.81792 Å, c = 14.06676 Å) (Fig.2A) are similar to that of commercial $LiCoO_2$ (a = 2.81664Å, c = 14.06165Å). Similarly, no YPO_4 diffraction peaks are observed up to a coating content of 20% (Fig.2B). This means that the

YPO_4 on $LiCoO_2$ is amorphous. Calculation shows that YPO_4 coating does not change the lattice parameters of $LiCoO_2$, either. These agree with the suggestion of Kweon *et al*[30, 31] who believed that few Mg^{2+} ions can be diffused into fairly- or well-crystallized core materials such as $Li_xNi_{1-y}Co_yO_2$ and $LiSr_{0.002}Ni_{0.9}Co_{0.1}O_2$ even if they were heat-treated over 10 hours at 750°C and 600°C, respectively. Therefore, the species on the particle surface in this work is supposed to be MgO and YPO_4, respectively.

Figure 3 shows the cycling profiles of commercial $LiCoO_2$. Obviously commercial $LiCoO_2$ shows specific capacities as high as *ca.*155, 190 and 265 mAh/g when charged to 4.3V, 4.5V and 4.7V respectively in the initial cycle. However, its cycling performance rapidly degrades in the subsequent cycles, especially for the cells charged to 4.5 and 4.7V. In less than 20 cycles, their capacities fade to half and one third of their initial values respectively. The charge plateau below *ca.* 4.2V ($x \geq 0.5$ in Li_xCoO_2) in these profiles represents the coexistence of two hexagonal phases ($x \geq 0.8$) and the growth of the second hexagonal phase ($0.8 \geq x \geq 0.5$) while the one above 4.2V ($x \leq 0.5$) is due to the M→H phase transition, consistent with the reports of Amatucci *et al*[36] on delithiating from $LiCoO_2$ and that of Pouillerie *et al*[16] on delithiating from $LiNiO_2$. Most authors agree that the rapid capacity fading of $LiCoO_2$ cathode material is due to the presence of this M→H phase transition when the material is heavily delithiated (over-charged). During this transition, the lattice parameter *c* shrinks significantly while the *a* value changes slightly. This inhomogeneous dimensional change induces a differential stress within the particle and causes a fracture event in the material.

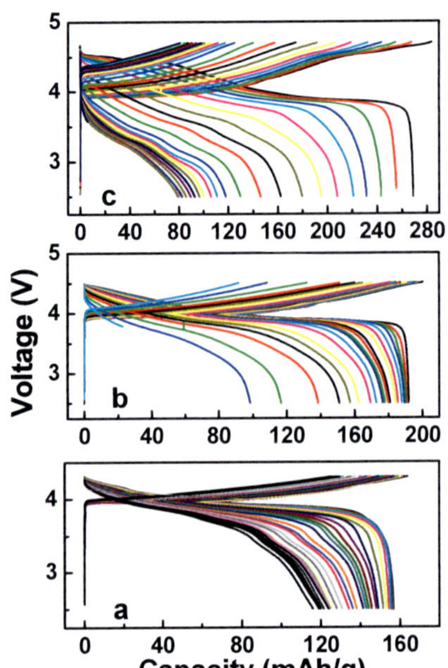

Figure 3. Charge/discharge profiles of commercial $LiCoO_2$ cycled between 2.5V and various charge cutoff voltages: (a) 4.3, (b) 4.5, and (c) 4.7 V.

The electrochemical signature of the above H→M→H transformation characterized with two oxidations peaks in the dx/dV profiles[36] is not obvious in the cyclic voltammetry (CV) of commercial $LiCoO_2$ in this work (Figure 4), probably due to overlapping or being covered up with the strong and broad 4.11V peak. However, their corresponding reduction peaks can be seen clearly at 4.13 and 4.03V in the CV plot. The strong oxidation peak at 4.11V indicates the good structural stability and electrochemical reversibility of the material when cycled below 4.2V. The strong 4.58V peak is attributed to the formation of Jahn-Teller low spin d^5 ions, according to Ohzuku and Ueda.[37] These two strong oxidation peaks have their counterparts at 4.30 and 3.73V in the discharge segment of the CV plot in the initial cycle. However, the 4.58V oxidation peak almost disappears in the second cycle while the 4.11V oxidation peak only varies slightly. In addition, the position of the reduction peaks moves from 4.30, 4.13, 4.03 and 3.73V in the initial cycle to 4.12, 4.02, 3.82 and 3.63V, respectively in the second cycle. This shifting is probably due to the migration of the Co^{4+} ions from their CoO_2 slab position into the interslab space at over-delithiated state. Amatucci *et al*[36] reported that only *ca.* 0.8Li can be reintercalated into CoO_2, the end member of $LiCoO_2$ charged to 5.2V. The presence of Co^{4+} in the interslab space can stabilize the hexagonal structure of $LiCoO_2$ and suppress the M→H phase transition in the subsequent cycles. However, the Co^{4+} ions in the interslab space hinder the transport of the Li^+ ions due to their difference in ionic radius during intercalation and deintercalation. Therefore, the polarization of the material increases. The suppression of the M→H phase transition is conformed with the disappearance of the higher plateau after a few cycles and the "straight" charge curves in Fig.3. After about 20 galvanostatic cycles, the 4.11V oxidation peak moves up to 4.6V and the reduction peaks shift down to 3.2 in the CV of the cell. The reduction peak at *ca.* 2.85V comes from another phase created during cycling.

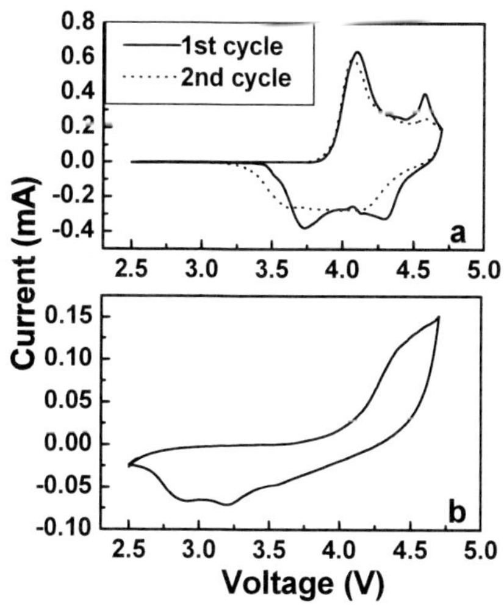

Figure 4. Comparison of slow scanned CV (0.04 mV/s) of commercial $LiCoO_2$ (a) in the first two cycles and (b) after 25 cycles between 2.5 and 4.7 V (Li^+ *vs.* Li).

The electrochemical performance of commercial $LiCoO_2$ is obviously improved with surface modification as discussed below. Figure 5 shows the cycling profiles of MgO-coated

$LiCoO_2$ cathode charged to various cutoff voltages. Compared with the cycling profiles of commercial $LiCoO_2$ and MgO-coated $LiCoO_2$ in the second and subsequent cycles, apparent polarization is observed in the initial cycle of MgO-coated cathode. This polarization is attributed to the insulation of MgO to both the electrons and the Li^+ ions. From the second cycle on, the polarization in the charge and discharge processes becomes negligible. This is probably due to the migration of the Mg^{2+} ions from the MgO shell into the core of the material. Mg^{2+} migration will result in two effects. On one hand, the diffusion of the Mg^{2+} ions makes the shell thinner and improves the ionic conduction of the surface layer by forming surface solid solution Li-Mg-(Co)-O. Therefore, the resistance of the shell to the electrons and ions becomes smaller. On the other hand, diffusion of Mg^{2+} into the bulk of $LiCoO_2$ enhances its conductivity by creating electronic holes, *i.e.* Co^{4+} ions.[14] As the charge/discharge plots in the other cycles are similar to each other, it is speculated that most of the Mg^{2+} ions finish migration in the initial cycle.

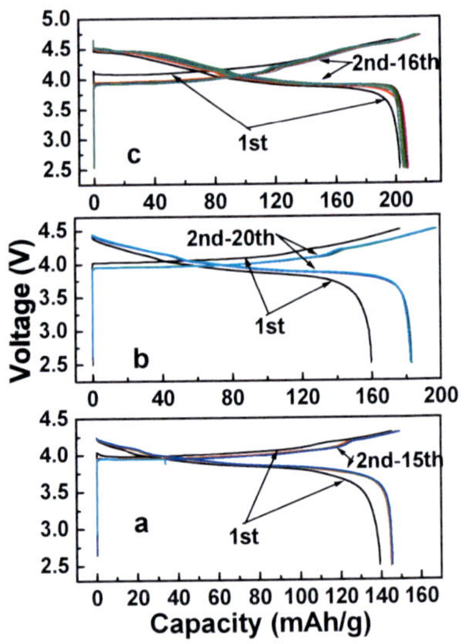

Figure 5. Cycling profiles of MgO-coated $LiCoO_2$ cathodes charged to 4.3, 4.5, and 4.7 V at 0.1 mA/cm^2

As shown in Fig.5, the polarization disappears after the initial cycle even in cells charged to 4.3V. This means that the migration of Mg^{2+} ions into $LiCoO_2$ bulk takes place at rather low potential. That is, the migration occurs when the material is still Li^+-abundant. In order to find out the exact procedure of the migration, cells composed of MgO-coated $LiCoO_2$ are cycled below 4.3V. It is observed that the cell potential increases quickly with time and reaches the summit of a hump in a few minutes. Then as the cell is further charged, its potential turns to decrease with time and reaches the bottom of a valley in another few minutes. Only after these processes will the cell potential increase monotonously with charging time. Figure 6 shows the dependence of the hump summit and the valley bottom potentials during charge in the first 20 cycles. It is seen that the potential values of these two points and the potential difference between them decrease with cycling. These two points unite into one and the potential value reaches its minimum after 15 cycles when the cell is cycled between 2.5 and 4.1V at 0.1mA/cm^2. Based on these experiments, it is determined that the Mg^{2+} migration begins around 4.1V and most of the Mg^{2+} ions finish migrating before 4.3V during charge or discharge.

There is still some argument as to the position of the doped Mg^{2+} ions in the lattice of $LiMg_xNi_{1-x}O_2$. Pouillerie *et al*[16] stated that Mg^{2+} ions migrate from the O-Ni-O slab to the O-Ni-O...O-Ni-O interslab space in $LiMg_xNi_{1-x}O_2$ (x = 0.05 and 0.10) at the end of the first discharge (2.7V) since they are destabilized in the covalent slab when almost all the nickel ions are in the tetravalent state. However, Chang *et al*[17] believed that the doped Mg cations occupy and remain on the Ni sites during cycling. Both groups of authors agree that the specific occupancy (in the interslab space or in the O-Ni-O slab) suppresses the phase transportation because this occupancy prevents Li vacancy ordering. The XRD pattern of the 60-cycled MgO-coated $LiCoO_2$ has been shown in Fig.2 at the discharge state. Calculations demonstrate that the *a* and *c* values of the material (*a* = 2.81152 Å and *c* = 14.03062 Å) are similar to that of commercial and fresh MgO-coated $LiCoO_2$. In addition, compared with the normal charge cutoff voltage for commercial $LiCoO_2$ or $LiNiO_2$ cathode materials and considering the severe polarization of the modified material in the initial cycle, the Li abundance is pretty high in the MgO-coated cathode at 4.1V (corresponding to ~$Li_{0.70}CoO_2$). This means that the Mg^{2+} migration takes place before obvious Li vacancy ordering begins. Most authors believe that the structure of the O-Co-O or O-Ni-O slab is very stable at this stage and the probability of migration of Co or Ni ions from their normal 3*a* sites to the 3*b* sites (usually observed in heavily delithiated states) is very small. The MgO-coated $LiCoO_2$ based cells were also discharged to 2.5V initially at a very slow rate (0.01mA/cm^2). It was found that very few Li^+ (and/or Mg^{2+}) ions can be inserted to the lattice (less than 0.001Li per formula of $LiCoO_2$). Therefore, the Mg^{2+} ions diffused into $LiCoO_2$ must share the interslab space with the Li^+ ions. Further evidence is needed to determine if the diffusion takes place during charge or discharge. In any case, the presence of the Mg^{2+} ions in the interslab space will hinder the ordering of the Li^+ ions and suppress the phase transition.

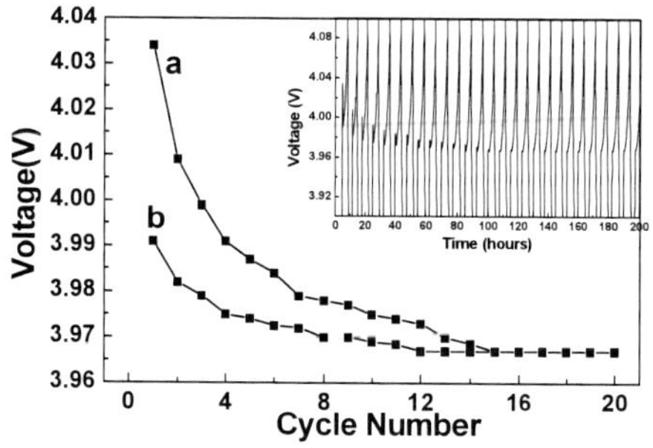

Figure 6. Evidence of Mg^{2+} migration into $LiCoO_2$ lattice, cycle number dependence of the starting voltages of Li^+ deintercalation at 0.1 mA/cm^2: (a) the hump summit value of the potential and (b) the valley bottom value of the potential during charge in every cycle.

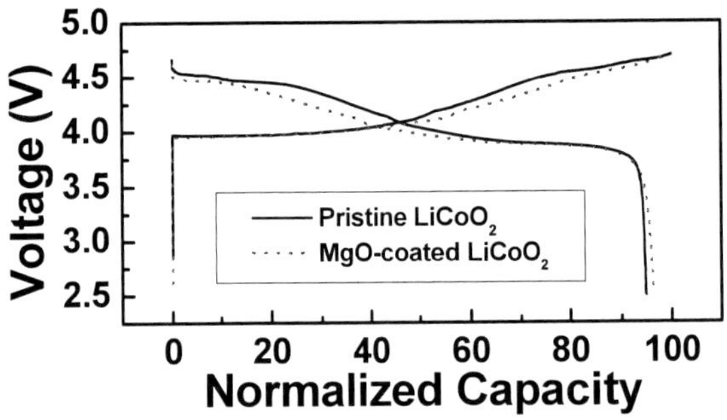

Figure 7. Comparison of cycling profile of commercial $LiCoO_2$ in the initial cycle and that of MgO-coated $LiCoO_2$ in the second cycle (at 0.1 mA/cm^2).

Due to the decrease of the polarization of the cathodes, the Li^+ deintercalation potential becomes comparable with that of commercial $LiCoO_2$. Therefore, the capacities of modified $LiCoO_2$ material are larger in the subsequent cycles than in the initial one. Figure 7 shows the charge/discharge profiles of commercial $LiCoO_2$ in the initial cycle and of MgO-coated $LiCoO_2$ in the second cycle. For the sake of comparing the shapes of the profiles, the charge capacity of each material is normalized to be 100%. Three features are observed in this figure. Firstly, the cycle profiles of these two materials overlap each other at low voltages (Li^+-rich states). This might result from two facts. On one hand, the migration of Mg^{2+} into $LiCoO_2$ and the formation of a solid solution lead to the increase of the bulk conductivity. On the other hand, the residual MgO on $LiCoO_2$ particle increases the surface resistance of the material. The voltage plots demonstrate the combined effects of these two factors. Secondly, the voltage plateau corresponding to the M→H phase transition is not as steep in commercial $LiCoO_2$ as in modified $LiCoO_2$. This slight difference brings about the disappearance of the oxidation peak (at 4.58V in commercial $LiCoO_2$) representative of the M→H phase transition in the CV plots in surface modified $LiCoO_2$ (Figure 8), indicating the suppression of phase transition at high potential. In fact, Pouillerie *et al*[16] named this potential plateau a pseudo-plateau as the potential increases continuously upon Li^+ deintercalation, even at very low cycling rate. Therefore, distinct from that in commercial $LiCoO_2$ cathode, the M→H phase transition is effectively suppressed in MgO-coated $LiCoO_2$. As seen in Fig.5, the capacity of the MgO-coated $LiCoO_2$ based cells keeps unchanged at 145, 175 and 210mAh/g for cells charged to 4.3V, 4.5V and 4.7V respectively in the first 15 to 20 cycles. Therefore, coating $LiCoO_2$ with amorphous MgO improves its cycling stability significantly. Thirdly, the discharge potential of the modified $LiCoO_2$ is lower than that of commercial $LiCoO_2$, consistent with the decrease of specific capacity and the suppression of the phase transition in the modified material.

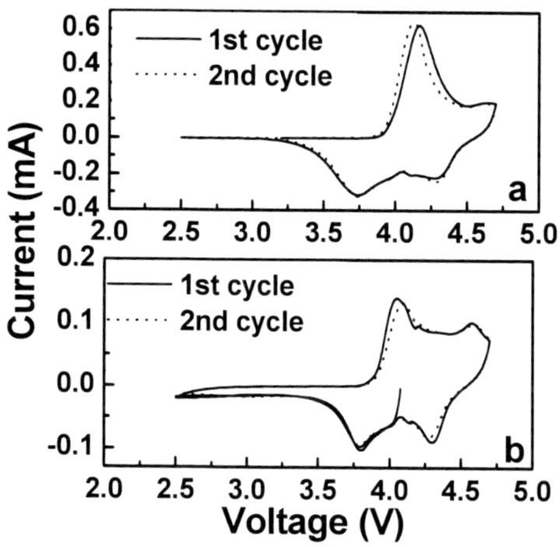

Figure 8. Slow scanned CV (0.04 mV/s) of MgO-coated $LiCoO_2$ (a) in the first two cycles and (b) after 20 cycles.

Figure 8 indicates that the CV plots of MgO-coated $LiCoO_2$ in the first two cycles overlap with each other completely except for the slight position shifting of the oxidation peak at *ca.* 4.11V, demonstrating the improved structural stability of the material. After about 20 cycles, capacity fading becomes obvious for all the MgO-coated $LiCoO_2$ based cells. The changes in the CV plots of the cells are as apparent. The CV plot of MgO-coated $LiCoO_2$ cathode becomes quite similar to that of commercial $LiCoO_2$ in the initial cycle. The 4.58V oxidation peak characteristic of commercial $LiCoO_2$ reappears, implying that the suppressed phase transition is partially recovered. The reason will be discussed in the following. In spite of that, the reversibility of the CV of modified $LiCoO_2$ is still much better than that of commercial $LiCoO_2$. As seen in Figs.3 and 5, the available capacities of modified $LiCoO_2$ charged to 4.5 and 4.7V are comparable to that of commercial $LiCoO_2$ charged to 4.3V and 4.5V, respectively. Figure 9 compares the cycling performances of commercial $LiCoO_2$ and modified $LiCoO_2$. Commercial $LiCoO_2$ charged to 4.7V is excluded here because it is obviously over-delithiated. Clearly the fading rate of the MgO-coated $LiCoO_2$ cathode is slower than that of commercial $LiCoO_2$ cathode (Fig.9A). This again demonstrates the effects of surface coating on improving the cycle stability of the materials. Similarly, though the initial discharge capacity of the 5%YPO_4-coated $LiCoO_2$ is only 177 mAh/g, its capacity retention is improved. After 17 cycles, its capacity becomes higher than that of the commercial $LiCoO_2$. In the subsequent cycles, its capacity remains stable. Within 80 cycles, its capacity fades only 26 mAh/g. Clearly, surface coating greatly improves the structural stability of $LiCoO_2$ at deep delithiation state.

Capacity fading in α-$NaFeO_2$ cathode materials is usually attributed to the side reactions[38, 39] such as the formation of inactive Co_3O_4, upon overcharging and the significant volume variations in the material accompanied with phase transition.[37] Chang *et al*[17] attributed the improved capacity stability exclusively to the prevention of material overcharging and therefore, prevention of the lithium vacancy ordering at high voltages due to the presence of inactive Mg^{2+} and Ni^{4+} species in the lattice. Pouillerie *et al*[15] also attributed the improved structural stability of $LiMg_xNi_{1-x}O_2$ to the pillaring effect of the Mg^{2+}

ions in the interslab space of the material. However, these alone cannot explain the improved cycling stability in the first 20 cycles and the capacity fading after that. Pouillerie *et al*[15] reported that 5% of Mg^{2+} is sufficient to suppress the phase transition of $LiNiO_2$ system. Nevertheless, the Mg^{2+} content in the bulk of $LiCoO_2$ should be less than 5% and might be insufficient to suppress the phase transition, considering that the initial total amount of MgO on the surface is only 1.5mol% in this work. Therefore, there must be some other reason(s) for the improvement and degradation of the cycle stability of the modified material. The MgO coating film on $LiCoO_2$ may take an important part in these processes.

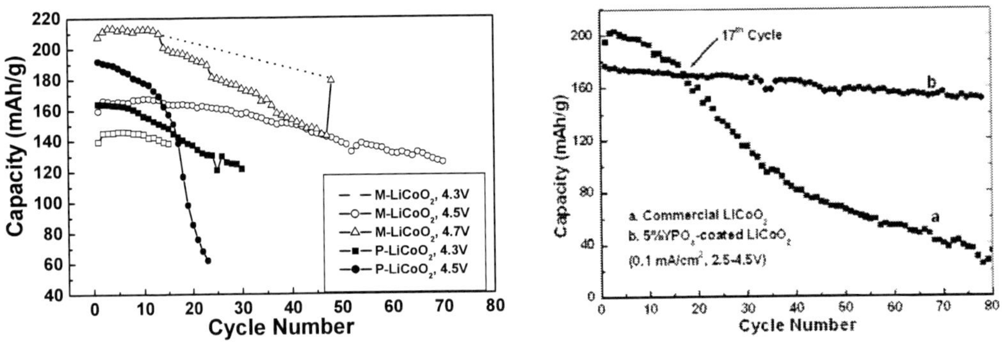

Figure 9. Cycle number dependence of the specific capacity of commercial and MgO-coated $LiCoO_2$ (A, left) charged to different voltages (M for modified $LiCoO_2$ and P for commercial $LiCoO_2$; the apparent stages for M-$LiCoO_2$ (4.7 V) are due to the different current densities applied, 0.1, 0.2, 0.4, and 0.8 mA/cm^2. The dashed line (---) is used to guide the eye to compare the recovered capacity when the current density changes back to 0.1 mA/cm^2). B (right) is for the cycling performances of (a) commercial and (b) 5%YPO_4-coated $LiCoO_2$ between 2.5 and 4.5V at a current density of 0.1mA/cm^2.

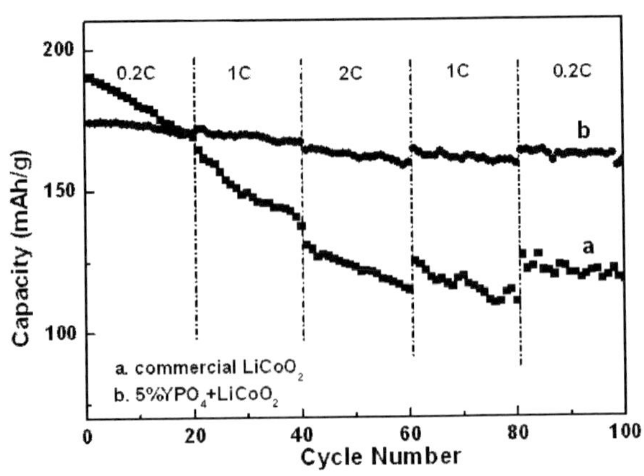

Figure 10. Comparison of rate performances of commercial (a) and 5%YPO_4-coated (b) $LiCoO_2$ at different current densities (1C =190 mAh/g).

Figure 10 exhibits the rate performance of commercial and 5%YPO_4-coated $LiCoO_2$ at different current densities at 25°C. The cells are charged galvanostatically to 4.5V at 0.2C rate but discharged to 2.5V at different current densities. It is seen that the capacity of commercial $LiCoO_2$ decreases sharply with increasing current density. In comparison, the capacity decrease of the 5%YPO_4-coated $LiCoO_2$ is much slower with increasing current density. This indicates that YPO_4 coating effectively improves the rate performance of commercial $LiCoO_2$.

The AC impedance spectra of the cell aged for 2 days were recorded in order to understand the improved rate performance of the YPO_4-coated $LiCoO_2$ (Figure 11). The semicircle in the high-frequency region of the Nyquist plot is mainly the contribution of the solid electrolyte interphase (SEI) film on the electrode.[15] The increasing diameter of the semicircle indicates that the impedance of the commercial $LiCoO_2$ electrode increases sharply with cycling after about 40 cycles while that of the YPO_4-coated $LiCoO_2$ increases very slowly.

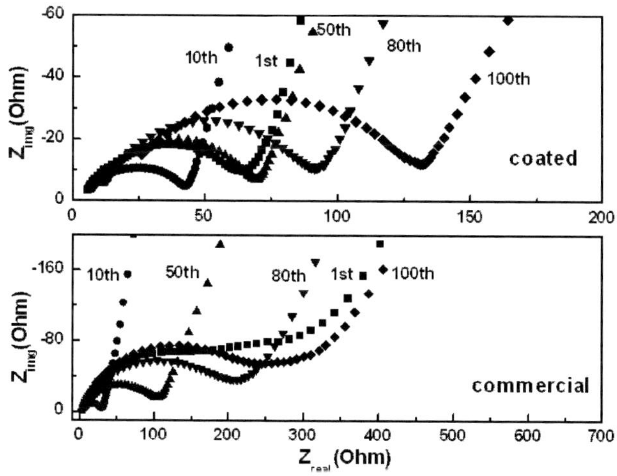

Figure 11. Nyquist plots of commercial and 5%YPO_4-coated $LiCoO_2$ after different cycles.

As the commercial $LiCoO_2$ and the YPO_4-coated $LiCoO_2$ are different electrodes, their impedances should not be quantitatively compared directly. Therefore, the impedance of each electrode after 10 cycles at discharge state is defined as 1 (normalized). The evolutions of the impedance of these two materials with cycling are compared in Figure 12. It indicates that the cell impedance of commercial $LiCoO_2$ is 5 times that of the 5%YPO_4-coated $LiCoO_2$ after 100 cycles. This variation implies that the impedance difference mainly comes from the cathode rather than the metallic Li electrode. The cell impedance of commercial $LiCoO_2$ at the 100th cycle is 10 times that of the 10th cycle while the cell impedance of the 5%YPO_4-coated $LiCoO_2$ increases very little. Chen and Dahn[40] suggested that impedance growth was responsible for the rapid capacity fading of $LiCoO_2$ cycled to 4.5V *vs.* Li^+/Li. These facts partially explain the good capacity retention of YPO_4-coated $LiCoO_2$ (Fig.9B) and insure the excellent rate performance of the materials (Fig.10). The reason for the suppressed increase of impedance will be discussed in Section 5 of this chapter.

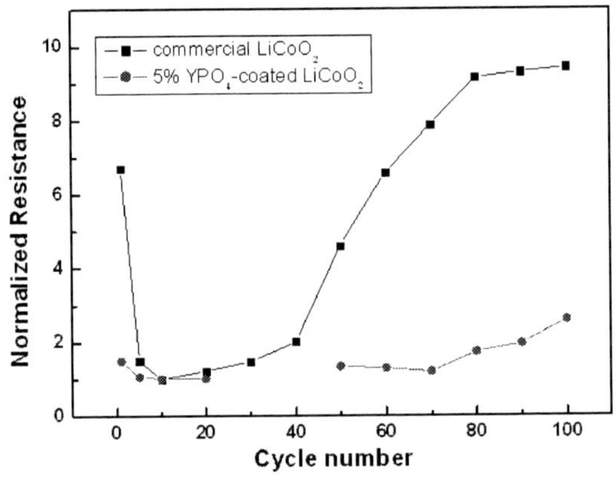

Figure 12. Comparison of normalized resistance of the cell after different cycles (the impedance after the first cycle of commercial $LiCoO_2$ electrode is due to experimental error because the semicircle after the 1st cycle is not perfect).

It has been reported that Co^{4+} ions are created at high potentials and prevent the possible decomposition of the material when charged to high potentials (above 4.3V). Amatucci *et al*[36] reported a strong and direct relationship between capacity loss and percentage of cobalt detected on the negative electrode for $LiCoO_2$-based lithium ion cells charged over 4.2V. For the MgO-coated $LiCoO_2$ cathode, the migration of the Mg^{2+} ions during cycling forms a uniform layer of Li-Mg-Co-O around the $LiCoO_2$ particle. This layer keeps the $LiCoO_2$ particles from direct contact with the electrolyte, and thereby prevents the escape of Co^{4+} species from the lattice of the core material and being dissolved in the electrolyte. Preventing the loss of the Co^{4+} ions may have two effects on keeping the reversibility of the material: (1) avoiding the formation of inactive substances and loss of active materials; (2) avoiding further formation of Co^{4+} ions and suppressing any reactions that create Co^{4+} ions in the material, which helps to avoid overcharging. Therefore the MgO coating layer is effective in protecting the cathode materials from Co^{4+} ion dissolution and ensures high cycling stability for the cathode material. However, if the Li-Mg-Co-O surface layer is damaged for whatever reasons (continuous Mg^{2+} migration into the core of $LiCoO_2$, for example) during cycling, it will lose its protective function to the core material and the Co^{4+} species can escape from the lattice of the core material. As impurities such as moisture will produce HF as by-product in the currently used liquid electrolyte, the thin Li-Mg-Co-O layer can be corroded during cycling. In this case the core material will be exposed to the electrolyte and the capacity fading is enhanced. Some experimental evidence is shown in Figure 13. In the SEM image, it is seen that most of the $LiCoO_2$ particles are covered with a layer of about 30 nm thick. In the cracks of the layer, some $LiCoO_2$ particles can be observed. The surface of the $LiCoO_2$ particle becomes rather smooth with many scale-like pits on it. The shape of the MgO beads becomes irregular, implying that part of the beads have been corroded during cycling. In addition, Mg and Co have been detected on the lithium foil and in the electrolyte solution by elemental analysis (ICP) to the sheets of the cell cycled 50 times. This indicates that the MgO coating layer has been corroded during cycling and lost its protective functions to $LiCoO_2$ after some cycles.

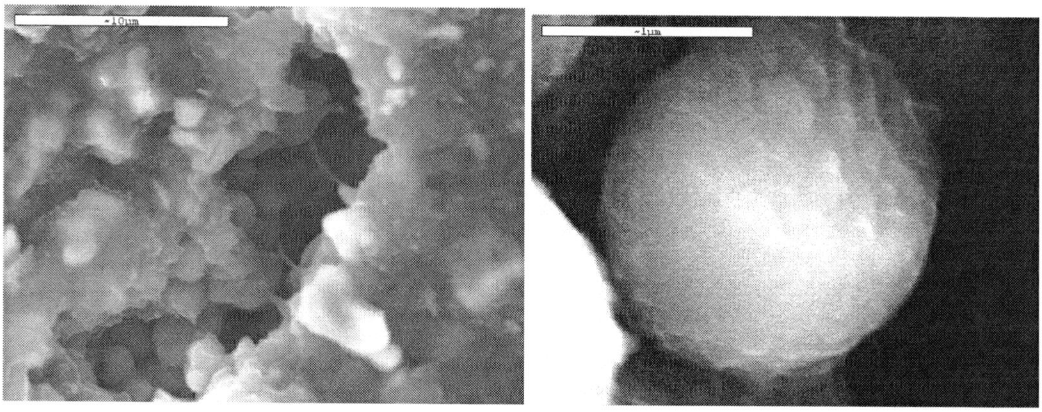

Figure 13. SEM imaging of (a, left) the surface of MgO-coated $LiCoO_2$ electrode after 70 cycles, and (b, right) an MgO-coated $LiCoO_2$ particle in the crack.

Usually the exothermic temperature becomes lower when the cathode material is charged to a higher voltage. This will lead to safety problems when the battery is overcharged or otherwise abused. Figure 14 compares the DSC traces of commercial and YPO_4-coated $LiCoO_2$ charged to 4.7V. It is seen that the exothermic reaction temperatures of the 1%, 5% and 20% YPO_4-coated $LiCoO_2$ charged to 4.7V are 6°C, 13°C and 18°C higher, respectively, than that of the commercial $LiCoO_2$ charged to the same voltages. In addition, the integrated areas of the exothermic peaks of the YPO_4-coated $LiCoO_2$ are much smaller than that of commercial $LiCoO_2$. Peak fitting to the DSC traces demonstrates that the integrated areas of the exothermic peaks are 95%, 47% and 18%, respectively, the integrated area of commercial $LiCoO_2$ for the 1%, 5% and 20% YPO_4-coated $LiCoO_2$. Therefore, surface modification with YPO_4 also improves the thermal stability of the material at charged state. However, as YPO_4 is insulating and electrochemical inactive, surface coating over 5%YPO_4 deteriorates the electrochemical performances of $LiCoO_2$.

Chen and Dahn improved the structural stability of $LiCoO_2$ by annealing the material at 550°C in air.[40] They attributed the performance improvement of surface-coated cathode materials to the essential heat treatment after surface coating. They believed that heat treatment removes the insulating surface impurities such as Li_2CO_3 and LiOH produced due to long-term storage of $LiCoO_2$ in (humid) air. However, as no heat treatment (except for sample drying at 120°C later) is necessary on coating $LiCoO_2$ with YPO_4 in this work, their suggestion may not be true for our case. In that review article,[40] they proposed three methods to change the surface chemistry of $LiCoO_2$, surface coating, particle grinding and heat-treatment. They admitted that none of those methods can improve the thermal stability of $LiCoO_2$. One has to modify the bulk structure of the material rather than its surface chemistry. However, our YPO_4 surface coating improves both the electrochemical performance and thermal stability of $LiCoO_2$ without any heat-treatment. With this, it seems that surface coating is not simply to create fresh $LiCoO_2$ surfaces. Its role should be much more complicated than we have currently understood. We believe that the interactions between the coating layer and the electrolyte and/or between the coating material and the active cathode material take more important part.

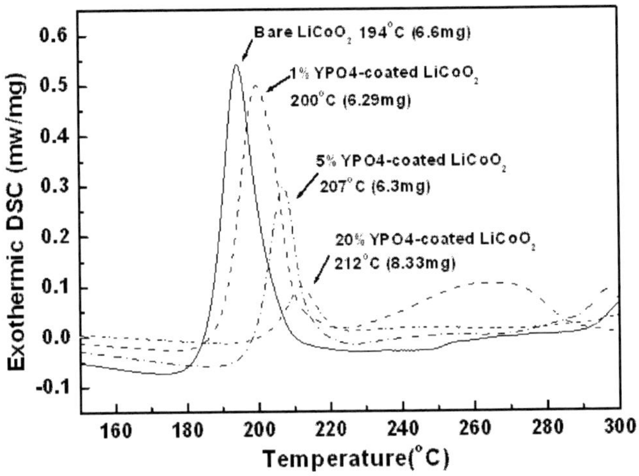

Figure 14. DSC traces of commercial and 1%, 5% and 20% YPO_4-coated $LiCoO_2$ charged to 4.7V at a heating rate of 5°C/min (the number in the parentheses is the net mass of the active material in each electrode sheet).

2.3. Summary

Surface modification with amorphous MgO or YPO_4 is effective to improve the structural stability of commercial $LiCoO_2$ cathode materials. Li/MgO-coated $LiCoO_2$ cells can be cycled between 2.5V and 4.7V and a high specific capacity of 210mAh/g be obtained without damaging the cycling stability of the material. These improvements are attributed to the formation of a surface solid solution on $LiCoO_2$. The diffusion of Mg^{2+} from the MgO coating layer into the core material occurs at rather low potentials. This helps to suppress the phase transition by occupying the Li vacancies and preventing the vacancy ordering at high charged potentials. The exothermic reaction temperature of the surface-modified $LiCoO_2$ is delayed by 6 to 18°C, depending on the amount of YPO_4 coated on $LiCoO_2$.

3. Spontaneous Reactions of $LiCoO_2$ with Electrolyte Solvent for Lithium Ion Batteries

3.1. Introduction

Surface modification can improve the electrochemical performances of the positive electrode materials for lithium ion batteries. However, the improvement mechanism has not been fully understood. Many authors believe that the modification layer separates the active material of the electrode from the electrolyte and prevents the escape of Li^+ ions at discharge states. In this and the following sections, the mechanism for the performance improvement by surface modifying the cathode materials will be comprehensively studied.

The importance of the surface of an electrode and its interface with the electrolyte cannot be overstated for the performance of a lithium ion battery. The nature of an electrode surface is critical for the electrochemical functionality of the material. Electrochemical (*e.g.*, charge transfer) and chemical (*e.g.*, Mn^{3+} disproportion in $LiMn_2O_4$) reactions occur at or near the surface, followed by mass transport into the bulk of the electrode, with structural changes as a result. Unwanted side reactions can take place as the electron meets the Li^+ ion at or near the surface of a cathode particle. Spontaneous reactions such as self-discharge and decomposition of the cathode material and electrolyte can also create a reactive surface, where solvent and salt can participate in reactions, resulting in further electrolyte decomposition. It is thus vital to obtain a basic understanding of the electrolyte/electrode interface during electrochemical storage and cycling.

Solid electrolyte interface (SEI) has long been known to play an important role in the electrochemical performance of the electrode material and of a battery.[41] Its formation starts upon the electrochemical cycling of the negative electrode of metals, metal oxides and various carbons. It is generally accepted that only in the case of alkali-metal electrodes may an SEI film appear just at contact with the electrolyte.[42] For all the other commonly used negative electrode materials, usually no signs of reduction could be observed solely due to contact of the electrolyte with the electrode material.

In contrast, investigation of the interfacial effects of electrolyte on the positive electrodes has been a rarity. Only recently has the formation of the SEI film been reported on positive electrodes.[19,43] Du Pasquier *et al.*[44,45] demonstrated that positive electrodes are covered with an organic SEI layer composed of decomposition products of alkyl carbonates. Aurbach[46] proposed that this layer may be either the re-precipitates of the reduction of carbonates on Li or Li-C negative electrode, or products of nucleophilic reactions between the oxides (negatively charged oxygen) and the highly electrophilic solvent molecules, ethylene carbonate (EC) and dimethyl carbonate (DMC), for example. Now more and more scientists realize that nucleophilicity of the positive electrode material plays important roles in the oxidation of electrophilic solvents. For instance, the intrinsic reactivity of highly nucleophilic $LiNiO_2$ with the solvent species has been found more pronounced than that of less nucleophilic $LiMn_2O_4$ spinel and Li_xMnO_2 (3V material) at some applied potentials. In addition, Ostrovskii *et al.*[47] reported that spontaneous reactions can occur on the surface of $LiNi_{0.8}Co_{0.2}O_2$ and $LiMn_2O_4$-based electrodes during storage in electrolyte of 1M $LiPF_6$ in EC/DMC (1:1 by volume) and 1M $LiClO_4$ in propylene carbonate (PC). That is, surface species can be formed in the absence of negative electrode materials and even without any applied voltage on the electrode. They believed that spontaneous electrode-electrolyte reaction occurs due to oxidation of the solvent molecules and salt anions, resulting in spontaneous lithium ion extraction from the active material. However, they failed to show the structural degradation of the electrode materials due to lithium ion extraction or identify the roles of the salt and the solvent on the reaction. McLarnon *et al.*[48] carried out further studies and compared the effects of storage on the structure of and surface film on $LiMn_2O_4$ thin film in pure solvent DMC and in electrolyte of 1mol/L $LiPF_6$ in EC/DMC at elevated temperatures. Thin electronically insulating surface layers were detected on all electrodes. The composition of the surface layer formed in DMC was found similar to that formed in the electrolyte solution. The surface layer in DMC can preserve the electrode material from further degradation but leads to complete electrode deactivation, probably due to loss of surface electronic conductivity and slow lithium ion transport rates through the surface layer. In contrast, the layer formed in the solution does not prevent $LiMn_2O_4$ decomposition or consequent electrode capacity loss. Earlier the same authors[49] reported spontaneous conversion from thin-film $Li_2Mn_4O_9$ to λ–MnO_2 in 1M $LiPF_6$ in EC/DMC solution. The original material was also found covered with a thin nonconductive layer, most likely Li_2O, after interaction with the electrolyte. These studies are helpful to understand the compatibility of the electrode material with the electrolyte. Nevertheless, they failed to illustrate the driving force for the dissolution of lithium ions from the material and whether the electrolyte or some of its components are decomposed in the surface reactions.

Therefore, some basic questions remain unanswered although much progress has been made in understanding the surface films on the positive electrode materials. First, as lithium salts were used in the above studies, the source of lithium for the well-known SEI (or more strictly, the surface layer, because its influence on the electrochemical performance of the positive electrode has not been clear yet) components such as $ROCO_2Li$ and Li_2CO_3 is unknown. Both the Li^+-containing positive electrode material and the lithium salt may be possible lithium sources for the generated species. In the latter case, a second question arises: if and how the salt concentration influences the properties of the surface layer and the structure of the positive electrode material? Third, it seems that most authors focused their attention on the composition of the solid surface layer. However, the mechanisms of the electrolyte decomposition and formation of surface film on positive electrode cannot be understood without knowledge of the other reaction products, such as the liquid and gas species. Fourth, it has been proved that surface modification can improve the electrochemical performance of $LiCoO_2$ and other positive electrode materials for lithium ion batteries. However, the mechanism remains unclear of why surface modification can improve the structural stability of these materials though some fundamental work has been done. Fifth, $LiCoO_2$ is the most successfully commercialized positive electrode material though other positive electrode materials are also promising. It has been believed stable at delithiated ($x<0.5$ in $Li_{1-x}CoO_2$) state and at high temperatures (above 55°C). Meanwhile, $LiPF_6$ dissolved in EC/DMC is one of the most popular electrolytes for both experimental lithium cells and commercial lithium ion batteries. However, the stability of $LiCoO_2$ towards the electrolyte at discharged (lithiated) state and at room temperature is rarely reported. Actually important criteria to the performance of a lithium ion battery include its shelf life and cycling life. The former is dependent on the compatibility of the material with the electrolyte at static (non-working) state while the latter is dependent on the endurance of the material to lattice distortion due to repeated lithiation and delithiation. The compatibility of the electrode material with the electrolyte components is crucial for a battery because its electrodes are immersed in the electrolyte from its assembly until the end of its life. Such compatibility is especially important for power-type lithium ion batteries in which the specific surface area of the electrode material is very large. However, most of the previous studies are focused on the electrochemical performances of the electrode material including its specific capacity and cycling stability or these behaviors at elevated temperatures. Very few reports have been made on the structural stability of the positive electrode material for lithium ion batteries towards the electrolyte or its solvent.[48,49]

The above questions are important because their answers can help to find ways to improve the chemical compatibility of the positive electrode materials with, and their electrochemical stability in, the electrolyte by modifying the surface of the electrode material and/or the molecular structure or composition of the electrolyte.

Before reporting the experimental results, we must point out that the purpose of this study is to find out some fundamental aspects of the solvent/$LiCoO_2$ system. Commercial $LiCoO_2$ is milled to "enlarge" its specific surface and, as a result, the surface reactions is promoted, rather than to emphasize the severity of the surface reactions or to improve its electrochemical performance this way. Al_2O_3 instead of MgO or other materials is coated on $LiCoO_2$ because Al_2O_3 was found easier to form a compact nano-film on $LiCoO_2$ than the latter.[50] Finally, we are trying to reach a better understanding to the mechanism of performance improvement by surface-coating $LiCoO_2$ rather than to deny the well-known experimental facts.

In this section, we will collect and characterize the gas, liquid and solid products so as to draw a clear picture about the effects of solvent and electrolyte soakage on the generated species and on the structural and electrochemical properties of the positive electrode material. Another aspect of the mechanism of performance improvement by modifying the surface of commercial $LiCoO_2$ will also be revealed based on these investigations.

3.2. Experimental

The same commercial $LiCoO_2$ used in the above section was high-energy ball-milled in ethanol and nanosized $LiCoO_2$ (nano-$LiCoO_2$) particles (*ca.* 200nm) were obtained. Part of the nano-$LiCoO_2$ powder was annealed at 500°C for 4 hours in air to eliminate most of the microstructural defects generated during ball-milling and to make the powder sufficiently dry. The rest of the powder was mixed with NH_4OH in distilled water. $Al(OH)_3$ was formed and deposited on the surface of $LiCoO_2$ particles when $Al(NO_3)_3$ aqueous solution was slowly added into the stirred mixture. The mixture was then filtered and flushed with distilled water three times and stored in an oven of 130°C for 12 hours. Later the dried $Al(OH)_3$-coated $LiCoO_2$ was heated at 500°C for 4 hours. This is expected to dehydrate $Al(OH)_3$ and to obtain dry Al_2O_3-coated nano-$LiCoO_2$. Transmission electron microscopy analysis indicated that the thickness of the Al_2O_3 coating layer on nano-$LiCoO_2$ particles was *ca.* 10 nm.

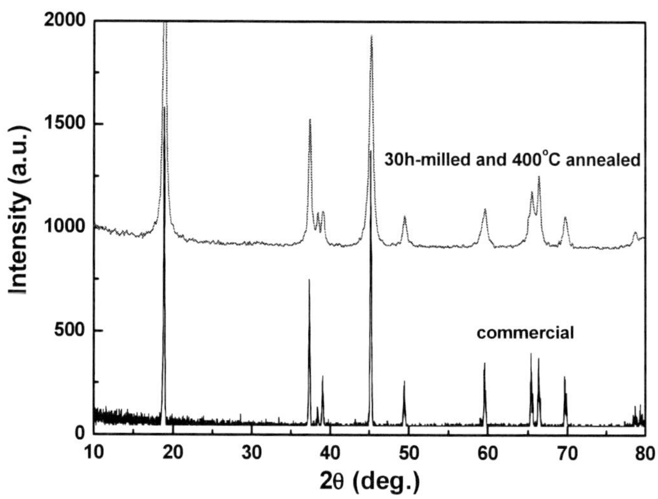

Figure 15. Comparison of XRD patterns of commercial and high-energy ball-milled $LiCoO_2$ heat-treated at 400°C for 4h in air.

Transmission electron microscopy analysis indicated that the thickness of the Al_2O_3 coating layer on nano-$LiCoO_2$ particles was *ca.* 10 nm. Both the bare and Al_2O_3-coated nano-$LiCoO_2$ were heated for 4 hours at 400°C in air prior to soakage. Such treatment was intended to eliminate the possible organic groups on the particle surface because organic media was employed and some structural defects might be created during ball milling, and to make sure that the powder was sufficiently dry. X-ray diffraction (XRD) (Figure 15) analysis indicates that high-energy ball-milling does not damage the structure of commercial $LiCoO_2$ or induce obvious lattice parameter variation of $LiCoO_2$ or generate new species, Li_2CO_3, Co_2O_3 or Co_3O_4, for example. However high-energy ball-milling led to obviously broadened diffraction lines because the $LiCoO_2$ particles were nanosized.

The nano-$LiCoO_2$ powder (300.0mg), a magnetic bar and battery-grade solvent EC/DMC (1:1 by volume; 4.0ml in total) with (1mol/L) or without $LiPF_6$ were sealed in a stainless steel (SS) container with a PTFE gasket (Figure 16) in the glove box. Both the oxygen and moisture contents in the glove box were below 10 ppm. Container made of SS could be used here because previous study[51] indicated that the same reactions took place in a glass container (replacing the SS with PTFE as the body material of the container did not influence the experimental results). The mixture was mechanically stirred for 7 days at room temperature (25°C). Then by puncturing the rubber window (2.5mm thick) of the container with an injector, 100μl of the gas in the sealed container was transferred into an HP-Plot Q fused silica capillary column (length=30m, inner diameter=0.32mm and film thickness=20μm) for gas chromatography & mass spectroscopy (GC-MS). Later the liquid and the solid in the SS container were separated after the container was opened in the glove box. The liquid was sealed in an optical test tube for the Raman and GC-MS tests. The solid was then rinsed with DMC and filtered three times (repeated rinsing and filtering took 2 days) and dried in the vacuum chamber of the glove box (for about 12h) for infrared and Raman spectroscopy, XRD and electrochemical evaluation. The purities of all the organic solvents and electrolyte were of battery grade.

The Raman spectra were recorded by radiating the liquid sample sealed in a test tube or the glass-sandwiched solid sample sealed in the glove box with 1064nm laser (50 mW output) of an FRA 106/S FT Raman spectrometer. The resolution for both the Raman and FTIR spectra was set at $4cm^{-1}$. The structure of the soaked solid was determined on MacScience M18A-HF (Cu Kα radiation) X-ray diffractometer after masking the solid sample on concaved glass (as the sample holder) with Mylar^{TM} film (as the X-ray window) in the glove box. The Raman spectra of some solid samples were recorded on Renishaw R-1000 Raman spectromcter with Ar^+ laser (633nm, 200mW output) as specified in the corresponding figures. The instrumental resolution was set at $3cm^{-1}$.

Fourier transformed infrared (FTIR) spectroscopic measurements were conducted on EQUINOX 55 (Bruker Instruments) by spreading a droplet of the liquid sample into a thin film on a KBr pellet and pressed together with another KBr pellet (sandwiched) or by pressing the mixture of KBr and the solid into pellets in the glove box. The pellet was sealed in a glass container in the glove box. When everything was ready for the FTIR instrument, the container was opened and the pellet was transferred into the vacuum chamber of the instrument and evacuated immediately. The exposure time of the sample in air was less than 10 seconds.

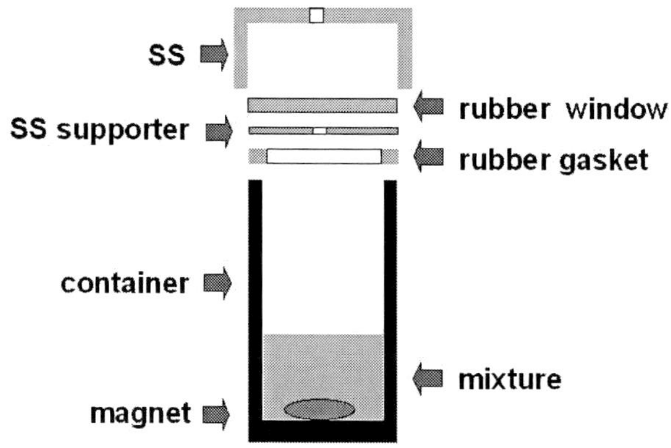

Figure 16. Scheme for the container.

GC was carried out on Trance 2000 series GC instrument (Finnigan Inc.). Part of the liquid sample was properly diluted with methanol and injected into a DB5 column (L = 30m, I.D. = 0.25mm, F.T. = 0.25μm) for GC measurement. The temperature of the GC oven was set at 50°C for the gas samples or heated gradually to 250°C in 25 min for the liquid samples. The inlet temperature was 50°C for both the gas and liquid samples. In both cases, helium (He) was used as the carrier gas. The species from the column outlet were continuously ionized and analyzed with Trance MS mass spectrometer (Finnigan). The ionization voltage and emission current were set at 70V and 350μA, respectively. The range of mass collection was set to be 10-190*amu* (atomic mass unit) for the gas and 17-257*amu* for the liquid samples. The range of retention time (RT) was 0 to 15min for the gas and 2.5 to 26min for the liquid. The setup of the retention time range was determined by extending the retention time to 30 min and making sure that no interesting species came into being after 15min for the gas or 26min for the liquid samples.

3.3.1. Spontaneous Solvent Decomposition on $LiCoO_2$

Figure 17 shows the GC and MS spectra of the generated gas of Al_2O_3-coated nano-$LiCoO_2$ soaked in solvent EC/DMC. The strongest line in the GC spectrum is from Ar+N_2 (at RT = *ca.* 1.80min in Fig.17a) because $LiCoO_2$ and the solvent were sealed in the Ar-filled glove box and some N_2 in air could enter the injector during sample transference from the sealed container into the GC column. Besides that, carbon dioxide (CO_2), carbon monoxide (CO), methane (CH_4), ethylene (C_2H_4), water (H_2O), ethane (C_2H_6) and oxygen (O_2), in the order of their maximum counts per second (cps) from high to low in MS, are also detected in the gas sample (Fig.17b). In order to make sure that the detected H_2O and O_2 are from the gas in the SS container rather than from air during sample transference, GC-MS spectra of air were recorded by simply injecting similar volume of air (100μl) in the room into the column. It is found that the relative contents of O_2 and H_2O (with N_2 from air as the reference) in air are about 30% lower than in the sealed container. We did not specifically analyze the gas in the glove box for possible trace water and oxygen.

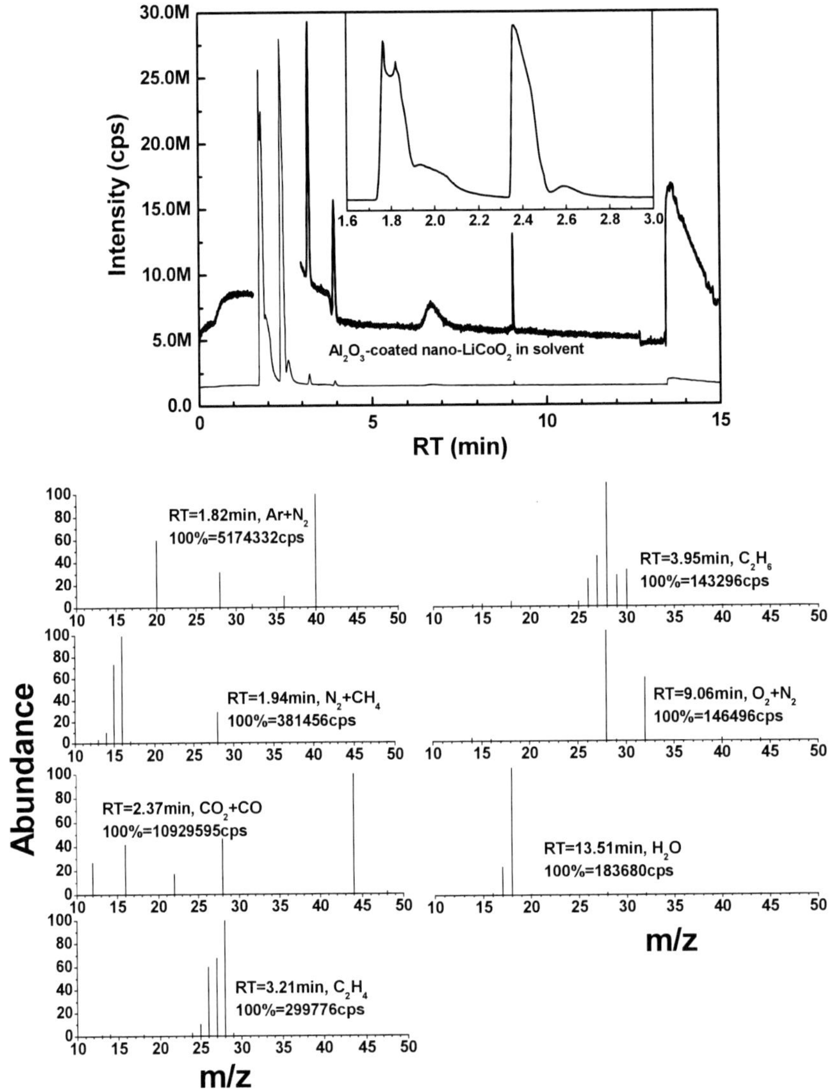

(** (1). the third spectrum on the left column: the line at 16 *amu* may also be explained as originated from atomic oxygen; this is principally possible. However, the line at 22 *amu* is mysterious to the authors; (2). theoretically one substance should have only a single retention time in a GC spectrum. However, N_2 reappears at different *RT*s in this work)

Figure 17. Typical GC (a) and MS (b**) spectra of gas species generated on $LiCoO_2$ soaked in EC/DMC (Al_2O_3-coated nano-$LiCoO_2$ here; the MS spectrum of Ar from the glove box and N_2 from air is shown for reference. The vertical scale is only for the bottom GC plot while the horizontal scale is for both the bottom and the middle plots; the inset is a selected part of the middle plot).

However, as will be seen in the other GC-MS spectra, the detected contents of oxygen and water are dependent on the specific sample. As a result, the above gases are believed to originate mainly from the reaction between the solvent and Al_2O_3-coated nano-$LiCoO_2$ in the SS container though we cannot exclude the contribution of moisture in the glove box to the results at the moment. GC-MS analysis is carried out to the gas from the soaked commercial and bare nano-$LiCoO_2$ as well. It is found that the main generated species in these three samples are the same but their contents are lower than that of Al_2O_3-coated $LiCoO_2$. In addition, the relative content of O_2 in Al_2O_3-coated nano-$LiCoO_2$ is lower than in the other two samples. Yang *et al*[52] examined the products of carbon electrode cycled in different electrolytes for lithium ion batteries. According to their MS analysis, CO_2 is the main product of EC decomposition. DMC is decomposed into CO and CH_4 while diethyl carbonate is decomposed into CO and C_2H_6. Our results show that CO and CH_3-CH_3 can also be generated from EC and/or DMC decomposition on $LiCoO_2$. In addition, no DMC trace is detected in the GC spectra of the gas samples within 30min. However, DMC was detected at RT = 36min. by elevating the temperature of the inlet and GC oven to 100°C in a separate experiment.

Identification of the generated species in the liquid was conducted by GC-MS, FTIR and Raman spectroscopy. Organic species with -C-O-C- units, most probably CH_3-O-$(CH_2)_2$-O-CH_3, is the only recognizable generated substance in these three liquid samples (Figure 18). Other detected substances include EC, DMC, H_2O and CH_3OH as well as some contaminant species. This is understandable because (1) EC and DMC as well as methanol are the dominant species in the liquid while the generated liquid species are only traces in the mixture; (2) it is very difficult to wash out the contaminant species in the long DB5 column for GC-MS spectra. The GC-MS spectra of these contaminants make it difficult for the identification of the trace generated species.

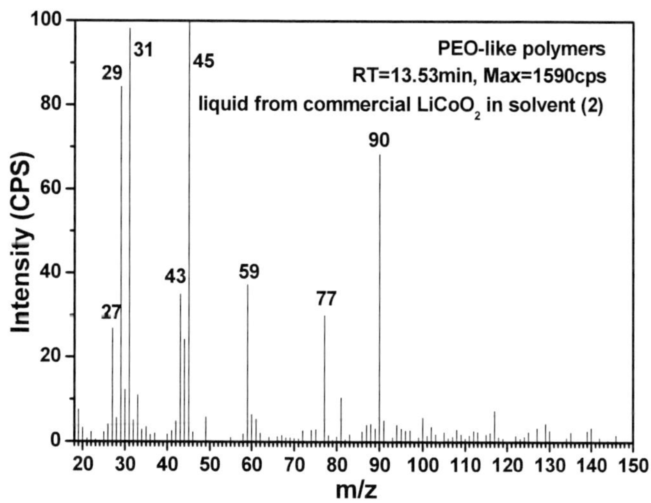

Figure 18. A typical MS spectrum of the liquid from solvent-soaked $LiCoO_2$ (bare nano-$LiCoO_2$ here).

As FTIR and Raman signals can be accumulated as many scans as required (200 scans for the FTIR and 400 scans for the Raman spectra in this work), a much higher signal/noise (S/N) ratio than GC-MS can be reached upon detecting trace species in the liquid. It is expectable that the signals from the generated liquid species are very weak and some of them might be submerged with strong signals from the solvent molecules. However, as listed in Table 1, some new bands can be recognized by carefully comparing the high S/N-ratio Raman and FTIR spectra of the liquid (not shown) with that of pure EC and DMC. The recognized bands are the characteristic lines of different functional groups. Based on Table 1, the liquid species generated in these three samples are actually the same though fewer bands are recognizable in the liquid of soaked commercial $LiCoO_2$. The main generated species have C-O-C and O=C structures. The former, consistent with that of liquid GC-MS results, confirms the existence of poly(ethylene oxide) (PEO)-like polymeric species while the latter demonstrates the formation of carbonates and other carbonyl groups in the liquid. These carbonates might include various solvable (lithium) alkyl carbonates. It is worth pointing out that the vibration bands close (with position difference $\leq \pm 6cm^{-1}$ *vs* the instrumental resolution of $4cm^{-1}$) to that of EC or DMC are not considered in this chapter. In addition, no traces of water are detected in the liquid, indicating that the solvent is sufficiently "dry" and the content of generated water (if any) should be low in the liquid.

Table 1. Comparison of the recognized FTIR absorption and Raman bands of liquid of solvent soaked $LiCoO_2$ and their assignments.

FTIR			Raman			Assignment
Commercial	Bare	Coated	Commercial	Bare	Coated	
1943	1942,1920	1943,1921	1912	1904	1947,1934 1916	overtone of CH_2 wag of aryl[53]
1698,1624 1617,1603 1573	1698,1664, 1624,1616 1576	1698,1624 1616,1575	1688,1671 1642,1632	1891,1695 1678,1668 1632	1723,1698 1680,1672 1662,1648 1628	carbonyl stretch [54,55]
1520,1339 1316	1533,1520 1339,1315, 1289	1518,1339 1316	1540,1523 1370,1344 1322,1283 1270	1545,1526 1368,1340 1325,1315 1288	1534,1515 1370,1347 1321,1304 1280	CH_3, CH_2 [54,55]
			1180,1142 1044	1184,1136, 1040	1184,1033 1017	C-O-C, C-OH [54,55]
			822,742	838,821 780	834,760 738	olefinic CH wag[53]
695,669 634,613	694,670 635,626 617	693b,668 636,627 617	679,660 624,613	650,626 612	664,635	C=O in-plane deform.[56]; CO_2 asym. Bend[53]
597	600	600				C-C-C deform [55]
	592 582 549	592 580 548	595,558 545	575,547		Li-O stretch;[53] C=O out of plane deform[55]
521,513	518	518				CH_2-CO-O deform;[55] Li-O stretch[53]
	503	503				Li-O stretch[53]
	492	492	495	499	489	
	485	485	483	483	483, 477	
454	458	457	468,456	463		
437,414			397,311 295,281 246,226 209	432,401 312,270 208	436,417 317,289 255,213	"chain expansion" of *n*-alkanes[54]

Figure 19 shows the IR absorption spectra of solvent-soaked $LiCoO_2$ rinsed with DMC (the sharp lines at 1385cm^{-1} with positive or negative intensities are due to instrumental error). Table 2 lists the main recognized bands and their assignments in the fresh and aged solid samples rinsed with DMC and dried in the glove box at room temperature (Note: the "fresh" samples were obtained right after the solid was dried in the vacuum chamber of the glove box, the "aged" samples were stored in the glove box for another 7 days after the fresh samples had been obtained). Clearly the strongest lines of EC become very weak in Fig.19 (the obvious water in the spectrum is from KBr rather than from the glove-box because no obvious water is detected in the aged samples).

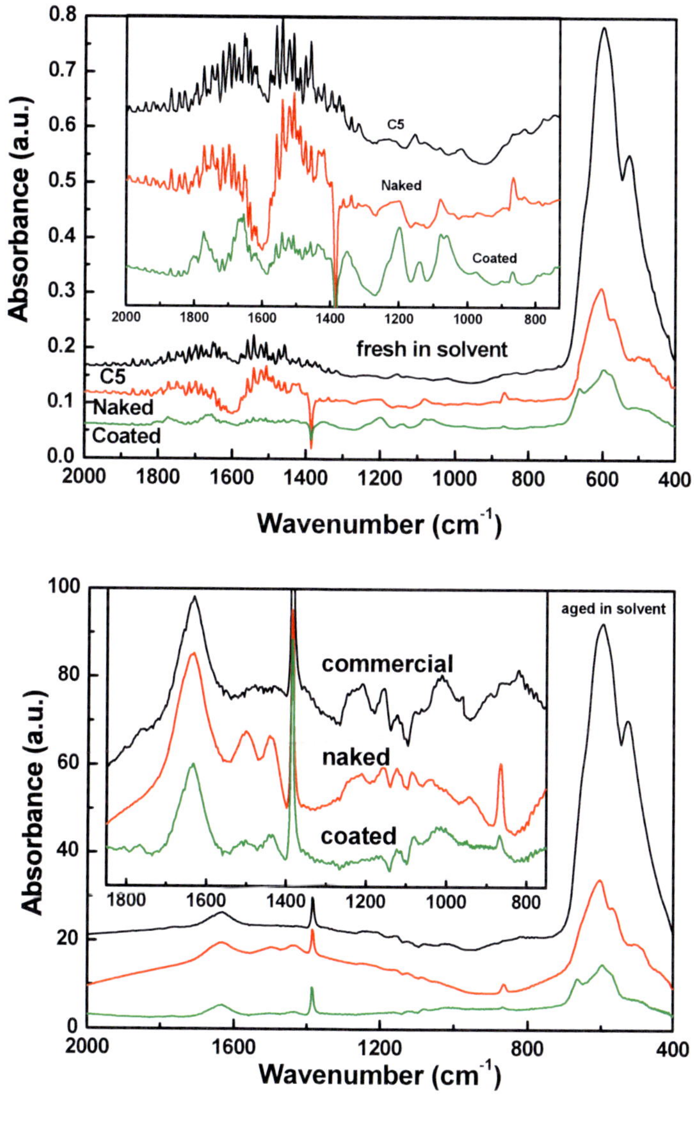

Figure 19. IR spectra of the fresh (a) and the aged (b) solids from solvent-soaked $LiCoO_2$.

This indicates that the DMC rinsing/filtering process is rather effective. Most of the observed bands belong to the species generated on the material surface. Comparing the IR spectra of the fresh samples (Fig.19a) with that of the aged (Fig.19b), it is seen that the corresponding IR spectra of the three samples are very similar to each other, indicating that the precipitates on $LiCoO_2$ are rather stable in dry Ar atmosphere at room temperature. Analysis to Fig.19 and Table 2 shows that the surface film is composed of lithium alkyl carbonates ($(ROCO_2Li)_2$, C=O stretching at 1635cm^{-1}) and Li_2CO_3 (868cm^{-1}). The similarity of the observed bands on different solids shown in Fig.19 and Table 2 also indicates that Al_2O_3 coating does not influence the composition of the surface film on $LiCoO_2$.

Table 2. Comparison and assignments of the recognized IR bands on solvent-soaked $LiCoO_2$.

Stored in solvent				Assignment
C5 (aged)	**B (aged)**	**C (aged)**	**C (fresh)**	
1635vs	1636vs	1635vs	1661	C=O stretch[54,55]
1499w	1499m	1501m		CH_3, CH_2 [54,55]
1429	1435s	1432m	1435	Li_2CO_3, CH, CH_3 asym. bend of SEI;[59] CH_3 asym. bend of SEI[60] and/or CH_2 scissors deform.[53]
	1245	1247		CH_3, CH_2 [54,55]
1236bs	1236w	1234		CH_3, CH_2 [54,55]
1226	1224	1226	1228	CH_3, CH_2 [54,55]
1209bs	1211	1208		CH_3, CH_2 [54,55]
1171w	1173	1173		SEI C-O stretch;[53] C-O-C stretch in formate esters[46]
	1 163	1161w		C-O-C, C-OH [54,55]
1156m	1153m	1152w		C-O-C, C-OH [54,55]
1130	1130	1130	1139	C=O stretch of SEI[61]; asym C-O-C stretch in aliphatic ethers (ROR');[53,61] C-O-C, C-OH [54,55]
1123m	1123m	1124		C-O-C, C-OH (1300-1000);[54,55] characteristic peak of PEO[56]
1113	1113	1112		C-O-C, C-OH [54,55]
1078m	1085m	1079m	1079	C-O stretch in primary (RCH_2OH) or secondary (RR'CHOH) alcohols;[62] C-O-C, C-OH [54,55]
1064w	1064	1062	1062	C-O-C, C-OH.[54]
1032	1037w	1033w		C-O-C, C-OH;[54] =C-H [54]
1013w	1013	1013		C-O-C, C-OH;[54] =C-H [63]
975	976	986	974	vinyl compounds;[53] out-of-plane C-H bending of CH_2 [59]
	941m			=C-H;[54] characteristic of PEO at 945cm^{-1} [56]
			883	CH_2 bend[64]
866	866vs	868m	868	Li_2CO_3
719	720	721	719	CO_2 asym of R in $R\text{-}CO_2Li$[59]
649	659	666	662	=C-H [54]
598	603	598	595	C=O out-of-plane deform.[55]
	565	568	568	Li-O stretch[53]
528				CH_2-CO-O deformation[55]
	492	506	510	CH_2-CO-O deformation[55]

Note: C5 is for Commercial, B for bare and C for Al_2O_3-coated

Ogumi *et al.*[57] and Yazami[58] detected polymer-like substances on carbon negative electrodes respectively. Ogumi *et al.*[57] further determined that these substances have repeated oxyethylene units, similar to PEO according to their pyrolysis/GC-MS analysis. These polymeric species are also detected in the current work with characteristic bands around 1120 and 940cm^{-1} (Table 2) as strong evidence for the presence of C-O-C structured polymeric substances on the solid,[56] in agreement with the observed C-O-C units in the liquid samples by GC-MS and FTIR spectroscopy.

In addition, the relative intensities of the bands around 600cm^{-1} (Co-O vibration[65]) are much stronger than those between 750 and 2000cm^{-1} after the solvent in the sample is well removed (Fig.19b). These indicate that the content of surface species is larger on Al_2O_3-coated $LiCoO_2$ than on bare $LiCoO_2$. These spectral features are also reflected in the following Raman spectra of DMC-rinsed solids.

The Raman spectra of the solid samples without rinsing are shown in Figure 20a. The two characteristic Raman bands of $LiCoO_2$ can be well recognized at 592 and 483cm^{-1} on commercial and nanosized $LiCoO_2$. However, these two bands become unrecognizable on solvent-soaked bare and Al_2O_3-coated nano-$LiCoO_2$. This indicates that the thickness of the surface film on them is much bigger than on commercial $LiCoO_2$. The particle size or the specific surface area of the material is important to the interaction between the solvent and the particle. The weak band at 894cm^{-1} is from the intense ring breathing band of residual EC on the solid.

Two strong bands are observed at *ca.*1090cm^{-1} and 560cm^{-1} on solvent-soaked nano-$LiCoO_2$. It is reported that the strongest Raman band of Li_2CO_3 is located at 1090cm^{-1} [63] while the Li-O bonding band of ROLi is around 560cm^{-1}.[53] However, the glass for sample sealing happens to have strong peaks at these two positions (the inset of Fig.20a). As the intensity of the second-to-strongest band of Li_2CO_3 (at 748cm^{-1}) is only 5% that of the strongest one (at 1090cm^{-1}),[63] the existence of Li_2CO_3 cannot be determined by Raman spectroscopy here. However, considering the variation of the intensity difference of these two bands, the species from the solid material should have some contribution to the 560cm^{-1} band because the intensity difference of these two bands becomes much smaller than in the glass alone. The 560cm^{-1} band is tentatively attributed to the Li-O bonding in ROLi.[53] The strong 560cm^{-1} band overlaps with the 592cm^{-1} band of $LiCoO_2$ and makes it unrecognizable for the soaked samples.

Fig.20b shows the Raman spectra of DMC-rinsed $LiCoO_2$ stored in the solvent. The two strong bands of pure $LiCoO_2$ at 592 and 483cm^{-1} are well defined in solvent-soaked commercial $LiCoO_2$. In nanosized $LiCoO_2$, however, the 592cm^{-1} band becomes difficult to recognize due to the presence of strong 560cm^{-1} band. When nano-$LiCoO_2$ is coated with nano-Al_2O_3, the 592cm^{-1} band overlaps with the 560cm^{-1} band and becomes unrecognizable. Meanwhile, the 483cm^{-1} band becomes very weak. These changes indicate that the thickness of the surface films or the surface reaction between the solvent and $LiCoO_2$ depends on the specific surface area of the substrate. That is, the surface layer is thicker on nano-$LiCoO_2$ than on commercial $LiCoO_2$. This is understandable. However, it is interesting to notice that the thickness of the surface layer on Al_2O_3-coated nano-$LiCoO_2$ is even thicker than on bare nano-$LiCoO_2$. This is consistent with the above observation by FTIR (Fig.19). Therefore, it seems that modifying the surface of nano-$LiCoO_2$ enhances the decomposition of the solvent and/or deposition of surface layer on it. Aurbach *et al.*[66] believed that the presence of activated Al_2O_3 promotes the decomposition of carbonate solvent and the formation of CO_2 and Li_2CO_3 on lithium electrode.

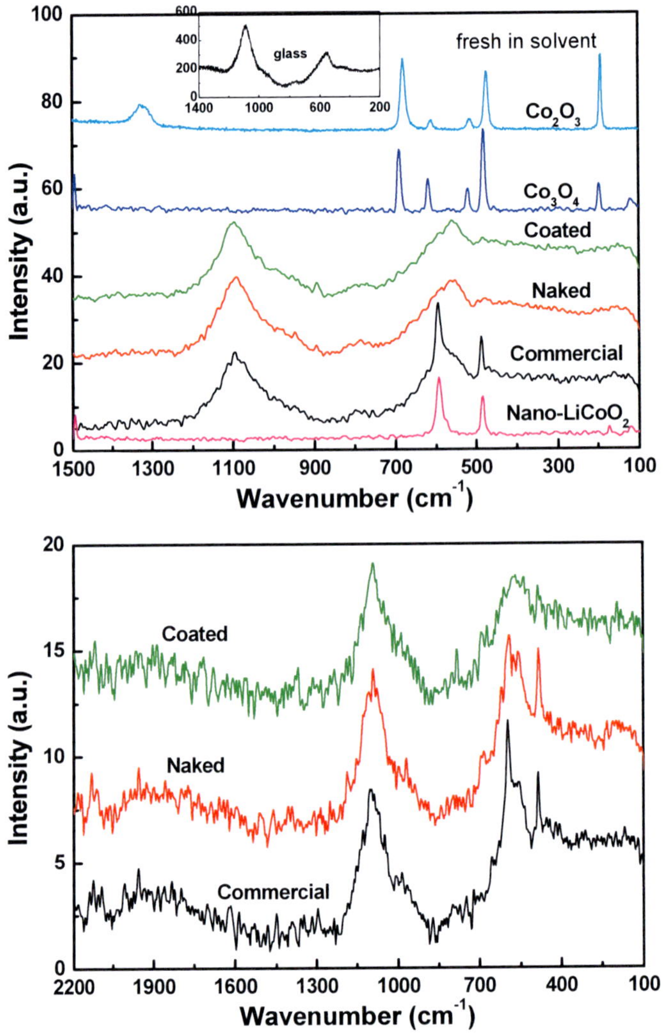

Figure 20. Raman spectra of the deposits on solvent-soaked $LiCoO_2$ before (a) and after (b) rinsed with DMC and then stored in the glove box for 7 days.

It was reported that the surface layer is only 1-2 nm thick on $LiNi_xCo_{1-x}O_2$ electrode.[47,67] This is also the average thickness of the SEI film on most of the studied negative electrodes. Considering that the penetration depth of the laser for the Raman spectrum is usually well over 10 nm (*ca.* 100nm in most cases) and that the Raman signals from $LiCoO_2$ or its derivatives should be masked with the surface layer, it is estimated that the thickness of the surface film on the soaked $LiCoO_2$ particles should also be well over 10 nm (Figure 21). It is known that the growth of the SEI film terminates on the negative electrodes when its thickness reaches the distance limit of electron tunneling effect.[42] As the thickness of the surface film on the soaked $LiCoO_2$ particles is well over 10nm, the mechanism of its growth and growth termination must be different from that on the negative electrodes. This will be further discussed in Section 5 of this chapter.

3.3.2. Structural Degradation of $LiCoO_2$ Soaked in EC/DMC Solvent

As there is no lithium source other than $LiCoO_2$ in the system, the above results indicate that simply soaking $LiCoO_2$ in the solvent can result in obvious lithium extraction from the lattice and structural degradation. As shown in Fig.1, the surface of commercial $LiCoO_2$ is very smooth and "clean" prior to soakage. However, it becomes very rough and covered with surface layer and many straight stripes or furrows after soakage in the mixed solvent EC/DMC (Fig.21). The size of the randomly distributed fragments is over 100nm. The straight stripes correspond to the 2-D layered structure of hexagonal $LiCoO_2$. The bumpy surface demonstrates the formation of thick and rough surface films on the particles. These features are typical of corroded surfaces.

Figure 21. Surface morphology of commercial $LiCoO_2$ soaked in EC/DMC solvent.

It was once believed that an important role of surface coating was to avoid direct contact of the active material with the electrolyte, prevent lithium escape from the positive electrode material and sustain its good electrochemical stability upon cycling.[47,68,69] However, Table 3 and the above spectroscopic studies demonstrate the opposite results: the Al_2O_3 coating layer on $LiCoO_2$ does not prevent the escape of lithium ions. Rather, the Co/Li atomic ratio in the solid of Al_2O_3-coated $LiCoO_2$ is even higher than in bare nano-$LiCoO_2$. These results are well reproducible, demonstrating that the performance improvement mechanism of surface modification is more complicated than what has been currently understood. There must be some other reasons for the improved electrochemical performance.

Table 3. Comparison of Co/Li atomic ratios in the high-energy ball-milled commercial $LiCoO_2$ under various conditions.

Condition	Co/Li ratio in solid	Co/Li ratio in liquid
Non-soaked	1:1.00	×
Bare and soaked	1:0.78	1:12.4
Coated and soaked	1:0.61	1:18.4

The above FTIR and Raman spectroscopic studies demonstrate that the surface film is mainly composed of Li-containing compounds and the presence of Al_2O_3 enhances the deposition of surface film on nano-$LiCoO_2$. This means that the contribution of the precipitated Li^+ ions to the Co/Li atomic ratio in Table 3 is more for the Al_2O_3-coated nano-$LiCoO_2$ than for the bare nano-$LiCoO_2$. In this case, a lower Co/Li ratio is expected for the Al_2O_3-coated $LiCoO_2$. However, this is opposite to the ICP results shown in Table 3. A reasonable explanation is that fewer Co^{2+} ions are converted from Co^{3+} or lost for the Al_2O_3-coated nano-$LiCoO_2$ than for the bare nano-$LiCoO_2$. As Co^{2+} is stable in liquid while Co^{3+} is stable in solid, it is proposed that Li^+ ion extraction from $LiCoO_2$ and the oxidation of the solvent result in the formation of Co_2O_3 in Al_2O_3-coated $LiCoO_2$ and Co_3O_4 in bare $LiCoO_2$, respectively. This proposal is verified with the following structural characterization of the generated solids by XRD and Raman spectroscopy.

Lithium loss by solvent soakage will definitely lead to structural degradation of $LiCoO_2$. Figure 22 compares the XRD patterns of commercial $LiCoO_2$, high-energy ball-milled $LiCoO_2$, solvent-soaked bare and Al_2O_3-coated nano-$LiCoO_2$. Comparison of the XRD patterns of commercial and nanosized $LiCoO_2$ indicates that high-energy ball-milling does not damage the structure of commercial $LiCoO_2$ or induce obvious variation in its lattice parameter or generate new species. However high-energy ball-milling leads to obvious broadening of diffraction lines because the $LiCoO_2$ particles have been nanosized. Besides the residual $LiCoO_2$, three species at least can be recognized in the solvent-soaked $LiCoO_2$ by comparing their XRD patterns with that of Co_3O_4 (JCPDS card # 9-418), Co_2O_3 (JCPDS card # 2-770) and Li_2CO_3 (JCPDS card #22-1141). The main reaction products in the soaked (bare or coated) $LiCoO_2$ are Co_3O_4 and Co_2O_3. Li_2CO_3 and lithium alkyl carbonate species have been observed on solvent- and electrolyte-soaked $LiNi_{1/3}Co_{1/3}Mn_{1/3}O_2$ particles in our previous report.[70] However, it seems that no authors have reported the formation of Co_2O_3 and Co_3O_4 in solvent- or electrolyte-soaked $LiCoO_2$ or electrochemically cycled $LiCoO_2$ electrodes. Based on the relative intensity of the strongest diffraction peaks of Co_2O_3 and Co_3O_4, their contents are comparable to that of the residual $LiCoO_2$ in the soaked material though the exact content of each cannot be determined by XRD. This agrees well with the above ICP analysis.

Calculation to the XRD patterns of the soaked materials indicates that the a values of the bare and Al_2O_3-coated nano-$LiCoO_2$ are 2.8188Å and 2.8197Å, respectively while their c values are 14.0364Å and 14.0780Å, respectively. Compared with the lattice parameters of nano-$LiCoO_2$ prior to soakage (a=2.8214Å, c=14.0379Å), these parameters do not show obvious changes. This implies that a *core-and-shell* structure is formed in the $LiCoO_2$ particle. The shell is Li-rare while the core is still Li-rich. The lithium content in the core is similar to that of commercial $LiCoO_2$. Suppose that there is no lithium ion in the shell (composed of Co_2O_3 and Co_3O_4) and the core is still $LiCoO_2$, the thickness of the shell should be over 15% of the particle diameter for the coated $LiCoO_2$. In comparison the thickness of the shell of bare $LiCoO_2$ is much smaller.

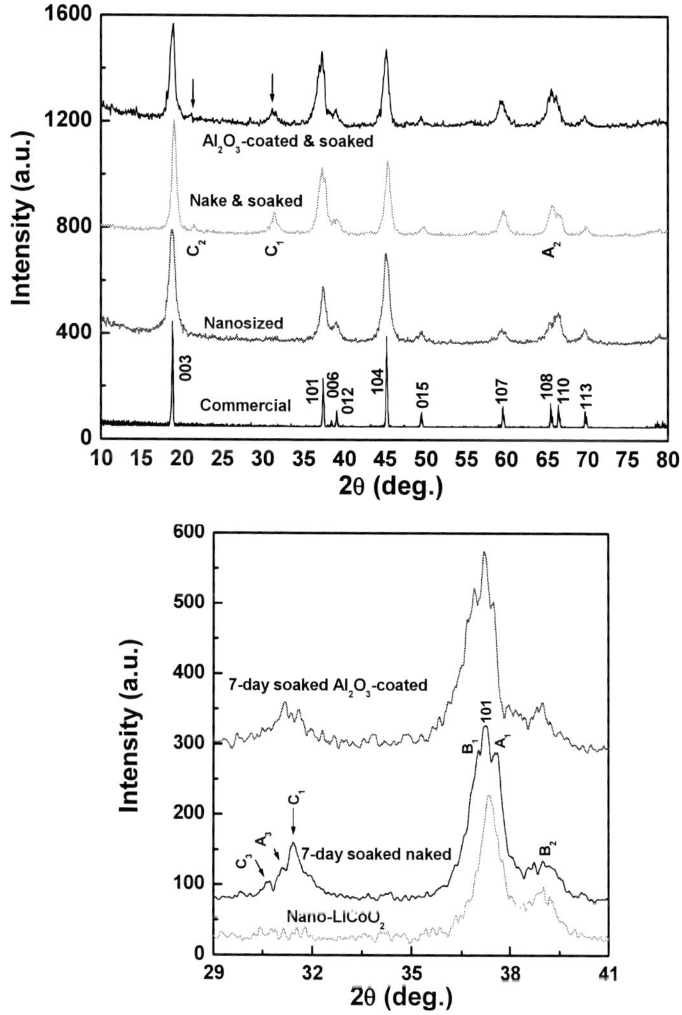

Figure 22. Comparison of (a) the XRD patterns of commercial $LiCoO_2$, high-energy ball-milled $LiCoO_2$, solvent-soaked nano-$LiCoO_2$ and Al_2O_3-coated nano-$LiCoO_2$ and (b) selected XRD patterns of (a): *A* for Co_3O_4, *B* for Co_2O_3 and *C* for Li_2CO_3. The footnote (1, 2 or 3) is for the first three strongest diffraction lines of each product in the order of their intensity from high to low.

It might be argued that the signal/noise (S/N) ratio of the above XRD patterns is not high enough. Therefore laser Raman spectra of the solvent-soaked $LiCoO_2$ with high S/N ratios are recorded (Figure 23) to help recognize the generated species in the material. Here the Raman spectra of pure $LiCoO_2$, Co_3O_4 and Co_2O_3 are compared with that of the solvent-soaked materials. Again, all the characteristic bands of Co_2O_3 and Co_3O_4 are observed in the soaked nano-$LiCoO_2$, verifying clearly that Co_2O_3 and Co_3O_4 are the main products in the material. Comparing the relative intensities of the spectral peaks of Co_2O_3 and Co_3O_4 in the material, it is seen that the content of $LiCoO_2$ is lower in Al_2O_3-coated $LiCoO_2$ than in bare $LiCoO_2$, consistent with the ICP analysis. Meanwhile, it is also found that the content ratio of Co_2O_3/Co_3O_4 is higher in Al_2O_3-coated $LiCoO_2$ than in bare $LiCoO_2$. No Li_2CO_3-related Raman peaks are observed (the characteristic Raman peak of pure Li_2CO_3 is at $1090 cm^{-1}$), implying that the content of Li_2CO_3 is much lower than that of Co_2O_3 and Co_3O_4 in the soaked material, consistent with the XRD results.

Some humps are observed in the Raman spectrum of soaked nano-$LiCoO_2$ between 800 and 1400cm^{-1}. They are related to the species in the surface film on nano-$LiCoO_2$ particles. However, as the samples have been exposed to air for a long time during which the air- and/or moisture-sensitive surface film might have been damaged, no further analysis is carried out on its composition.

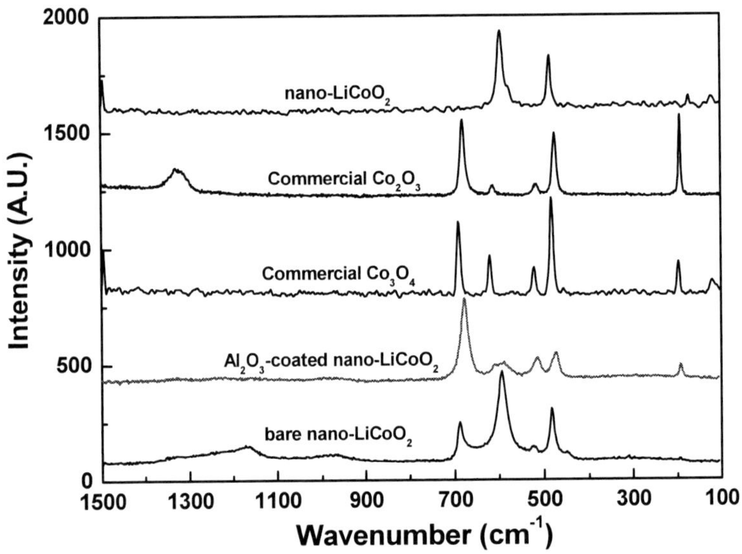

Figure 23. Comparison of the Raman spectra of commercial, nanoscaled and solvent-soaked nano-$LiCoO_2$ and other cobalt oxides.

It is difficult to determine the thickness of the surface film based on the above data. As seen in Fig.23, the relative intensities of the humps (in respect to that of residual $LiCoO_2$) are higher on Al_2O_3-coated nano-$LiCoO_2$ than on bare nano-$LiCoO_2$, indicating that the surface film on the Al_2O_3-coated $LiCoO_2$ is thicker than on the bare $LiCoO_2$. In this sense, the surface film on $LiCoO_2$ should not be called a passivation film because it cannot protect $LiCoO_2$ from further structural degradation. In addition, as will be seen at the end of this section (Figure 30), the average thickness of the surface film on commercial $LiCoO_2$ soaked in EC/DMC mixture is well over 10 nm. The surface film on nano-$LiCoO_2$ should be thicker than on commercial $LiCoO_2$ because the reactivity of nano-$LiCoO_2$ is higher than that of micro-$LiCoO_2$.

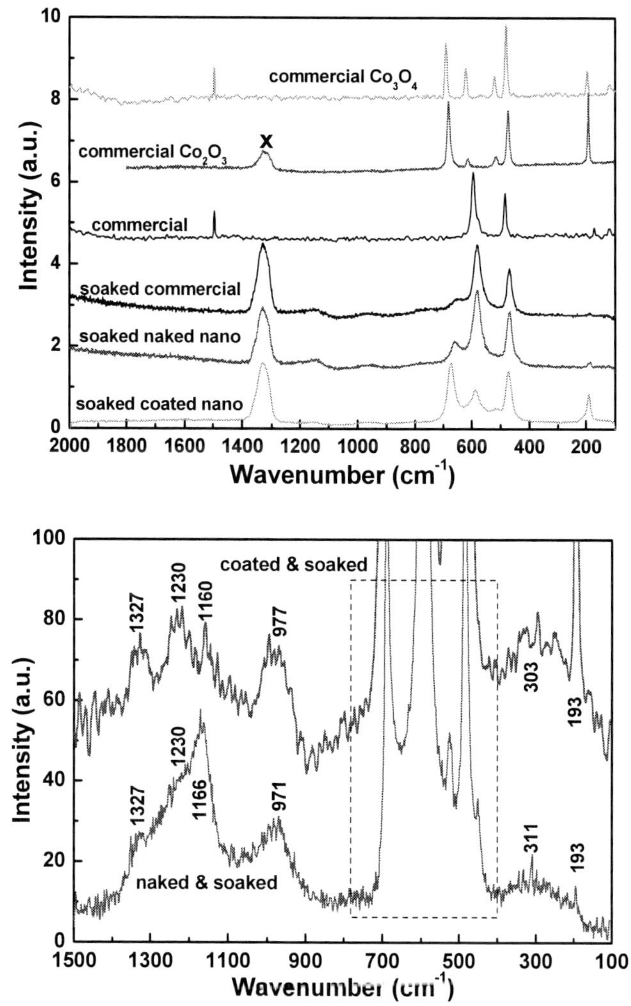

Figure 24. Comparison of the laser Renishaw Raman spectra of the solvent-soaked $LiCoO_2$ exposed to air for 14 days and pure $LiCoO_2$, Co_3O_4 and Co_2O_3.

The Raman spectra of solvent-soaked $LiCoO_2$ exposed to air are shown in Figure 24 to help identify the produced species in the material. Exposing the generated solid species to air will probably damage the surface film and render the $LiCoO_2$ particles to reappear to the laser. Here the Raman spectra of commercial $LiCoO_2$, Co_3O_4 and Co_2O_3 are recorded and compared with that of the solvent-soaked materials (Fig.24a). Again, all the characteristic bands of Co_2O_3 and Co_3O_4 are observed in the soaked materials, indicating clearly that Co_2O_3 and Co_3O_4 are the main produced components in the material, consistent with the results of XRD analysis. Comparing the relative intensities of the spectral peaks of Co_2O_3 and Co_3O_4 in the material, we can see that the content of $LiCoO_2$ is lower in Al_2O_3-coated $LiCoO_2$ than in bare $LiCoO_2$. Meanwhile, it is also found that the content ratio of Co_2O_3/Co_3O_4 is higher in Al_2O_3-coated $LiCoO_2$ than in bare $LiCoO_2$, verifying the above proposal.

The Raman spectra of soaked $LiCoO_2$ are magnified so that their weak bands can be analyzed (Fig.24b). The contours of these two spectra look quite different from each other.

However, considering that the positions of most of the observed bands in one spectrum are similar to their counterparts in the other, the composition of the chemical species on these two solvent-soaked nano-$LiCoO_2$ materials are qualitatively the same. Neither EC nor DMC is detected in the samples because their characteristic Raman bands (the strongest 893 and 712cm^{-1} bands of EC, for example) are not observed in Fig.24b.

Li_2CO_3 is an important component of the SEI film on different electrodes charged or discharged to various voltages. It is evidently detected in all the above solid samples by FTIR spectroscopy and also detected in the above XRD patterns before exposed to air. However, as Li_2CO_3 is known stable, it is amazing that no traces of Li_2CO_3 are detected in the solid samples exposed to air. Nor is ROLi detected in the Raman spectrum. The reason is not clear at present.

$ROCO_2Li$ actually represents a series of lithium alkyl carbonates and has been regarded as the main component of the SEI film. These species are not detected by XRD because they are usually amorphous. It seems that the soakage product contains oxygen-involved ring-structured (ring breathing at 939cm^{-1} and ring stretching at 1230cm^{-1}) and carboxylic (C-O stretching at 1160cm^{-1}) groups as well as C-C skeletons (skeleton deformation at 311cm^{-1} and CH deformation at 1327cm^{-1}).[71] However, it is not of much help to recognize the composition of the surface film in Fig.24 because the film might have been damaged due to exposure to air during which the air- and/or moisture-sensitive components in the surface film have been changed and cannot reflect the real composition of the surface layer any longer.

With the above proposed *core-and-shell* structure of the soaked $LiCoO_2$ particles, the formation of a very thick surface film on the particle can be further discussed. Suppose that the surface film formed by spontaneous reaction is porous, the solvent can reach the surface of $LiCoO_2$ particle and get decomposed continuously. With the increase of the thickness of the shell composed of Co_2O_3/Co_3O_4, the thickness of the surface film also increases. On the other hand, growth of the surface layer and the shell makes it difficult for the Li^+ ions to transport from the *Li*-rich core to the surface and for the solvent molecules to penetrate the surface film and reach the particle surface, the rate of the solvent-electrode reaction becomes slow and finally the reaction terminates. This process does not apply to the electron tunneling effect. Therefore the thickness of the surface film can be well over that of the distance limit of electron tunneling effect.

3.3.3. Degradation of Electrochemical Performance of $LiCoO_2$ Soaked in EC/DMC

Figure 25 shows the charge/discharge profiles of the non-soaked nano-$LiCoO_2$ and solvent-soaked materials, bare and Al_2O_3-coated. Only a capacity of about 60mAh/g is obtained for bare $LiCoO_2$ and 40 mAh/g for Al_2O_3-coated nano-$LiCoO_2$ in their first cycles, much smaller than that of the non-soaked nano-$LiCoO_2$ (80mAh/g). This is understandable as neither Co_2O_3 nor Co_3O_4 is electrochemically active for lithium storage above 2.5V. In the subsequent cycles, the capacity of the soaked material decreases with cycling. An "extra" discharge plateau is observed in bare $LiCoO_2$ at *ca.* 3.4V. This plateau has been observed in commercial $LiCoO_2$ experiencing many deep charge/discharge cycles (between 4.7V and 2.5V).[21] It is interesting that the capacity corresponding to this plateau (*ca.* 20 mAh/g) roughly equals to the capacity difference between the bare and Al_2O_3-coated nano-$LiCoO_2$. Its origin needs to be further studied.

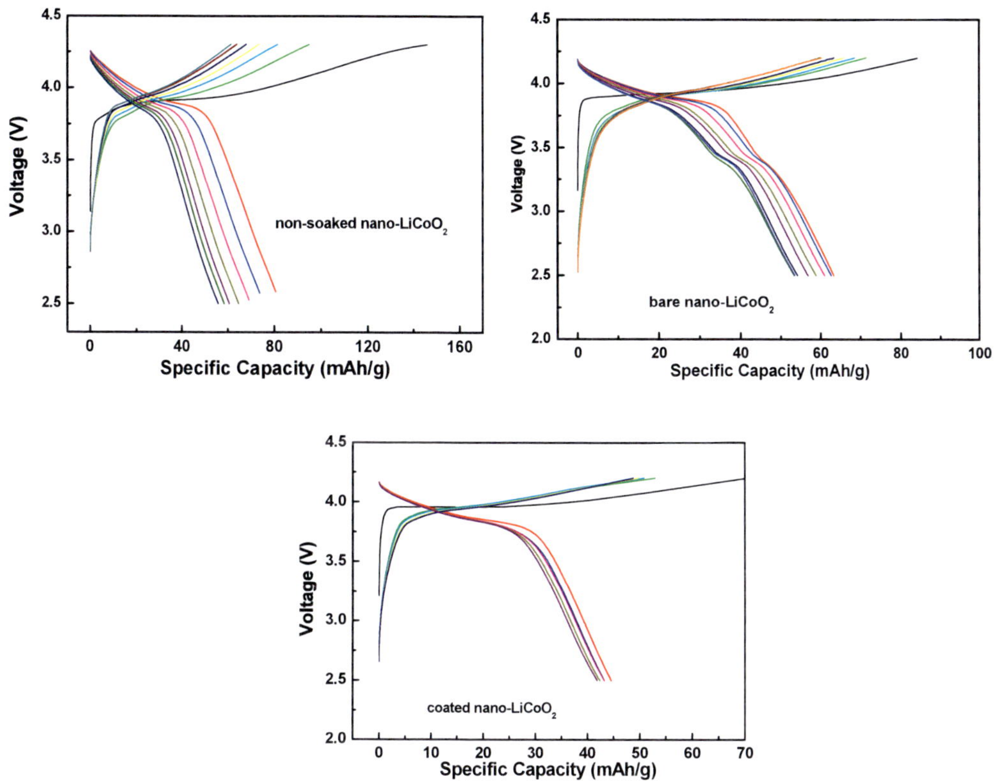

Figure 25. Charge/discharge profiles of as-prepared (non-soaked bare) nano-$LiCoO_2$ (a), solvent-soaked bare nano-$LiCoO_2$, (b) and Al_2O_3-coated nano-$LiCoO_2$ (c).

Lithium ion loss is severe in the above solvent-soaked nano-$LiCoO_2$. However, it is worth pointing out that lithium escape might be not as obvious in a practical lithium ion battery or lithiated positive electrode material as shown here for the following reasons. Their specific surface areas are usually smaller than that of the nano-$LiCoO_2$. Therefore the contact area of the solvent or electrolyte with the material is much smaller, leading to less obvious reaction between $LiCoO_2$ and the solvent. This is evident in Fig.20. However, for some positive electrodes using nanoscaled active materials, especially those obtained from soft chemistry for power-type batteries, the spontaneous reactions have to be considered upon designing a battery. In addition, lithium dissolution might be less significant due to the less difference of Li^+ ion contents in the material and in the electrolyte. This will be further discussed in the following paragraphs. Finally the volume ratio of $LiCoO_2$ to the solvent is much bigger than in a practical battery and the mixture is continuously stirred. Anyway, the above phenomena demonstrate clearly that $LiCoO_2$ is not stable in EC/DMC-based electrolyte and the extensively employed solvent for the electrolyte is not completely without problem. In addition, surface coating cannot prevent lithium escape from the $LiCoO_2$ lattice. These problems can irreversibly degrade the performance (especially the calendar life) of a battery with nanosized electrode materials.

3.3.4. $LiCoO_2$ Soaked in Electrolyte

In order to determine the possible decomposition of the lithium salt and the impact of Li^+ ion content in the liquid (solvent and electrolyte) on the surface reactions and on the dissolution of Li^+ ions from various $LiCoO_2$, commercial and nano-$LiCoO_2$, bare and Al_2O_3-coated, are soaked in 1mol/L $LiPF_6$ in EC/DMC (1:1 by volume). Figure 26 shows the MS spectra of the gas species generated on commercial $LiCoO_2$. Clearly CH_4, CH_3-CH_3, CH_2=CH_2, CO_2, CO and H_2O are the main products. The same gas species are detected in (Al_2O_3-coated) nano-$LiCoO_2$ soaked in the electrolyte. However, the corresponding contents of the generated gases are lower than in the solvent. This can be seen by comparing the maximum *cps* of each species with that of the corresponding species generated in solvent-soaked samples. Oxygen is not detected in the electrolyte-soaked samples, probably due to its low content in the gas. GC-MS spectra of the liquids are also recorded and again substances with C-O-C skeleton are detected.

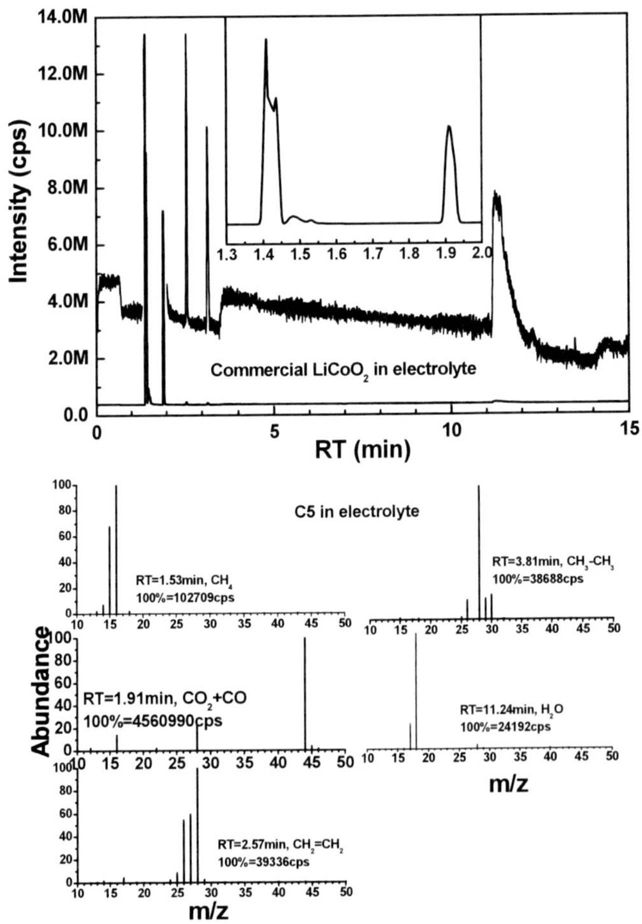

Figure 26. Typical MS spectra of the gases generated in electrolyte-soaked commercial $LiCoO_2$.

Figure 27 compares the FTIR spectra of fresh electrolyte and the liquid obtained by soaking Al_2O_3-coated $LiCoO_2$ in the electrolyte. Different from the FTIR spectral results of $LiCoO_2$ soaked in the solvent, broad and strong bands are observed at 3550 and 1635cm^{-1}. As the water content in the fresh electrolyte is rather low and the solid was heat-treated prior to being transferred into the container, these two bands are attributed to the water as a reaction product in the electrolyte. Strong bands around 3000cm^{-1} are from CH_2 or CH_3 groups of esters.[53] Weak bands between 2800 and 2200cm^{-1} are due to the CH vibrations of the SEI components.[53] A group of new bands are observed below 1500cm^{-1} in comparison with that listed in Table 1. According to literature, the 1406cm^{-1} band is from $(ROCO_2Li)_2$.[53] The 1196cm^{-1} band is assigned to C-O stretch[53] or C-O-C stretch in formate esters in the SEI film.[46] The band at 559 cm^{-1} is due to the overlapping of $LiPF_6$ band (see the spectrum of fresh electrolyte in Fig.27) and the Li-O vibration of ROLi.[53] No bands related to P-F bonding[72] are detected in the liquid other than that of $LiPF_6$ itself.

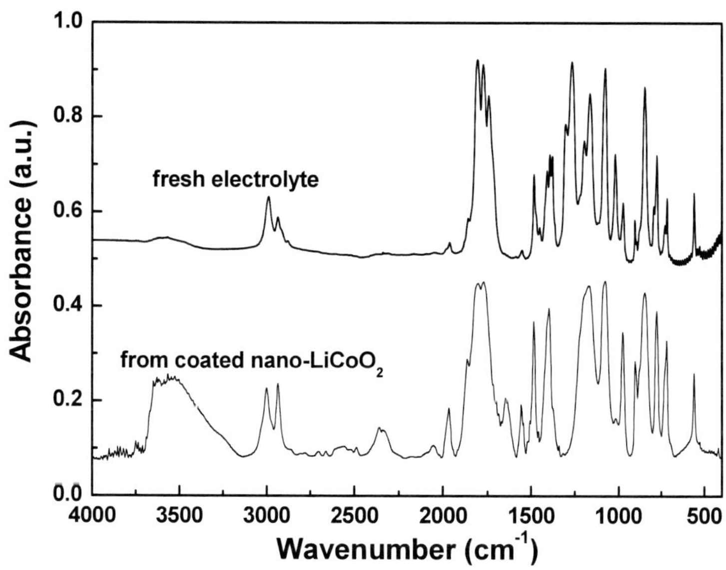

Figure 27. Comparison of the FTIR spectra of fresh electrolyte and the liquid from electrolyte-soaked Al_2O_3-coated nano-$LiCoO_2$.

Figure 28 shows the FTIR spectra of fresh solid species on various $LiCoO_2$ soaked in the electrolyte. It is seen that the IR spectra of the surface species on different substrates are similar to each other, indicating that surface modification does not influence the composition of the surface film on $LiCoO_2$. This is consistent with the observation of the surface film on $LiCoO_2$ soaked in the solvent. In addition, LiF is detected in the solid (the 775cm^{-1} band), indicating that $LiPF_6$ is decomposed during storage. However, no vibration bands related to P-F bonding are detected.

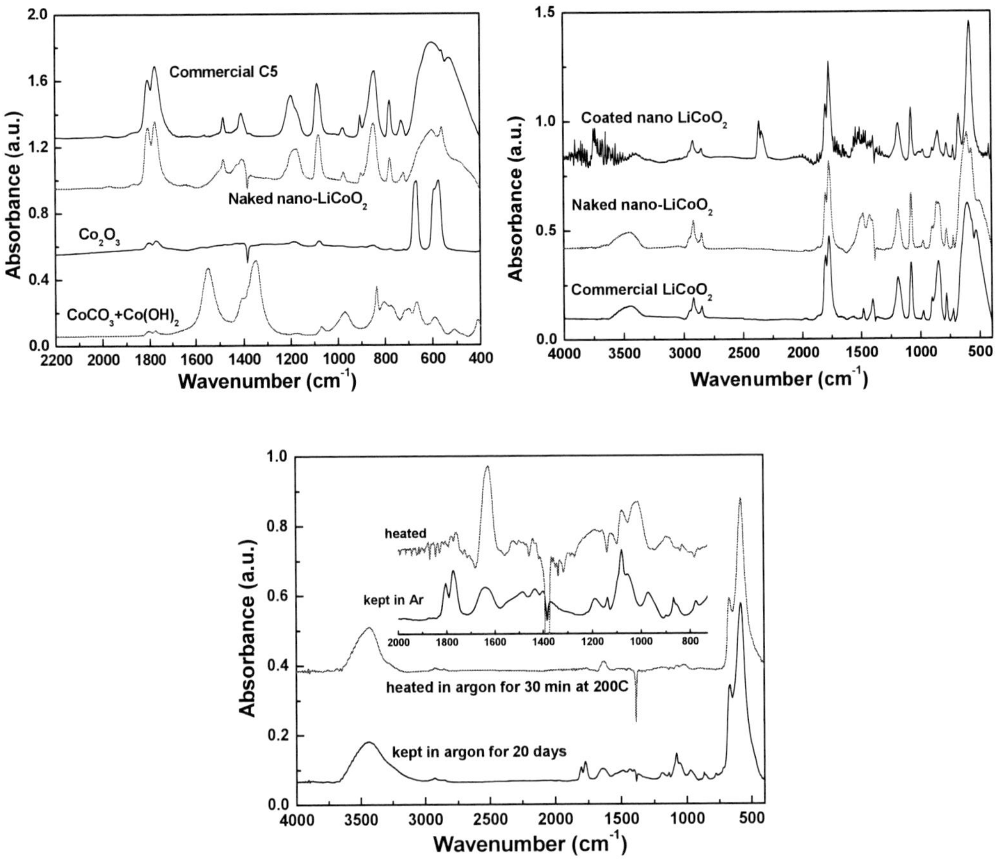

Figure 28. FTIR spectra of fresh solid species on various $LiCoO_2$ soaked in the electrolyte.

Figure 29 shows the Raman spectra of $LiCoO_2$ materials covered with surface films. The dependence of the thickness of the surface film on the specific surface area of the material in the electrolyte is less obvious than in the corresponding $LiCoO_2$ materials soaked in the solvent (Fig.29a). In addition, the thickness of the surface film on corresponding $LiCoO_2$ in electrolyte is estimated to be smaller than on $LiCoO_2$ soaked in the solvent because the characteristic bands of $LiCoO_2$ can easily be recognized at 592 and 483cm^{-1}, respectively. Different from solvent-soaked $LiCoO_2$, lithium dissolution is significantly suppressed in the electrolyte, implying that the concentration difference of the Li^+ ions in and out of the $LiCoO_2$ particle is one of the driving forces for lithium dissolution. As a result, the structural degradation of $LiCoO_2$ is less severe in the electrolyte than in the solvent. Similar to the case of $LiCoO_2$ soaked in the solvent, the bare $LiCoO_2$ is mainly degraded to Co_3O_4 while the Al_2O_3-coated nano-$LiCoO_2$ is mainly degraded to Co_2O_3. Finally, it seems that the influence of $LiCoO_2$ particle size on the structural degradation is also less pronounced in the electrolyte than in the solvent.

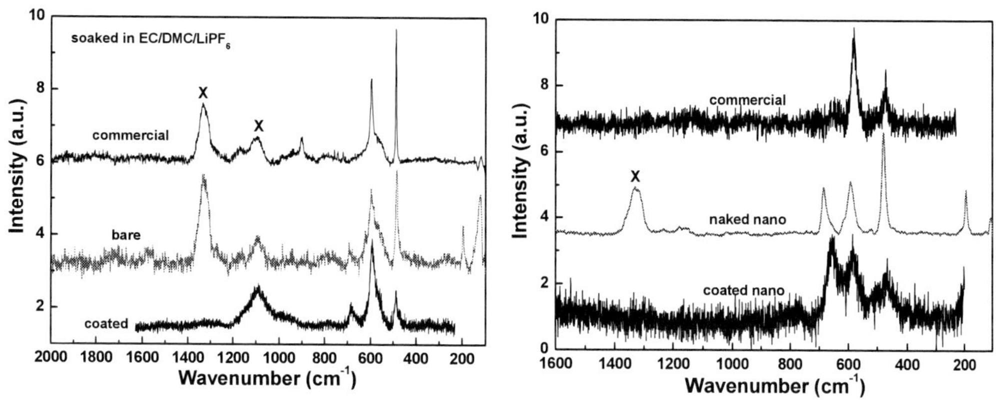

Figure 29. Renishaw Raman spectra of $LiCoO_2$ materials covered with surface films kept in dry Ar for 7 days.

Figure 30 shows the surface morphology of commercial $LiCoO_2$ soaked in the electrolyte. Clearly the soakage-induced corrosion is still obvious but significantly suppressed. Grid-like as well as stripe-like structures are observed (Fig.30a). Meanwhile, a thick surface film can be observed on the particle surface. Amazingly, layer cleavage is observed in the particle as shown in Fig.30b. This is direct evidence of Li^+ extraction (lithium escape leads to strong repulsive forces between the neighboring O-Co-O slabs). Due to severe aggregation and small size of the nano-$LiCoO_2$ particles, the surface morphology of bare and coated nano-$LiCoO_2$ is not available. However, considering the strong spontaneous reactions between the positive electrode material and the solvent/electrolyte, the structural degradation of nano-$LiCoO_2$ might be more severe.

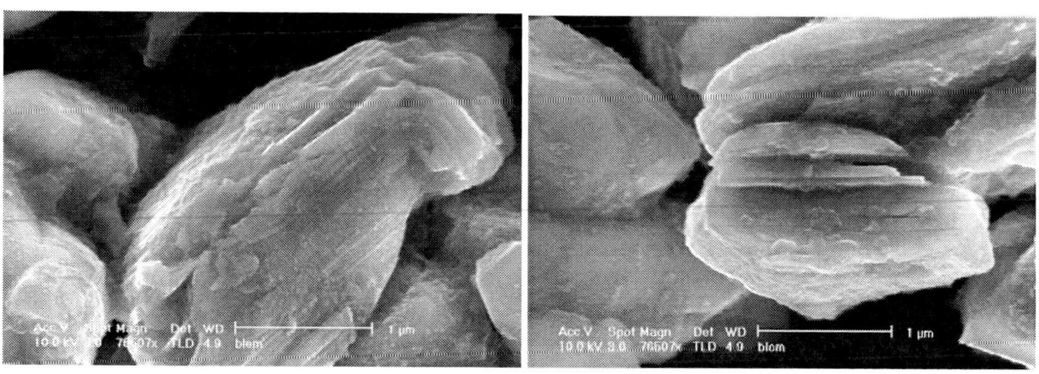

Figure 30. Surface morphology (a) and layer cleavage (b) of commercial $LiCoO_2$ particles soaked in 1M $LiPF_6$ dissolved in EC/DMC for 7 days.

3.3.5. Solvent decomposition and Li^+ dissolution from $LiCoO_2$

The following reactions have been proposed and confirmed by previous authors upon soaking $LiCoO_2$ in the solvent and electrolyte.

$2H_2O + 2e^- + 2Li^+ \rightarrow 2LiOH\downarrow + H_2\uparrow$[73]

$2EC+2e^{-}+2Li^{+}\rightarrow (CH_2OCO_2Li)_2\downarrow+CH_2CH_2\uparrow$[74]

$2EC+2e^{-}+2Li^{+}\rightarrow LiCH_2CH_2OCO_2Li\downarrow$[46]

$ROLi+CO_2\rightarrow ROCO_2Li\downarrow$[75]

$2RCO_3Li+H_2O\rightarrow Li_2CO_3\downarrow+2ROH+CO_2\uparrow$[76]

$2CO_2+2Li^{+}+2e^{-}\rightarrow CO\uparrow+Li_2CO_3\downarrow$[73]

$LiPF_6\rightarrow LiF\downarrow+PF_5$[76]

$PF_5+H_2O\rightarrow 2HF+PF_3O$[76]

$PF_3O+Li^{+}+e^{-}\rightarrow LiF\downarrow+Li_xPOF_y\downarrow$[76]

As for the current system, the Co^{3+} ion in $LiCoO_2$ attacks and oxidizes the carbonate groups of the solvent molecules due to its acidic/nucleophilic properties and is reduced to Co^{2+}. This will bring about two effects. On one hand, the reduction of Co^{3+} leads to the degradation of $LiCoO_2$ to Co_2O_3 and Co_3O_4, and the dissolution of some Li^{+} and Co^{2+} ions into the liquid. On the other hand, the Li^{+} ions in the liquid are solvated and combine with the Co^{3+}-oxidized solvent molecules to form RCO_2OLi. The great difference of the Li^{+} concentrations in $LiCoO_2$ and in the solvent facilitates the above reaction. As the Li^{+} ions leave the lattice, it is understandable that a *core-and-shell* structure is the most possible for the soaked $LiCoO_2$. These processes will lead to the formation of $(ROCO_2Li)_2$, Li_2CO_3 and other small molecules. This is quite similar to the formation of SEI film on the positive and negative electrodes. McLarnon *et al.*[48] studied the spontaneous degradation of $LiMn_2O_4$ thin films in pure DMC and in EC-DMC-$LiPF_6$ electrolyte, respectively. They attributed the formation of surface films on $LiMn_2O_4$ to surface oxidation of DMC by Mn^{4+}. Therefore the SEI film formation was accompanied by conversion of the original $LiMn_2O_4$ into Mn_2O_3 and λ-MnO_2 in pure DMC. This might also be true of the present case.

The presence of Al_2O_3 on the surface of coated nano-$LiCoO_2$ enhances the solvent oxidation and suppresses the $LiCoO_2$ oxidation induced by Li^{+} extraction. As a result, Co_2O_3 is the dominant degradation product on Al_2O_3-coated $LiCoO_2$. As for the bare $LiCoO_2$, spontaneous redox reaction occurs: Co^{3+} is reduced to Co^{2+} while oxygen (O_2) is released. Therefore, the main degradation product of bare $LiCoO_2$ is Co_3O_4. With the presence of lithium salt in the solvent, the Li^{+} concentration difference is reduced in and out of the $LiCoO_2$ particle. Therefore the above reactions are suppressed.

Based on the above analysis, the following reactions are supposed to occur between the carbonate solvent and $LiCoO_2$. However, it is worth emphasizing that these reactions may take place without any electrochemical treatment to the samples.

$$DMC-2e^{-}+2Li^{+}\rightarrow CH_3CH_3\uparrow+Li_2CO_3\downarrow \tag{1}$$

$$DMC-e^{-}+Li^{+}\rightarrow \dot{C}H_3+CH_3OCO_2Li\downarrow \tag{2}$$

$$DMC-e^{-}+Li^{+}\rightarrow CH_3O\dot{C}O+CH_3OLi\downarrow \tag{3}$$

$$3RCO_3 + 3LiCo^{3+}O_2 \rightarrow 3ROCO_2Li + Co_2^{3+}Co^{2+}O_4 + O_2\uparrow \text{ (on bare } LiCoO_2) \quad (4)$$

$$2LiCo^{3+}O_2 \rightarrow Co_2^{3+}O_3 + Li_2O \text{ (on } Al_2O_3\text{-coated } LiCoO_2) \quad (5)$$

$$Li_2O + H_2O \rightarrow 2LiOH \quad (6)$$

$$LiOH + HF \rightarrow LiF\downarrow + H_2O \quad (7)$$

$$Li_2CO_3 + HF \rightarrow LiF\downarrow + CO_2\uparrow + H_2O \quad (8)$$

$$Li_2O + HF \rightarrow LiF\downarrow + H_2O \quad (9)$$

*Eqs.*1 to 3 explain the spontaneous decomposition of the solvent on the surface of the positive electrode particles. *Eq.*4 actually explains why Al_2O_3 surface modification can improve the electrochemical performance of $LiCoO_2$. This suggestion is based on the findings of the GC-MS and the X-ray photoelectron spectroscopic study as shown in Section 4 of this chapter.[77] There, it was found that surface modification suppresses the formation of oxygen with high oxidizing power. Actually lithium extraction from lattice by chemical (as it is here) or electrochemical (during charging) driving force leads to the same effects: solvent decomposition, structural degradation and the formation of SEI film and oxygen. When there is no Al_2O_3 on $LiCoO_2$, it degrades to Co_2O_3 and Li_2O (*Eq.*5). *Eqs.* 6 to 9 are based on the fact that $LiPF_6$ is decomposed and HF is formed due to trace water in the electrolyte. Reactions of HF with various metal oxides will produce more water in the electrolyte.

The following reactions are supposed to take place upon soaking nano-$LiCoO_2$ in the solvent. The Co^{3+} ion in $LiCoO_2$ attack and oxidize the carbonate groups of the solvent molecules due to its acidic/nucleophilic properties and is reduced to Co^{2+}. This will bring about two effects. On one hand, the reduction of Co^{3+} leads to the degradation of $LiCoO_2$ to Co_2O_3 and/or Co_3O_4 and the dissolution of some Li^+ and Co^{2+} ions into the liquid. On the other hand, the Li^+ ions in the liquid are solvated and combine with the Co^{3+}-oxidized solvent molecules to form RCO_2OLi. The great difference of the Li^+ concentrations in $LiCoO_2$ and in the solvent facilitates the above reaction. As the Li^+ ions leave the lattice due to their large concentration gradient in and out of the $LiCoO_2$ particle, it is understandable that a *core-and-shell* structure is the most possible for the soaked $LiCoO_2$. Out of the shell is the surface film composed of various Li-containing compounds.

3.4. Summary

The above studies show that storage of commercial and nanosized $LiCoO_2$ in the electrolyte or its solvent leads to Li^+ ion dissolution from the lattice and degradations of $LiCoO_2$ in structure. The content of generated Co_2O_3 on solvent-soaked Al_2O_3-coated $LiCoO_2$ is higher than on bare nano-$LiCoO_2$. Dissolution of lithium also leads to degradation in specific capacity, reversibility and cyclability of the material. Lithium dissolution and $LiCoO_2$ structural degradation are obviously suppressed in 1mol/L $LiPF_6$/EC/DMC electrolyte. The driving force for lithium dissolution is attributed to the great concentration difference in and out of a $LiCoO_2$ particle.

Meanwhile, direct contact of $LiCoO_2$ with the non-aqueous electrolyte and the solvent results in obvious decomposition of the latter. The precipitates from the decomposed solvent form a thick layer of surface film. The thickness of the surface film, estimated to be well over 10 nm, is much larger than previously proposed. This is an interesting finding because it was previously believed that the growth of the surface film on the electrode terminates when the film thickness reaches the distance limit of electron tunneling. As the electron tunneling distance is only a few anstroms (Å), it will be interesting to find out the real mechanism of the surface film growth on the positive electrode material. Based on the experimental results, kinetics effect is believed to be responsible for the termination of the surface film growth.

Accompanied with the precipitation of surface film on the electrode material, different gases are generated due to solvent decomposition, such as CH_4, CH_3-CH_3, CH_2=CH_2, CO, CO_2, O_2 and H_2O. The strong reaction of $LiCoO_2$ with the solvent molecules are attributed to the strong electrophilic effect of the carbonate solvent and the nucleophilic effects of the Li^+ ions in $LiCoO_2$. Surface modification to $LiCoO_2$ cannot prevent the dissolution of Li^+ ions from its lattice. However, it suppresses the reduction of $LiCoO_2$ to Co_3O_4, the escape of some Co^{2+} ions and the formation of oxygen. Suppression of oxygen formation will also be beneficial to enhance the structural stability of the material and reduce the level of safety concerns of a lithium ion battery. This might be an important reason for the improved structural stability of surface-modified $LiCoO_2$ as positive electrode materials at high charge voltage or deep delithiation.

As surface films have been detected on a number of positive electrode materials while the decomposition of the lithium salt are mostly reported on electrode with applied potentials, it is supposed that the decomposition of the solvent and the decomposition of the lithium salt are two separate processes in a dry $LiPF_6$-based electrolyte. The decomposition of carbonate-based solvent molecules can take place spontaneously at any potential on $LiCoO_2$ while $LiPF_6$ can be decomposed due to trace water in the electrolyte and in the glove box. In order to alleviate the spontaneous reaction between the solvent molecules and the positive electrode materials, non-nucleophilic atoms should be employed to modify the surface of the positive electrode materials. Meanwhile, atoms that help to suppress the formation of oxygen with high oxidizing power and stabilize the crystalline structure of the material should also be used. That is, the surface modification should be multi-layered or composite. The decomposition of $LiPF_6$ and the formation of HF can induce a series of severe problems including corrosions to the active electrode material, the surface coating metal oxide layer and some of the components in the SEI itself. Therefore, the electrolyte in the glove box and the electrode materials should be sufficiently dry so as to suppress the decomposition of $LiPF_6$ and the formation of strongly corrosive HF.

Finally before ending this section, we emphasize that the purpose of this work is to make a clearer picture to some basic questions put forward in the *Introduction* of this section rather than to demonstrate how severe the spontaneous reaction is between commercial $LiCoO_2$ and electrolyte. Although high-energy ball-milling will definitely create some microstructural defects in $LiCoO_2$ and its reactivity with the electrolyte/solvent is “magnified”, answers to the basic questions are still important and will be seen in the 5th section of this chapter on nano-$LiCoO_2$ prepared by molten-salt method.

4. Evolution of Electronic Structure of Cathode Materials with Charge Voltages

4.1. Introduction

With decades of extensive study, it is realized that the surface chemistry, morphology and surface species of the cathode and its interface with the electrolyte have significant influence on the electrochemical performances of a lithium ion battery, such as its reversibility and safety. Surface modification has proved effective in improving the electrochemical performances of the cathode materials.[30,32,33] However, investigations of why the coating layer and the interface of coating/coated materials can improve the electrochemical performance of the cathode material have been rare. The above study shows that surface coating cannot prevent the dissolution of Li and Co from $LiCoO_2$ and that the presence of coating material promotes the corrosion of $LiCoO_2$ to some extent. Previous studies on commercial and Al_2O_3-coated $LiCoO_2$ by *in situ* synchrotron XRD during cycling in this Laboratory indicated that surface coating does not suppress the phase transition during cycling. Rather it makes the phase transition easier.[78] Therefore, the reason for the performance improvement of surface coated $LiCoO_2$ is still unclear.

This section will compare the electronic structures of commercial and nano-MgO coated $LiCoO_2$ charged to various voltages and make some basic explanations to the improved electrochemical performances.

4.2. Experimental

Test cells were assembled as in the above sections and charged to various voltages. In order to have the electrode charged to the required open circuit voltages (OCVs), the cells were first charged to preset voltages at a constant current (<0.1mA/cm^2) and then potentiostatically charged until the nominal charge current faded below 0.1μA. The voltage values shown in this section were picked up after the cell was disconnected from the tester and stored for more than 24 hours.

The test cell was then dissected and the working electrode was removed in the glove box. The electrode was soaked in and then rinsed repeatedly with pure DMC to remove the electrolyte. The washed electrode was stored in a vacuumed mini-chamber of the glove box to have DMC evaporated at room temperature. Nano-MgO powder was prepared by filtering and washing the reaction product ($Mg(OH)_2$ gel) of $MgCl_2$ with NH_4OH in distilled water three times. Then the gel was heated at 600°C for 2 hours in air to get the nano-MgO powder. The washing and drying processes for nano-MgO powder, commercial $LiCoO_2$ and MgO surface-modified $LiCoO_2$ ($MgO/LiCoO_2$) electrodes were the same as for the charged electrodes, after they were soaked in the electrolyte for an hour.

All the above operations were carried out in argon atmosphere. When everything was ready for the XPS instruments, the containers were opened and the samples were transferred into the vacuum chambers of the instruments and the chambers were vacuumed immediately. The exposure time of the sample to air was less than 10 seconds. XPS spectra were collected on an ESCALAB5 (VG Scientific; energy resolution: 0.1eV) with a non-monochromatic Mg

Kα radiation (1253.6eV). Before measurements, the XPS samples were sputtered with Ar^+ beam (2 KeV, 40μA) for 10 minutes to remove the SEI layer.

4.3. Electronic Structures of Cathode Materials at Different Voltages

Thomas[79] proposed that the electrolyte decomposition on charged $LiCoO_2$ surface is induced by the strong oxidizing power of Co^{4+} cations. However, Montoro *et al*[80] reported that the Co ions are in a trivalent Co^{3+} low-spin state in $LiCoO_2$ and remain mostly unaffected by Li^+ extraction in chemically delithiated $LiCoO_2$. On the other hand, the O^{2-} ions in $LiCoO_2$ are partially reduced upon Li deintercalation. Figure 31 shows the XPS spectra of commercial $LiCoO_2$ and MgO/$LiCoO_2$ charged to various voltages. The spectra correspond to Co 2p →Co 3p transitions and are dominated by multiplet effects. The Co 2p spectrum is split by the spin-orbital interaction into Co $2p_{1/2}$ and Co $2p_{3/2}$ regions. In turn these regions are further split by Co 2p-3d interaction and crystal field effects. The shapes of the spectra are directly related to the ground state of the Co ions.[81] Only Co 2p bands corresponding to trivalent Co^{3+} are observed in commercial and MgO/$LiCoO_2$ before electrochemical treatment, representative of the Co $2p_{3/2}$ at 780.5 eV and Co$2p_{1/2}$ at 795.7 eV. Another two Co 2p bands are observed at 786.6 and 803.2eV on both the commercial and the MgO/$LiCoO_2$ cathodes after charging the cathodes to a certain voltage. These two new bands might be responsible for the presence of tetravalent Co^{4+} ions in the charged electrodes. Interestingly these two bands appear at rather low charge potential (4.1V, corresponding to the extraction of ca. 0.3mole of Li from $LiCoO_2$) and remain there until most of the Li^+ ions are extracted (4.8V *vs* Li/Li^+). It is surprising that the detected oxidation states of Co cations in commercial $LiCoO_2$ and MgO/$LiCoO_2$ is independent of the charge voltage, consistent with the investigation results of Montoro *et al.*[80]

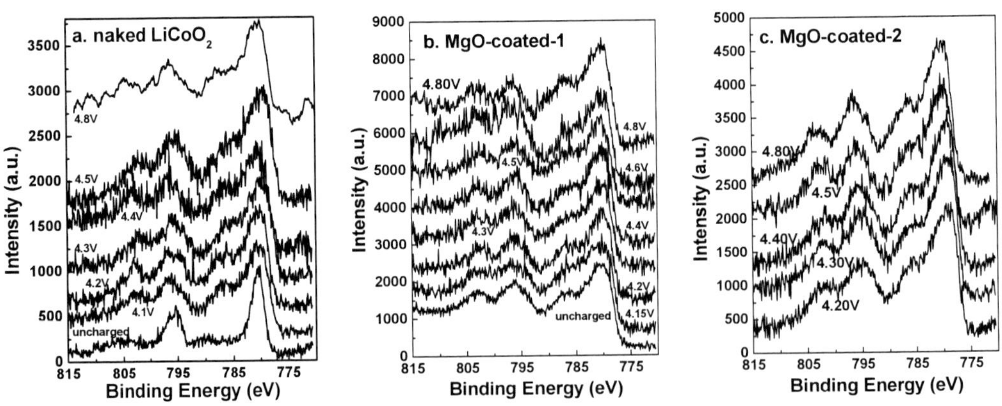

Figure 31. Comparison of the Co2p spectra of commercial (a) and MgO-coated (b) $LiCoO_2$ charged to various voltages. (c) MgO-coated $LiCoO_2$ after sputtering.

An explanation is that the XPS spectra are only sensitive to a very small depth (about 20 Å at most) under the surface of a $LiCoO_2$ particle. Further sputtering the electrode surface is expected to remove the surface species and detect the species deeper under the surface. Fig.31c demonstrates that the relative intensity of the Co^{4+} band slightly increases with the charge voltage compared with that of Co^{3+}. Notice that the $LiCoO_2$ particles are not in the same geometric plane and sputtering the electrode surface can only remove the surface species on some of the $LiCoO_2$ particles. It is understandable that the variation of the relative intensity of Co^{4+} to Co^{3+} is not as significant as expected, with the charge voltage.

The difference of the O 1s spectra of commercial and coated $LiCoO_2$ and their variation with the charge voltage are much more obvious than in the Co 2p spectra. As shown in Figure 32a, the electronic structure of oxygen varies steadily (at 529.4 eV in uncharged commercial $LiCoO_2$ electrode) with the increase of the charge voltage and a new component appears at 532.6eV, corresponding to oxygen atoms with stronger oxidizing power. The content of such oxygen atoms increases with the charge voltage and becomes dominant in the electrode at high voltages. This is opposite to the report of Montoro *et al.*[80] Oxygen atoms with higher binding energy (at 531.6eV) in the coated cathode, however, appear at lower charge voltage but their content increases more slowly than in commercial $LiCoO_2$ cathode (Figure 32b). These oxygen atoms cannot dominate the cathode material up to 4.8V. Moreover as the binding energy of these oxygen atoms is lower in charged MgO/$LiCoO_2$ cathode than in charged commercial $LiCoO_2$, their oxidizing power should also be weaker than those in commercial $LiCoO_2$ cathode. These results indicate that modifying the surface of $LiCoO_2$ particles can help to suppress the formation of oxygen atoms with stronger_oxidizing power and the decomposition of electrolyte. These will both increase the cycling efficiency of the cell and prolong its cycle life especially when it is deeply delithiated. In addition, suppression of the formation of oxygen atoms with high oxidizing power will improve the safety of these cell, especially at deeply delithiated state.

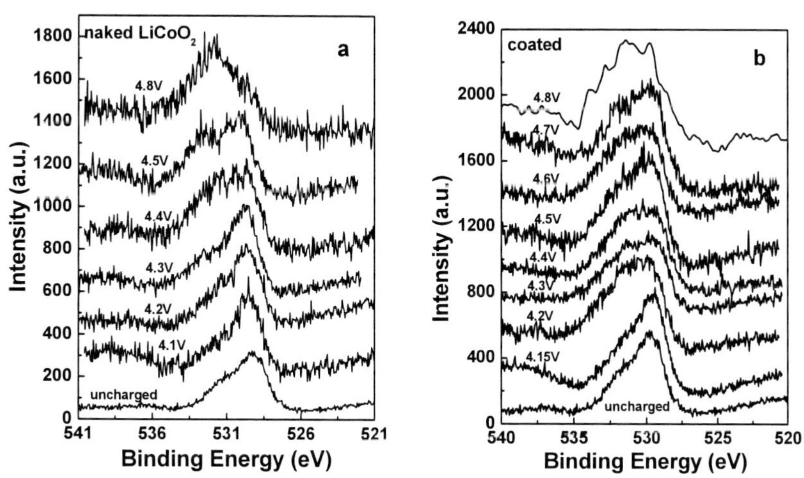

Figure 32. Comparison of O 1s spectra of commercial (a) and MgO-coated (b) $LiCoO_2$ charged to various voltages.

The Mg 1s spectra of the MgO/$LiCoO_2$ cathode charged to various voltages are also recorded and shown in Figure 33. The Mg1s peak is observed at about 50.4eV and does not shift with the charge voltage. The other peak at 60.0eV is assigned to Co3p.[82] As the Co cations cannot be expected to accumulate at the surface of a $LiCoO_2$ particle upon charging, it can be suggested that the content of the Mg cations or the thickness of the MgO coating decreases with increasing charge voltage. It would be interesting to find out whether the electrolyte decomposition on $MgO/LiCoO_2$ at high voltages is related to the decreasing thickness of the coating layer. Indeed it has been observed that the MgO coating is corroded on cycled $LiCoO_2$ cathodes in our previous study.[21]

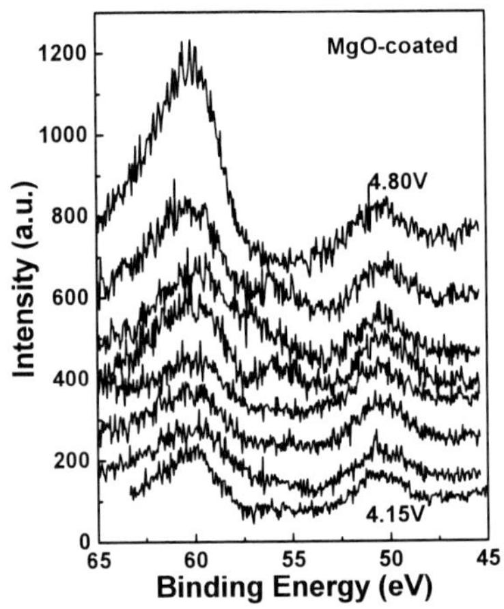

Figure 33. Mg1s and Co3p spectra of MgO coated $LiCoO_2$ charged to various voltages (from bottom to top: a. 4.15V; 4.2V; 4.3V; 4.4V; 4.5V; 4.6V; 4.7V and 4.8V).

5. New Understanding and New Methods of Performance Improvement of Cathode Materials

5.1. Introduction

Surface coating is an effective method to improve the cycling stability of cathode materials. Some authors reported that surface coating can even improve the rate performance [83,84,85] and thermal safety of the $LiCoO_2$ cathode material.[35,84] Various metal oxides such as MgO,[21] SnO_2,[23] ZnO,[25] Al_2O_3,[28] ZrO_2,[22, 40] TiO_2,[86] SiO_2[86] and La_2O_3[87] have been used as coating materials and have shown obvious improvement effects. However, the coating process is rather complicated and the mechanisms are in controversy. For example, Cho *et al*[28] attributed the improved performance to the mechanical constraint effect of the coating layer to the lattice expansion/contraction during cycling. Dahn *et al*,[40] however, pointed out that this is not true because it contravenes the basic physics principium.

Protection to the cathode materials against the corrosion of HF in $LiPF_6$-based electrolyte was supposed to be another important function of surface coating. In this sense, the coating layer separates the active material from the moisture-containing electrolyte and scavenges traces of HF in commercial electrolyte, *e.g.*[88]

$$ZrO_2 + 4HF \longrightarrow ZrF_4{\cdot}2H_2O \quad \text{and} \quad ZrO_2 + 5HF \longrightarrow ZrF_4{\cdot}HF{\cdot}3H_2O$$

Clearly physical separation requires that the coating layer be compact. However, some recent reports claimed that loose[89] or incomplete [90,91] coating can still improve the performance of the cathode materials. In these cases a coverage as low as 13.7% was sufficient in improving the performances of $LiCoO_2$.[90] These observations are great challenges to the protection model because the coating layer cannot effectively separate the active material from the electrolyte. The above comparative study on the spontaneous reaction between the electrolyte and the (surface-coated) $LiCoO_2$ indicated that surface coating does not hinder the dissolution of Li^+ and Co^{3+} ions from the active material into the electrolyte solvent, with or without $LiPF_6$ in it. However, the Al_2O_3 coating indeed suppressed the generation of O_2 and H_2O in the liquid and more Co_3O_4 was detected in the Al_2O_3-coated $LiCoO_2$ than in bare $LiCoO_2$ when they were soaked in the same electrolyte. Actually the Degussa AG (Company) in Germany claimed that their Separion™ separator (porous polypropylene (PP) or polyethylenterephthalate (PET) nonwoven cloth impregnated with alumina and/or silica particles) for lithium ion batteries can delay the exothermic reaction of bare $LiCoO_2$ charged to 4.3V by more than 10°C. These facts give us enough revelation that it is not essential for the coating layer to be compact or to be on the surface of the active material. On the other hand, even compact surface coating cannot hinder the dissolution of Li^+ and Co^{3+} ions into the electrolyte. In both cases, interaction between the coating material and the electrolyte takes a more important role than the morphology (surface coverage ratio) of the coating layer.

In this section, we present experimental evidence to show that the electrochemical performances and the thermal safety of commercial $LiCoO_2$ can be improved by using commercial electrolyte where nano-oxides have been soaked in and then removed. Based on extended and comprehensive analysis, we propose a solid super-acid model to explain how the surface coating (or to be more exact, surface modification) improves the performances of $LiCoO_2$. This model does not agree to the traditional improvement mechanisms. Based on this model, it is predicted that some other metal oxides can also be used as additives in the active electrode material or in the electrolyte.

5.2. Experimental

Nano-$LiCoO_2$ was prepared by firing a mixture of commercial Co_3O_4, $LiOH{\cdot}H_2O$ and KNO_3 at 600°C for 8h in air by the molten salt method.[91] KNO_3 was removed later by dissolving the mixture in distilled water and then alcohol. Commercial $LiCoO_2$ was the product of Japan Chemicals Co. Ltd..

Both α- and γ-Al_2O_3 with average diameters around 30nm (Nanjing Hitech Nano-Materials Co. Ltd., China) were used in this study. The nano-alumina powders were dried at 120°C for 12h before added into the commercial electrolyte. 1g nano-Al_2O_3 was soaked in 50mL commercial electrolyte, 1 mol/L $LiPF_6$ dissolved EC/DMC (1:1 v/v). The mixture was sealed into a PTFE container (see Section 2) in an Ar-filled glove box. After 10 days, the rubber window of the PTFE container was penetrated through with an injector and the gas in it was injected into the column of a GC-MS instrument. Then the upper liquid of the mixture was centrifugally separated from the solid and used as the new electrolyte or for other analyses. The solid was repeatedly rinsed with DMC and then dried under vacuum in the mini-chamber of the glove box. Similar to this, nano-$LiCoO_2$ was soaked in the commercial electrolyte or the new electrolyte for 7 days and then separated from the liquid.

The structure of the samples was analyzed on a JEM 2010 transmission electron microscope at 200 KeV. The X-ray photoelectron spectroscopic (XPS) analysis was performed on PHI Quantera SXM using an Al target and the binding energies (BE) were calibrated with the BE of the C1s (284.5eV) as reference.

The H_2O contamination in the electrolytes was determined on Mettler DL32D by the Karl-Fisher method. For determining the content of HF and other acids in the commercial and the new electrolytes by NaOH titration, 5mL of the electrolyte was mixed with 10mL of water/ice mixture (0°C) for the titration test. The test is carried out at 0°C to prevent the decomposition of $LiPF_6$ in water. Then some NaOH solution (0.01-0.03 mol/L) was added into the electrolyte. Some known amount of 0.1wt% bromthymol blue was dripped into the mixture as indicator. The color of the electrolyte changed from light blue to colorless with increasing amount of added NaOH. The volume (V) of the NaOH and the weight (W) of the tested electrolyte consumed were recorded. The acidity (assume all the acids exist as HF) of the electrolyte can be calculated as follows:

$$Acidity = V_{NaOH} \times C_{NaOH} \times M_{HF} \times 1{,}000{,}000 \div W_{electrolyte} (ppm) \tag{10}$$

5.3. Results and Discussion

5.3.1. Addition of nano-Al_2O_3 in Commercial Electrolyte Improves the Performances of $LiCoO_2$

Fig.34 shows the galvanostatic voltage profile of a Li/$LiCoO_2$ cell using the new electrolyte between 3.3 and 4.5V at 0.15mA/cm^2. Three interesting features are observed. Firstly, the capacity retention of commercial $LiCoO_2$ is significantly improved. The initial discharge capacity of $LiCoO_2$ in the new electrolyte reaches 192 mAh/g, 85% of which is remained after 100 cycles. In contrast, only half the initial capacity is retained after 40 cycles when the commercial electrolyte is used. These results are similar to that of commercial $LiCoO_2$ surface-coated with Al_2O_3 or ZrO_2.[22,28] However, a battery with such electrolyte is expected to have a higher specific capacity and energy density than one using surface-coated $LiCoO_2$ cathode and/or commercial electrolyte. In addition, separating Al_2O_3 from the mixture is easier than coating the surface of $LiCoO_2$ with Al_2O_3. Therefore, we discover a more practical method to improve the electrochemical performance of commercial $LiCoO_2$ than surface coating. With this method, the specific capacity of $LiCoO_2$ is remarkably increased while its cycling stability is maintained. Secondly, a sharp hump appears at the beginning of the initial charge (Fig.34). The cell voltage increases to 4.08V in a few minutes and then drops to 3.9V before it turns to rise with charging time. This hump disappears in the second cycle if the cell is initially charged to 4.1V or over, consistent with our previous observation in MgO-coated $LiCoO_2$ (Section 1 in this chapter) and will be further discussed in the following. Thirdly, the charge voltage is about 5mV lower while the discharge voltage is about 3mV higher in the second and later cycles than in the first cycle. This indicates that the polarization of the cathode is reduced after the first cycle.

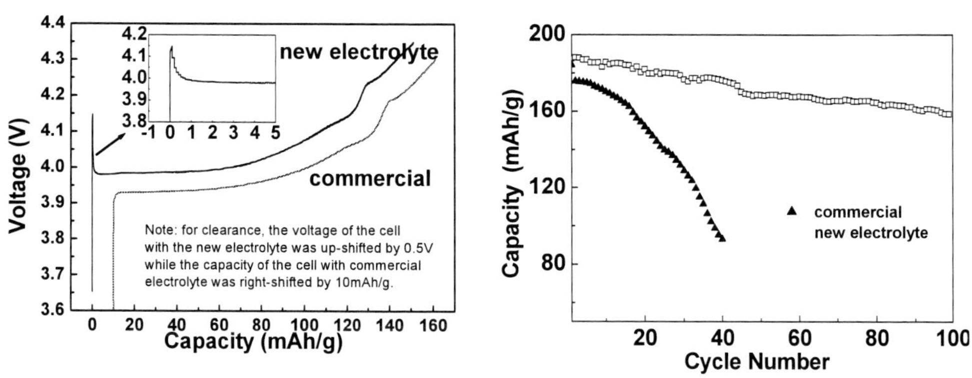

Figure 34. Comparison of the initial galvonostatic charge voltage profiles (A) and discharge capacity retention (B) Li/$LiCoO_2$ cell cycled between 3.3 and 4.5V at 0.15mA/cm^2 using different electrolytes.

Fig.35 shows the rate performance of one Li/$LiCoO_2$ cell at different current densities and cycled between 3.55 and 4.50V (*vs* Li^+/Li). The capacity decreases with increasing current density. At the current density of 0.8mA/cm^2 (~100mA/g), a capacity of 140mAh/g can still be obtained when the new electrolyte is used. When the current is restored to 0.1mA/cm^2 after a total of 160 cycles, the capacity recovers to 175mAh/g. On the contrary, the commercial $LiCoO_2$ material with commercial electrolyte decreases to 100mAh/g after 40 cycling at 0.1mA/cm^2. These demonstrate the improved rate performance and capacity retention of $LiCoO_2$ in the new electrolyte.

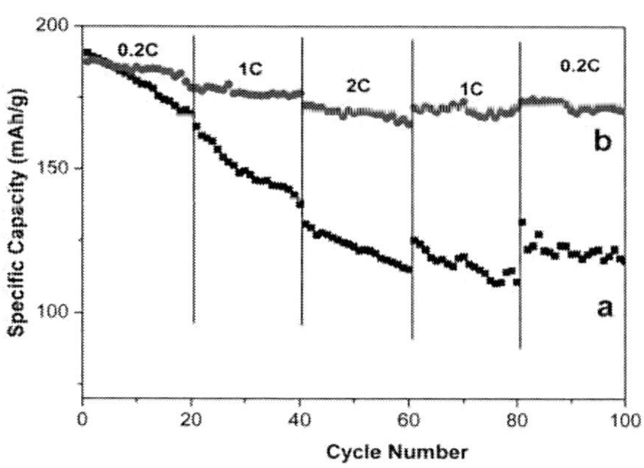

Figure 35. Comparison of rate performances of $LiCoO_2$/Li cells with (a) commercial electrolyte and (b) new electrolyte at different current densities.

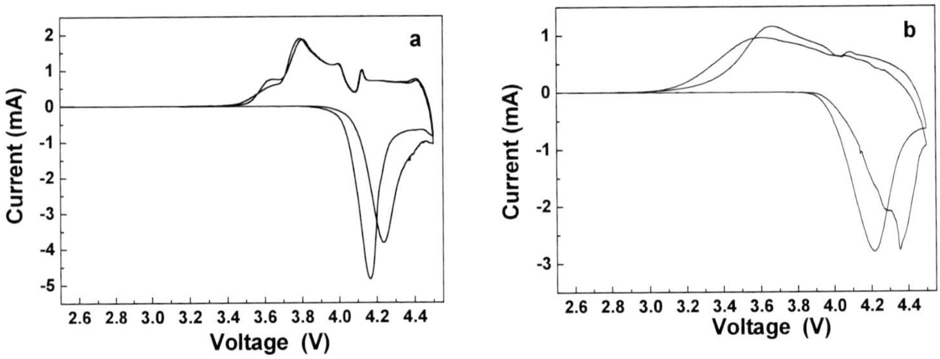

Figure 36. Comparison of cyclic voltammetry of $LiCoO_2$/Li cells between 2.5 and 4.5V with (a) new electrolyte and (b) commercial electrolytes at 0.1mV/s.

Fig.36 compares the cyclic voltammograms of Li/$LiCoO_2$ cells in the first two cycles between 2.5 and 4.5V with different electrolytes, respectively. Three obvious features are observed. Firstly, the differences between the oxidation potentials and the corresponding reduction potentials of the cell using the new electrolyte are smaller than that of the cell using commercial electrolyte. This means that the impedance is reduced and the polarization is decreased in the new electrolyte. Secondly, both the oxidation and reduction peaks in the new electrolyte are sharper than in the commercial electrolyte, indicating a quick electrode reaction in the new electrolyte than in the commercial electrolyte. This is consistent with the excellent rate performance of the Li/$LiCoO_2$ cell using the new electrolyte shown in Fig.35. Thirdly, both the position and intensity of the reduction peaks change very little with cycling in the new electrolyte. In contrast, the reduction peak of the cell using commercial electrolyte becomes weaker and shifts to lower potentials with cycling. These indicate that the Li/$LiCoO_2$ cell using the new electrolyte has better capacity retention than the one using commercial electrolyte. This feature supports the improved capacity retention (Fig.34) of the Li/$LiCoO_2$ cell using the new electrolyte.

Usually an SEI film is formed on the electrode and the impedance of the cell gradually increases during cycling. This is one of the main reasons of capacity decay of the lithium cells. Fig.37 demonstrates the impedance spectra of two Li/$LiCoO_2$ cells with the new electrolyte and commercial electrolyte, respectively, at the discharged state after 1 and 3 cycles, respectively. These two cells have comparable impedances at the open circuit state. It is seen that the charge transfer resistances of both cells at the discharged state increases with cycle number. However, the impedance of the cell using the new electrolyte increases more slowly than that of the cell using commercial electrolyte in the subsequent cycles. This indicates that the addition of nano-Al_2O_3 suppresses the increase of impedance of the SEI film on either the Li or $LiCoO_2$ electrodes or both. This explains the improved rate performance of the commercial $LiCoO_2$ cathode in the new electrolyte or the Al_2O_3-added $LiCoO_2$ cathode.

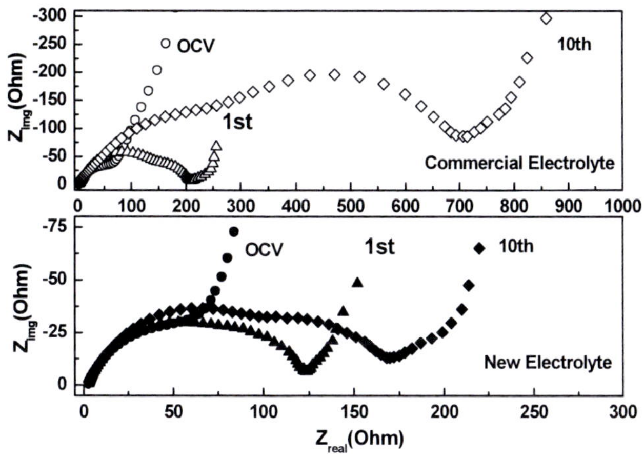

Figure 37. Evolution of the impedance spectra of a $LiCoO_2$/Li cells with the commercial and new electrolyte at discharged state after different cycles.

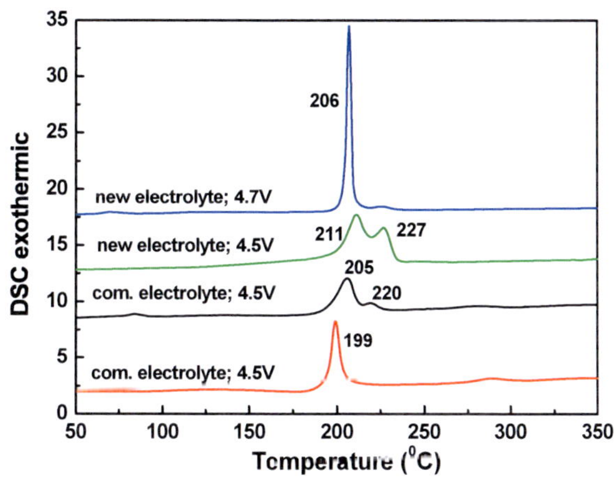

Figure 38. Comparison of the exothermic reaction temperatures of $LiCoO_2$ charged to 4.5V and 4.7V in commercial electrolyte and in the same commercial electrolyte where nano-Al_2O_3 (γ-phase) has been soaked for several days and then removed (*C* is for commercial electrolyte while *N* is for the new electrolyte in this figure).

The exothermic reaction between the electrode materials and the flammable organic solvent in the electrolyte brings about safety concerns to the battery. The temperature of the exothermic reaction of charged cathode becomes lower as the charge cutoff voltage rises. This problem has been harassing the scientists and engineers since the birth of lithium ion batteries. Fig.38 compares the differential scanning calorimetry (DSC) traces of cathode sheets of commercial $LiCoO_2$ in Li/$LiCoO_2$ cells charged to 4.5 and 4.7 V, respectively. It is seen that employing the new electrolyte delays the exothermic reaction by 6 to 7 °C. Therefore, the new electrolyte improves the thermal safety of the charged cathode material.

5.3.2. Analysis to Al_2O_3-induced Electrolyte Decomposition

In order to understand the reason for the remarkably improved electrochemical and thermal performances of the cathode materials, extended analysis was carried out to the products in the Al_2O_3-soaked electrolyte in the order of the gas, the liquid and the solid.

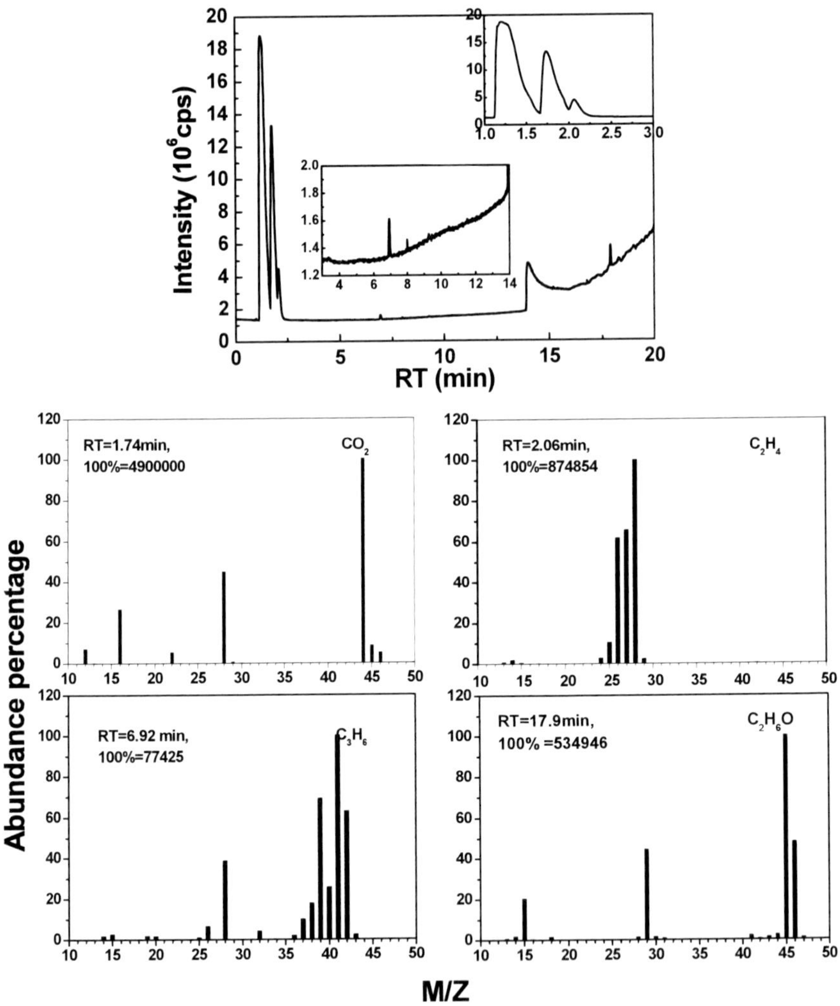

Figure 39. GC and MS spectra of the gas species in the PTFE container where nano-Al_2O_3 is soaked in commercial electrolyte.

By penetrating through the rubber window of the PTFE container, the gas in it was taken out and transferred into the column of the GC-MS instrument. The container was not opened in this progress. GC-MS measurement demonstrates the presence of CO_2, C_2H_4, C_2H_6O andC_3H_6 in the gas (in the order of their abundance from high to low) as well as argon and DMC (Fig.39). This indicates that the solvent (mainly the DMC therein) interacts with Al_2O_3 and is decomposed.

After GC-MS analysis, the container was opened and the liquid and solid were separated in the glove box for spectroscopic and other studies. It is found that the electrolyte became light yellow after soakage. New Raman peaks are observed in the liquid at 639, 777, 794,

1074, 1161 and 1382cm^{-1} as well as peaks belonging to the electrolyte (Fig.40a). The presence of -C-O-C- structures (peaks at 1161, 1074cm^{-1}) indicates the formation of PEO-like species in the electrolyte [53,54,55] The presence of low-molecular weight polymerized PEO-like molecules is confirmed with the FTIR spectra (Fig.40b). $ROCO_2Li$ species (at 583 and 703cm^{-1}) and Li_2CO_3 (at 1432cm^{-1}) are also detected in the new electrolyte with IR absorbance spectroscopy.[59, 60]

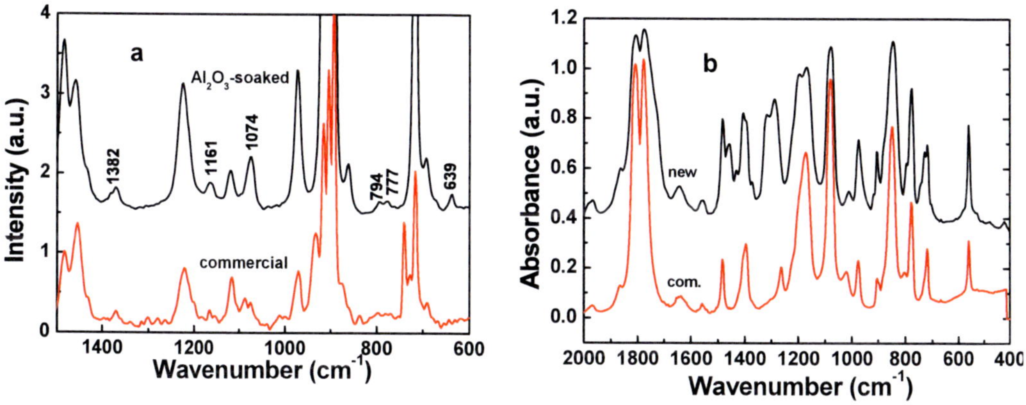

Figure 40. Comparison of the Raman (a) and FTIR (b) spectra of commercial electrolyte and the new electrolyte.

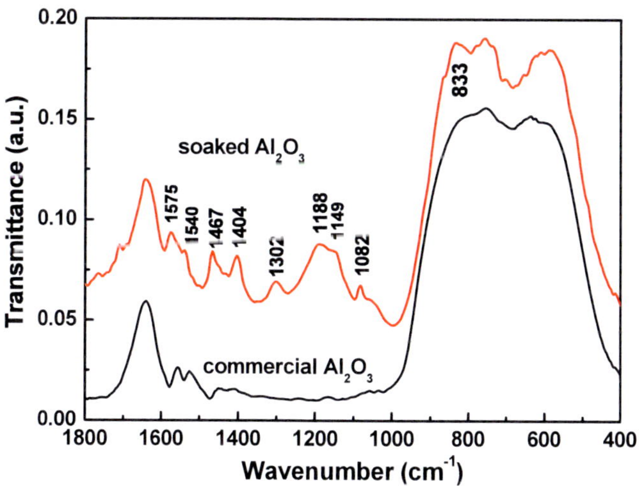

Figure 41. Comparison of selected FTIR spectra of commercial Al_2O_3 before and after soakage in the commercial electrolyte.

The solid was rinsed repeatedly with DMC to remove the residual EC and $LiPF_6$ prior to further analysis. Several new IR absorption peaks are observed in the solid (Fig.41). The presence of the -C-O-C- structures (at 1188 and 1149cm^{-1}) demonstrates the formation of PEO-like species on Al_2O_3 as well as in the electrolyte. In addition, the observation of the $ROCO_2Li$ species (at 833 and 1404cm^{-1}) indicates that the decomposition products of the electrolyte on Al_2O_3 are similar to that on $LiCoO_2$ as previously observed,[92] meaning that both the solvent and the salt take part in the reaction.

The above spectroscopic results agree with our previous observations in $LiCoO_2$-soaked electrolyte [92, 93] and are further supported with ICP analysis. The atomic ratio of Li:P is 0.89 in the liquid (the new electrolyte) but becomes 2.76 in the solid. This confirms the decomposition of both $LiPF_6$ and the solvent. It also helps to understand the origin of the above detected species on nano-Al_2O_3 by FTIR.

The above results indicate that dried nanoscaled Al_2O_3 induces decomposition of the electrolyte (solvent and salt). Low-molecular weight polymerized PEO and lithium alkyl carbonates are formed in both the solid and liquid. However, these products have also been observed when $LiCoO_2$ is soaked in the commercial electrolyte.[92,93] Therefore, these observations still cannot explain the performance improvements of $LiCoO_2$ in the new electrolyte.

5.3.3. Recognition of Solid Superacids on Electrolyte-soaked $LiCoO_2$

Fig.42 compares the surface morphology of commercial $LiCoO_2$ soaked in commercial electrolyte and in the new electrolyte. Obvious corrosion defects are observed on $LiCoO_2$ soaked in the new electrolyte. In contrast, the corrosion of the commercial electrolyte to $LiCoO_2$ is much slighter. The $LiCoO_2$ surface is still rather smooth except for some small irregular chips.

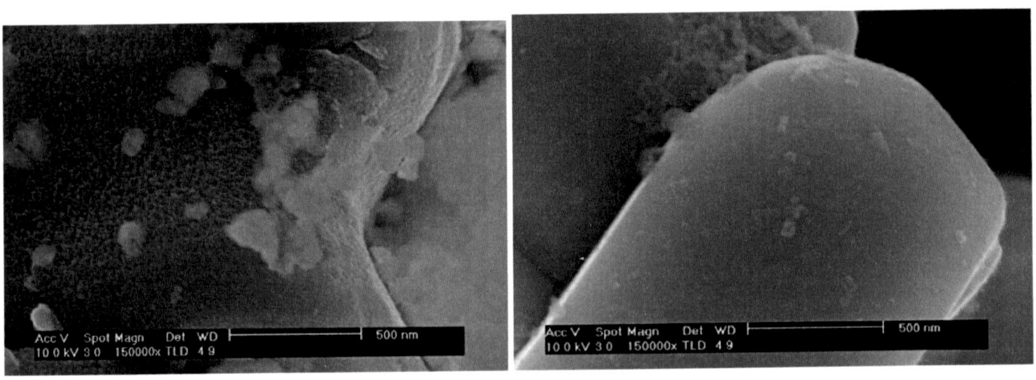

Figure 42. SEM images of commercial $LiCoO_2$ soaked in (a, left) new electrolyte and (b) commercial electrolyte for 2 weeks.

The structures of nano-$LiCoO_2$ soaked in commercial electrolyte and in the new electrolyte were analyzed with high-resolution transmission electron microscope (HRTEM). It is seen that the as-prepared nano-$LiCoO_2$ is well crystallized (Fig.43a). The particle surface is very "clean" and the arris of the particle is sharp; no amorphous species are observed there. After soakage in DMC or in the commercial electrolyte for 8 hours, the nano-$LiCoO_2$ particle is covered with a layer of amorphous species of about 5nm thick (Fig.43b). In both cases, the lattice distance between two (003) planes is homogeneous throughout the particle. Elongating the soakage time only slightly increases the thickness of the solid electrolyte interphase (SEI) layer but the particle surface remains well crystallized. In contrast, the surface of nano-$LiCoO_2$ is obviously corroded after it is soaked in the new electrolyte for 7 days. The thickness of this amorphous layer still ranges 3 to 5nm, but the lattice distances are remarkably different in the bulk and near the surface (Fig.43c). The variation of the lattice distance may be divided into three distinctive areas. The lattice distance of the most inner area (area I) is the same as that of commercial $LiCoO_2$, corresponding to the undisturbed $LiCoO_2$. Outside area I, the fringe is bent and the lattice distance becomes slightly bigger. It is well known that the *c* value of $LiCoO_2$ becomes bigger when some of its Li^+ ions are electrochemically extracted during charge process. As the variation is continuous, this area is recognized to be hexagonal $Li_{1-x}Co_{1-y}O_2$ (x, y ≤1), corresponding to $LiCoO_2$ whose Li^+ and/or Co^{3+} ions are partially extracted. Further outward the particle, the lattice distance becomes bigger and finally each fringe is split into two in the most outer part of the particle, area III. The lattice distance is *ca.* 0.24nm. This area is recognized to be Co_3O_4. Co_3O_4 was detected by XRD in commercial $LiCoO_2$ soaked in commercial electrolyte in Section 3.[93] As the thicknesses of areas II and III are very small, it is difficult to determine their structures with selected area electron diffraction (SAED) or determine their elemental composition by energy dispersion spectroscopy (EDX) technique.

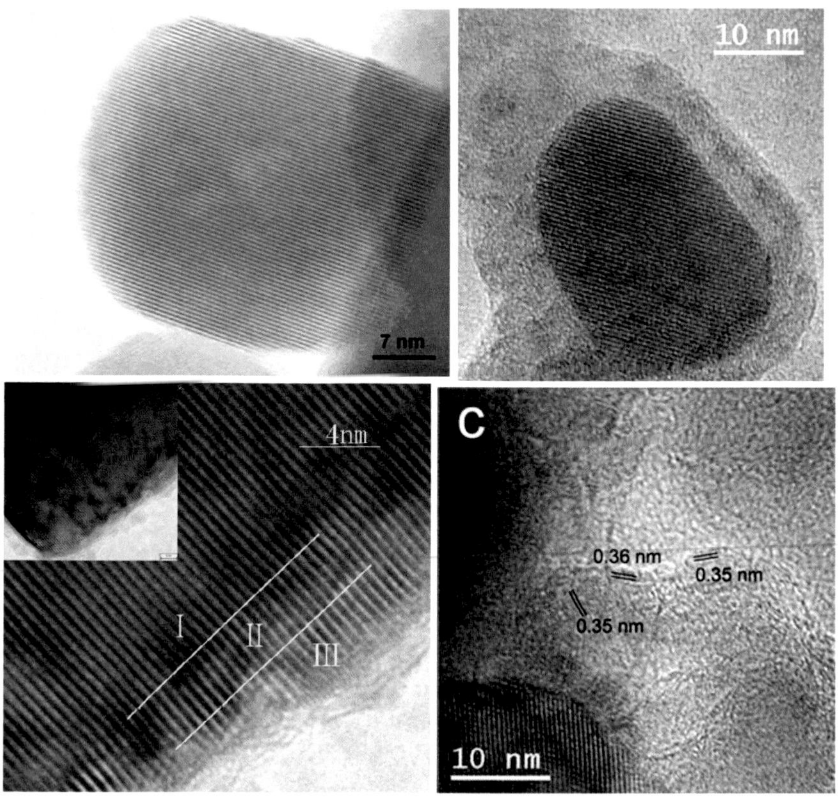

Figure 43. HRTEM imaging of nano-$LiCoO_2$ (a, upper left) soaked in commercial electrolyte (b, upper right) and soaked in the new electrolyte (c, lower left) emphasizing the variation of lattice distances from the inner to the edge of the particle (a selected area of the inset); (d, lower right) shows the hexagonal AlF_3 in the SEI layer on an $LiCoO_2$ particle.

The inset of Fig.43c also shows that some poorly crystallized particles adhere on the surface of $LiCoO_2$ soaked in the new electrolyte. These particles are recognized to be γ-Al_2O_3

according to the measured lattice distances. This means that it is difficult to remove all the Al_2O_3 particles completely, especially the very fine ones, from the new electrolyte. This is why the content of Al is not considered in the elemental analysis (ICP) in the above. Around these fine Al_2O_3 particles, some smaller crystallites are observed in the SEI layer on nano-$LiCoO_2$ (Fig.43d). The lattice distance of these crystallites is *ca.* 0.35nm. This value is similar to that of the (012) planes of hexagonal AlF_3. Therefore, these poorly crystallized fine particles are speculated to be hexagonal AlF_3.

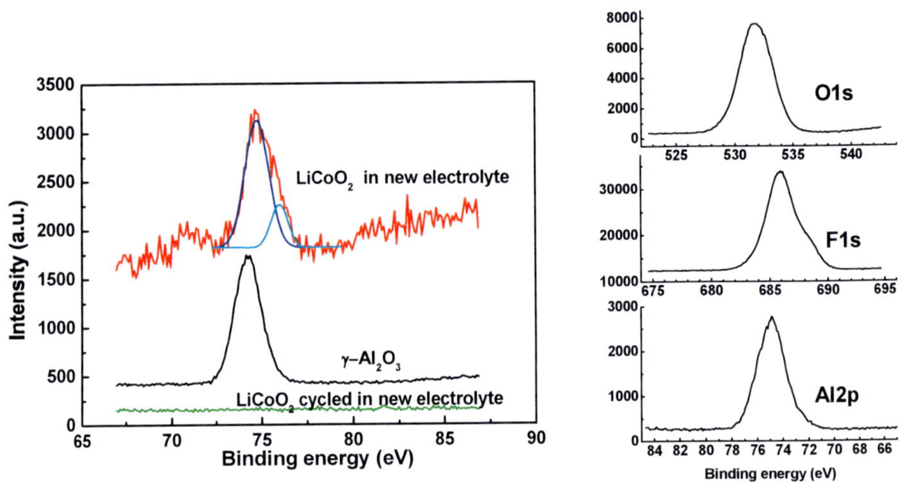

Figure 44. Comparison of (a, left) the Al 2p spectra in commercial nano-sized γ-Al_2O_3, γ-Al_2O_3 fine particles adhered on $LiCoO_2$ soaked in the new electrolyte and on commercial $LiCoO_2$ cycled electrolyte and (b, right) the binding energies of Al 2p, O 1s and F1s in Al_2O_3 soaked in the commercial electrolyte for 60 days.

It is true that the lattice fringe of the AlF_3 particles is fussy. However, considering the small size and poor crystallinity of these particles, it is really difficult to obtain a TEM image with higher quality. In this case, other techniques may help to recognize the products. XPS is an effective tool for surface analysis. The chemical statuses of Al on the Al_2O_3 surface before and after soakage in the new electrolyte were analyzed and compared with that in $LiCoO_2$ cycled between 3.35 and 4.5V. It is seen that the Al 2p peak shifts from 74.2eV in γ-Al_2O_3 to 74.7eV in $LiCoO_2$ soaked in the new electrolyte for 14 days (Fig.44a). This means that the status of Al_2O_3 adhered on $LiCoO_2$ is different from that in free Al_2O_3. Some of the electron around the Al nuclei is transferred elsewhere. It has been reported that the binding energy of Al 2p in γ-Al_2O_3 is 74.2eV while that in AlF_3 is 76.2eV.[94] Considering that the shape of the Al 2p peak is not symmetric in the precipitate, the Al 2p peak of the precipitate is actually a combination of Al 2p in Al_2O_3 and an Al-compound with chemical bonding stronger than Al-O. As a complementary to the above data, the Al 2p spectrum of the precipitate obtained by soaking nanoscale γ-Al_2O_3 in commercial electrolyte for 60 days was recorded (Fig.44b). It is found that its Al 2p binding energy is still at 74.7eV. The F 1s peak, however, is located at 686.0eV, well consistent with the reported binding energy of F 1s in AlF_3, 686.3 eV.[82] Therefore, the F 1s spectrum should be strong evidence for the formation of AlF_3 on the surface of $LiCoO_2$. That is, the nano-Al_2O_3 is partially fluorinated and becomes AlF_3. No Al

2p signal was detected in the $LiCoO_2$ electrode sheet charged to 2.5V after three 10-cycles in the new electrolyte, probably due to the formation of new SEI film covering the old one because of the electrochemical decomposition of the electrolyte or the electrochemical decomposition of AlF_3 during cycling of the Li/$LiCoO_2$ cell. Further evidence is required to find out the reason.

Fig.45 compares the XRD patterns of nanoscaled γ-Al_2O_3 and its precipitate on the bottom of the PTFE container after soakage in the commercial electrolyte for 21 days. The raw alumina is poorly crystallized, agreeing with the HRTEM results. New diffraction peaks are observed after soakage. These peaks are indexed to Li_3AlF_6 (JCPDS Card No.52-1151), suggesting that alumina can react with commercial electrolyte and is fluorinated.

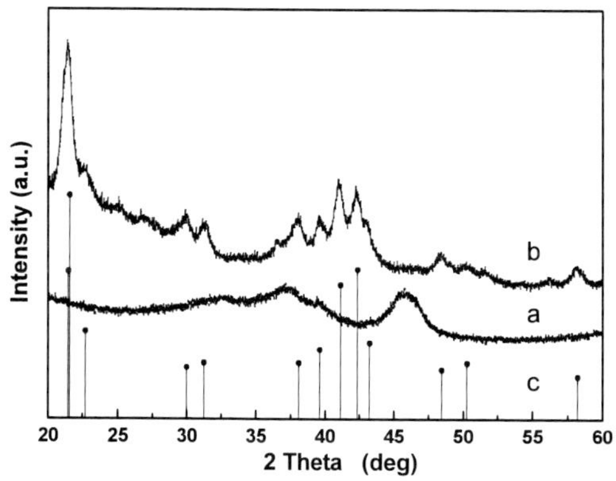

Figure 45. Comparison of the XRD patterns of commercial nano-Al_2O_3 before (a) and after (b) soaked in commercial electrolyte. The standard XRD pattern of Li_3AlF_6 (c; JCPDS Card No.52-1151) is shown for reference.

The following two reactions are well known in the field of lithium ion batteries:[95]

$$LiPF_6 \leftrightarrow PF_5 + LiF \quad (11)$$

$$PF_5 + H_2O \rightarrow 2HF + POF_3 \quad (12)$$

It is believed that HF is detrimental to the transition metal oxide cathode materials. Therefore, worldwide electrolyte manufacturers are trying to remove the water and acidic species in the electrolyte. Of the various mechanisms, the belief is popular that the coating layer separates the electrode material from the electrolyte and scavenges the detrimental HF in the $LiPF_6$-based electrolytes.[88,96] An important purpose of surface coating was also to scavenge HF and other acidic species in the electrolyte so as to suppress the dissolution of the cathode materials.[88, 96] However, it seems that no authors have actually analyzed the variation of the acidity of the electrolyte where the surface-coated material is soaked in or when the surface-coated materials are used in a lithium (ion) cell. To the surprise of most authors, our NaOH titration analysis with bromthymol blue as indicator at 0°C indicates that soakage of Al_2O_3 or SiO_2 in commercial electrolyte increases rather than decreases the acidity of the electrolyte (the acidity of the commercial electrolyte is 17.0 *ppm* while that of the new electrolyte is 1360 *ppm*) **(Table 4).** These results are well reproducible. For example, the measured acidity of the same commercial electrolyte increased over 2000 *ppm* after SiO_2 was soaked in it for 14 days. In addition, the readings of a digital pH-gauge became over-flowed as soon as its tip (the probe) was dipped into the new electrolyte. In contrast, the water content in the electrolyte has no obvious reduction (15.82 *ppm* in commercial electrolyte *vs* 14.09 *ppm* after Al_2O_3 soakage or 10.59 *ppm* after SiO_2 soakage). The increase of acidity disagrees with the previous expectation and eliminates the possibility of the acidity suppression effect of the coating materials but well explains the surface corrosion of $LiCoO_2$ in the new electrolyte.

Table 4. Acidity variation of the electrolyte determined by NaOH titration

Sample	Acid content as HF (ppm)
Commercial electrolyte	17.38
YPO_4 soaked in electrolyte for 7 days	2397.69
Commercial $LiCoO_2$ soaked in electrolyte for 7 days	978.44
5%YPO_4-coated $LiCoO_2$ soaked in electrolyte for 7 days	1307.85

Based on the above HRTEM, XPS and XRD results and *Eqs*. 11 and 12, the following reactions are supposed:

$$6HF+Al_2O_3 \rightarrow 2AlF_3+3H_2O \qquad (13)$$

$$3PF_5+Al_2O_3+6LiF \rightarrow 2\ Li_3AlF_6+3POF_3 \qquad (14)$$

AlF_3/Al_2O_3 is a well-known solid superacid[94] while Li_3AlF_6 can be regarded as a solid solution of LiF and AlF_3,

$$Li_3AlF_6 \rightarrow AlF_3 \cdot 3LiF \qquad (15)$$

Superacids are widely applied as catalysts because of their extremely high acidic strength and catalystic power.[97,98] For example, the acidity of superacid HF/SbF_5 (acidic strength Ho = -20 for HF:SbF_5 =1:0.03 mol/mol) is 10^8 times that of 100% H_2SO_4 (Ho = -11.93).[99] As the content of water in commercial electrolyte is only a few to tens of *ppm*'s, the acidity of commercial electrolyte is not high although some acidic products are present in the $LiPF_6$-based electrolyte (*Eqs*.11 and 12). However, solid superacids such as AlF_3/Al_2O_3 and Li_3AlF_6/Al_2O_3 are formed when HF and PF_5 react with nano-Al_2O_3 added in the electrolyte or in the cathode (see *Eq*.11 through *Eq*.14). As a result, the acidity and the reaction activity of the electrolyte are remarkably increased though the contents of HF, PF_5 and H_2O are very low. This explains the remarkable acidity increase of the electrolyte after nano-Al_2O_3 soakage. A scheme is drawn in Fig.46 to show the formation of solid superacids around a $LiCoO_2$ particle. Clearly these solid superacids can also be formed in the electrolyte because there are fine Al_2O_3 particles suspending in the electrolyte. Indeed our experiments indicate that the electrochemical and thermal properties of $LiCoO_2$ can also be improved by adding nano-Al_2O_3 into the electrode materials during preparation of the cathode sheets.

It is worth pointing out that no performance improvement is observed when the same nano-Al_2O_3 powder is added into $LiClO_4$-based electrolyte (1mol/L $LiClO_4$ in EC:DMC, 1:1 v/v) or into $LiCoO_2$ cathode for cells using $LiClO_4$-based electrolyte. NaOH titration analysis indicates that the acidity of the $LiClO_4$-based electrolyte changes very little before (20.00 *ppm*) and after (~50 *ppm*) Al_2O_3 is soaked in it for 30 days at room temperature. This is probably due to the higher stability of $LiClO_4$ than $LiPF_6$ at room temperature. This supports the above explanation from the opposite side.

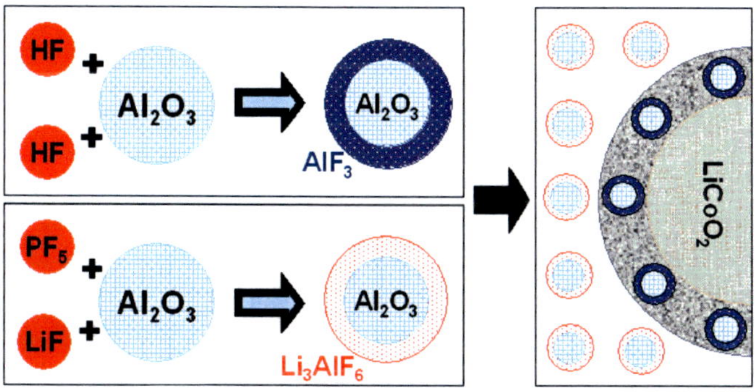

Figure 46. Scheme of the formation solid superacids on $LiCoO_2$ in $LiPF_6$-based electrolyte.

5.3.4 Influence of solid superacids on structure and performances of $LiCoO_2$

The solid superacids in the electrolyte and/or in the SEI layer have important effects on the structure, the surface impurities and the ion transport in the SEI film of the cathode material.

a. Removing insulating surface impurities on commercial $LiCoO_2$

Usually the $LiCoO_2$ surface is covered with a layer of insulating alkaline impurities such as Li_2CO_3, LiOH, etc. Dahn *et al*[40] believed that these impurities deteriorate the electrochemical performance of $LiCoO_2$ above 4.3V. By heat treatment at 500°C, these impurities can be removed and the fresh surface of $LiCoO_2$ is exposed. Fresh surfaces can also be created by ball-milling the material as well as by using newly prepared $LiCoO_2$. They attributed the improved performance of $LiCoO_2$ to the removal of the insulating impurities during the essential heat treatment after surface coating. Clearly these impurities can also be removed with the solid superacids induced with the additives in the electrolyte or in the cathode sheets. Therefore, the corrosive effects of the solid superacids to the impurities are responsible for the performance improvements of the cathode material. However, the acidity of the commercial electrolyte is not high enough to remove these impurities. In comparison to the significantly increased acidity, soaking nano-Al_2O_3 in commercial electrolyte does not influence the H_2O content in it. Aurbach[100] reported that the H_2O content in the electrolyte does not impact the electrochemical performances of $LiCoO_2$ until 200 *ppm*. Therefore, the significant increase of acidity takes much more important role than the slight decrease of water content. The fresh interface stabilizes the structure of $LiCoO_2$ and is beneficial for the ion transport during cycling above 4.3V.

b. Promoting formation of more stable surface solid solution

Many efforts have been taken to suppress the corrosive effect of the electrolyte by suppressing the decomposition of $LiPF_6$ and removal of the moisture in the electrolyte.[88,89] The solid superacids in the new electrolyte not only remove the insulating surface impurities on $LiCoO_2$ but also corrode $LiCoO_2$ itself. In this way, some Li^+ and Co^{3+} ions are dissolved from $LiCoO_2$ into the electrolyte and leave some Li and Co vacancies in $LiCoO_2$. Meanwhile, however, nano-Al_2O_3 and AlF_3 will also be partially dissolved and some of their Al^{3+} ions can migrate to fill the Li^+ and/or Co^{3+} vacancies during cycling, forming a layer of solid solution ($Li_xAl_yCoO_2$, $x+y\leq1$ or $Li_xAl_yCo_{1-y}O_2$) on $LiCoO_2$. The migration of the Al^{3+} ions is evidenced with the obvious hump in the voltage profile in the first few cycles (depending on the charge cutoff potential, this hump will appear in one to a few cycles) and the increased resistance at the open-circuit states. Similar migration process has been observed in $LiCoO_2$ surface-coated with MgO (Section 2 in this chapter). Jang *et al*[12] reported that Al substitution improves the structural stability of $LiCoO_2$ because the Al-O bonding is stronger than the Co-O bonding. This can explain the improved capacity retention and the thermal safety of $LiCoO_2$ at deep delithiation states. Although the $LiCoO_2$ surface can also be corroded to some extent with commercial $LiPF_6$-based electrolyte containing protonic species (purposively added or $LiPF_6$ decomposed), surface solid solutions cannot be formed during cycling. Therefore, the corrosive effect of the acidic commercial electrolyte only deteriorates the storage performance of the material and lead to capacity "leakage". In this sense, it may say that the added nano-Al_2O_3 scavenges the protonic acids (HF, for example) and converts to solid superacids, which attack the cathode surface in a less negative way than the protonic species.

Chen *et al*[101] reported that addition of Al_2O_3 in $LiPF_6$-based electrolyte does not help to improve the capacity retention of $LiCoO_2$ charged to 4.5V. They attributed the capacity fading of $LiCoO_2$ to the side reactions involving moisture-related chemical species on the $LiCoO_2$ surface and $LiPF_6$-based electrolyte. In contrast, they believed that SiO_2-contained electrolyte improves the electrochemical performance of $LiCoO_2$ because SiO_2 consumes the harmful species (mostly HF) in the electrolyte. Indeed we found that the improvement effect becomes obvious only after the oxide was soaked for 7 days or more and the electrolyte became light yellow. They failed to observe the improved performance of $LiCoO_2$ probably because the storage time of their cell containing Al_2O_3 was not long enough (usually a few hours). However, they did not show any experimental evidence for the HF-consuming reactions.

c. Enhancing ionic conduction and suppressing growth of SEI layer

Existence of solid superacids in the SEI is beneficial for improving the ionic conduction and the ion transference in the SEI film. Raman and FTIR spectroscopic study in Section 3 of this chapter has confirmed the presence of PEO-like low-molecular weight polymerized species in both the liquid- and solid-state soakage products. Xi *et al.*[102] reported that addition of solid superacid SO_4^{2-}/ZrO_2 can enhance the ionic conductivity and the lithium ion transference number of the polymer electrolyte (PEO_{12}-$LiClO_4$) by effectively reducing the crystallinity of PEO through the strong Lewis acid-base interaction with the PEO chains. Yang *et al*[103] reported similar observations. Our EIS study on Li/$LiCoO_2$ cells using the new electrolyte indicates that the charge transfer resistance of $LiCoO_2$ at discharged state increases with cycle number but remains much lower than that of the Li/$LiCoO_2$ cell using commercial electrolyte. This means that the SEI layer containing solid superacids is beneficial for the charge transfer on the surface of $LiCoO_2$ and ion transport in the new electrolyte. These facts indicate that addition of nano-Al_2O_3 can improve the ionic conduction of the SEI layer on the cathode material. This explains the improved rate performance of the Al_2O_3-added $LiCoO_2$ cathode or commercial $LiCoO_2$ cathode in the new electrolyte (not shown).

It might be argued that the presence of nano-Al_2O_3 rather than the formation of solid superacid AlF_3 enhances the ionic conduction of the polymer electrolyte-containing SEI layer because addition of the nanoscale ceramic powder can also improve the conduction of the polymer electrolyte. However, considering the experimental condition in this work (1g Al_2O_3 *vs* 50mL electrolyte and the centrifugally removal of the solid additives), the content of residual Al_2O_3 in the electrolyte should below 1wt%, much lower than the reported usual content (≥5wt%) of nanoscale additives in composite polymer electrolyte. Actually, we found that the Al content in the electrolyte is around the detection limit of ICP technique (0.1%). Therefore, the contribution of the solid superacids to the ionic conductivity of the SEI should not, at least, be negligible.

It is worth pointing out that no performance improvement was observed when the same nano-Al_2O_3 powder was added into $LiClO_4$-based electrolytes or into $LiCoO_2$ cathode for cells using $LiClO_4$-based electrolyte (1mol/L $LiClO_4$ in EC:DMC, 1:1 v/v). NaOH titration analysis shows that the acidity of the $LiClO_4$-based electrolyte changes very little before (20.00 *ppm*) and after (~50 *ppm*) Al_2O_3 is soaked in it for 30 days at room temperature, probably because $LiClO_4$ is more stable than $LiPF_6$ at room temperature. This supports the above solid superacid model from the opposite side.

With the above discussion, it is understandable that the formation of solid superacids is beneficial for improving both the structural (including cycling and thermal) stability and the rate performance of $LiCoO_2$.

5.3.5. Extending the electrolyte soakage to other metal compounds

We have extended the soakage of Al_2O_3 to other metal oxides or compounds and found that similar reactions can take place between commercial electrolyte and some other metal compounds. Obvious increase of acidity was also observed in TiO_2- and Y_2O_3-soaked commercial electrolyte in this Laboratory. $LiCoO_2$/Li cells using such electrolytes showed much better performances than those using commercial electrolyte. Our recent study indicates that the electrochemical and thermal performances of commercial $LiCoO_2$ can also be remarkably improved by coating its surface with YPO_4.[104] No annealing process is necessary during or after surface coating. NaOH titration demonstrates that the acidity of the electrolyte is significantly increased after soaking nano-YPO_4 powder or the YPO_4-coated $LiCoO_2$ in it. This should also be evidence for the importance of solid superacids. Similar to these, metal compounds were mixed with commercial $LiCoO_2$ as additive on preparing the cathode sheets. Similar improvement effects were observed as using the new electrolyte. Clearly these improvements are for the same reason, the added metal compound react with the electrolyte and convert HF to solid superacids.

With the above facts, we speculate that various superacids can be formed if the added compound can react with LiF, HF and PF_5 in the $LiPF_6$-based commercial electrolytes. Such a solid superacid model can explain why many metal oxides such as MgO, ZnO, TiO_2, ZrO_2,

SiO_2, *etc.* can also be used as the coating material to improve the performances of $LiCoO_2$ and show similar improvement effects. Loose or incomplete coating and the Al_2O_3- and/or SiO_2-impregnated Separion™ separator can still improve the performances of the cathode materials because the above solid superacid model does not require a compact coating or direct contact between the additive and the particles of the cathode materials. On the contrary, even compact coating cannot prevent the dissolution of Li and Co from $LiCoO_2$ due to the enhanced corrosive effect by the coating materials.

Studies of the impacts of solid superacids on other components (the porous polymer separator, the metallic current collector and the anode material) of a lithium (ion) battery are in progress. No negative effects have yet been found on the Al and Cu current collectors and the PE/PC polymer separators. The effects of the presence of superacids on the performances of the graphitic (natural graphite and MCMB), metal oxide (Co_3O_4) and alloy anode materials are being evaluated.

5.3.6 Nano-Al_2O_3 Additive Improves Performances of Cathode Materials for Lithium Ion Batteries

Now that nano-Al_2O_3 can be added into the commercial electrolyte and forms solid superacids via interaction with the components of the electrolyte, we believe that similar reactions can still occur when nano-Al_2O_3 is added into commercial $LiCoO_2$ and assembled into a battery. Here we show that the electrochemical performances of commercial $LiCoO_2$ powder can also be improved by adding very low content of nano-Al_2O_3 as additives. These results are not only a simplification of techniques of electrode or electrolyte preparation, rather they are based on comprehensive understanding to the surface modification mechanism. Applications of these methods will greatly improve the thermal safety, elevate the energy/power density, and prolong the cycle life of lithium ion batteries.

Figure 47 compares the cycling performances of commercial $LiCoO_2$ and commercial $LiCoO_2$ containing 10 wt% nano-Al_2O_3 (γ-phase). The initial discharge capacity of Al_2O_3-added $LiCoO_2$ is *ca.* 190 mAh/g, only slightly lower than the highest specific capacity of the additive-free $LiCoO_2$ cathode charged to the same voltage (4.5V). After 12 cycles, the reversible capacity of the Al_2O_3-added $LiCoO_2$ overruns that of the additive-free $LiCoO_2$. Obviously, addition of nano-Al_2O_3 in the active material improves the cycling stability of the latter, especially at deep delithiation states. This improvement is quite similar to that of surface-coated $LiCoO_2$ cathode materials.[28] More details for the performance improvement can be found in our new paper to be published in Journal of Power Sources.

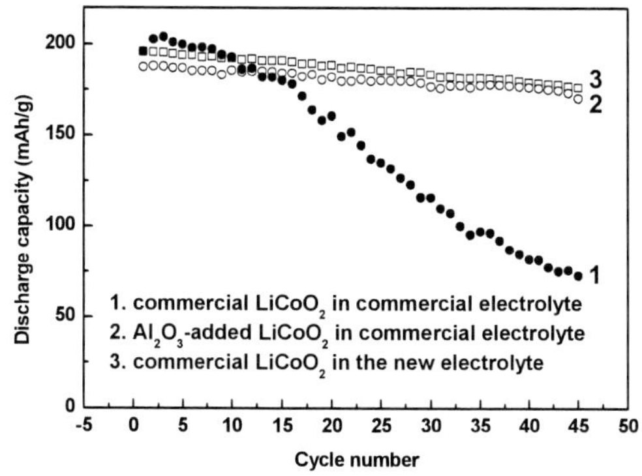

Figure 47. Comparison of the cycling stabilities of LiCoO2 with (10 wt%) and without nano-Al2O3 (γ-phase) additives in commercial electrolyte and LiCoO2 in the new electrolyte and cycled between 2.5 and 4.5V vs. Li/Li+ at a current density of 0.2 mA/cm2.

5.4 Summary

Based on the above discussion, we may summarize the roles of the solid superacids on improving the performances of $LiCoO_2$ cathode material and explain the previous observations (as mentioned in the *Introduction* in this section) as follows. Various solid superacids may be formed when the coating material, adhered on $LiCoO_2$ particle or separated from it in any form and for any reason, is in contact with the $LiPF_6$-based electrolyte. The solid superacids help to 1) remove the insulating impurities on $LiCoO_2$ particles and exposing its fresh surfaces to the electrolyte; 2) enhance the conduction of the SEI layer and suppress its growth with cycling; 3) dissolve Li^+ and/or Co^{3+} ions from $LiCoO_2$ and form a surface solid solution more stable than $LiCoO_2$ itself during cycling. Loose or incomplete coating can improve the performances of the cathode materials because the above solid superacid model does not require a compact coating. On the contrary, even compact coating cannot prevent the dissolution of Li and Co from $LiCoO_2$ due to the enhanced corrosive effect of the solid superacids converted from the coating materials.

Clearly soakage of oxides in the electrolyte or addition of the oxides into the electrode powder is much easier than coating the surface of the active cathode materials. Considering the extensive applications of lithium ion batteries and their importance in energy storage in our times short of fossil fuels, addition of nano-oxides into the active material or into the electrolyte is a very practical method on elevating the energy and power densities, prolonging the cycle life and improving the thermal safety of the lithium ion batteries. We believe that the above findings can be extended to other metal oxides using $LiPF_6$-based electrolytes (by the time this chapter is finished, we have confirmed the acidity increase of the electrolyte and performance improvement of $LiCoO_2$ material by adding SiO_2, TiO_2, Y_2O_3 and YPO_4 nanoscale powder into the commercial electrolyte). Finally the present finding also induces a number of interesting questions to the scientists and engineers in the field of lithium ion batteries.

References

[1] T. Ohzuku, A. Ueda, *J. Electrochem. Soc.*, 141, 2972(1994); J N Reimers, J R Dahn, *J Electrochem Soc.* 139, 2091(1992)
[2] M. Holzapfel, R. Schreiner, A. Ott, *Electrochem. Acta* 46, 1063(2001)
[3] M. Holzapfel, C. Haak, A. Ott, *J. Solid State Chem.*156, 470(2001)
[4] S. NietoRamos, M. S. Tomar, S. Hermandez, F. Aliev, *Thin Solid Films* 377, 745(2000)
[5] Z. L. Liu, A. S. Yu, J. Y. Lee, *J. Power Sources* 81-81,416(1999).
[6] G. Ceder, Y. M. Chiang, D. R. Sadoway, M. K. Aydinol, Y. I. Jang, B. Huang, *Nature* 392, 694(1998)
[7] S. T. Nyung, N. Kumagai, S. Komaba, H. T. Chung, *Solid State Ionics*, 139, 47(2001)
[8] W. S. Yoon, K. K. Lee, *Electrochem. Solid State Lett.* 4, A35(2001)
[9] P. S. Dobal, R. S. Katiyar, M. S. Tomar, A. Hidalgo, *J. Mater. Res.* 16, 1(2001)
[10] C. Julien, G. A. Nazri, A. Rougier, *Solid State Ionics* 135, 121(2000).
[11] W. S. Yoon, K. K. Lee, K. B. Kim, *J. Electrochem. Soc.* 147, 2023(2000)
[12] Y. I. Jang, B. Y. Huang, H. F. Wang, D. R. Sadoway, G. Ceder, Y. M. Chiang, H. Liu, H. Tamura, *J. Electrochem. Soc.* 146, 862(1999)
[13] J D. Perkins, C. S. Bahn, P. A. Parilla, J. M. McGraw, M. L. Fu, M. Duncan, H. Yu., D. S. Ginley, *J. Power Sources* 81-82, 675(1999)
[14] H. Tukamoto, A. R. West, J. Electrochem. Soc. 144, 3164(1997)
[15] C. Pouillerie, L. Croguennec, Ph. Biensan, P. Willmann, C. Delmas, *J. Electrochem.Soc.* 147, 2061(2000)
[16] C. Pouillerie, L. Croguennec, C. Delmas, *Solid State Ionics* 132, 15(2000)
[17] C. C. Chang, J. Y. Kim, P. N. Kumta, *J. Electrochem. Soc.* 147, 1722(2000)
[18] G. G. Amatucci, J. M. Tarascon, L. C. Klein, *Solid State Ionics* 83, 167(1996)
[19] D. Aurbach, K. Gamosky, B. Markovsky, G. Salitra, Y. Gofer, U. Heider, R. Oesten, M. Schmidt, *J. Electrochem Soc.* 147, 1322(2000)
[20] H.-J. Kweon, G.-B. Kim, D.-G. Park, *K. R. Patent Appl.* 0012005(1998).
[21] Z. X. Wang, C. Wu, L. J. Liu, F. Wu, L. Q. Chen, X. J. Huang, *J. Electrochem. Soc.* 149, A466(2002).
[22] J. Cho, Y. J. Kim, T.-J. Kim, B. Park, *Angew. Chem. Int. Ed.* 40, 3367(2001).
[23] J. Cho, C. S. Kim, S. I. Yoo, *Electrochem. Solid State Lett* 3, 362(2000)
[24] H. W. Ha, K. H. Jeong, N. J. Yun, M. Z. Hong, K. Kim, *Electrochim. Acta* 50, 3764(2005).
[25] T. Fang, J. G. Duh, S. R. Sheen, *J. Electrochem. Soc.* 152, A1701(2005).
[26] J. Cho, J. G. Lee, B. Kim, B. Park, *Chem. Mater.* 15, 3190 (2003).
[27] H. Omanda, T. Brousse, C. Marhic, D. M. Schleich, *J. Electrochem. Soc.* 151, A922(2004).
[28] J. Cho, Y. J. Kim, B. Park, *Chem. Mater.* 12, 3788(2000)
[29] E. Endo, T. Yasuda, A. Kita, K. Yamaura, K. Sekai, *J. Electrochem. Soc.* 147, 1291(2000)
[30] H. J. Kweon, S. J. Kim, D. G. Park, *J. Power Sources* 88, 255(2000)
[31] H. J. Kweon, D. G. Park, *Electrochem Solid State Lett.* 3, 128(2000)
[32] J. Cho, G. B. Kim, H. S. Lim, C. S. Kim, and S. I. Yoo, *Electrochem. Solid State Lett.* 2, 607(1999).
[33] J. Cho, Y. J. Kim, T. J. Kim, B. Park, *Chem. Mater.*, 13, 18(2001).

[34] J. Cho, G. Kim, *Electrochem. Solid State Lett.* 2, 253(1999).
[35] J. Cho, *Electrochim. Acta* 48, 2807(2003).
[36] G. G. Amatucci, J. M. Tarascon, L. C. Klein, *J. Electrochem. Soc.* 143, 1114(1996).
[37] T. Ohzuku and A. Ueda, *Solid State Ionics* 69, 201(1994)
[38] J. R. Dahn, E. W. Fuller, M. Obrovac, U. von Sacken, *Solid State Ionics* 69,265(1994).
[39] Z. Zhang, D. Fouchard, J. R. Ren, *J. Power Sources* 70, 16(1998).
[40] Z. Chen, J. R. Dahn, *Electrochim. Acta* 49, 1079(2004).
[41] N. Dey, B. P. Sullivan, *J. Electrochem. Soc.* 117, 222(1970).
[42] E. Peled, *J. Electrochem. Soc.*, 126, 2047(1979).
[43] D. Aurbach, B. Markovsky, M. D. Levi, A. Schechter, M. Moshkovich, Y. Cohen, *J. Power Sources* 81/82, 95(1999);
[44] du Pasquier, A. Blyr, P. Courjal, A. Cressent, C. Lenain, G. Amatucci, B. Gerand, and J. M. Tarascon, *J.Electrochem.Soc.*146, 428(1999);
[45] du Pasquier, A. Blyr, A. Cressent, C. Lenain, G. Amatucci, and J. M. Tarascon, *J. Power Sources,* 81/82, 54(1999).
[46] D. Aurbach, *J. Power Sources* 89, 206(2000).
[47] D. Ostrovskii, F. Ronci, B. Scrosati, P. Jacobsson, *J. Power Sources* 103, 10(2001).
[48] Y. Matsuo, R. Kostecki, F. McLarnon, *J. Electrochem. Soc.* 148, A687(2001).
[49] R. Kostecki, F. Kong, Y. Matsuo, F. McLarnon, *Electrochim. Acta* 45, 225(1999).
[50] L.J. Liu, Z.X. Wang, H. Li, L.Q. Chen, X.J. Huang, *Solid State Ionics* 152, 341(2002).
[51] Z. X. Wang, H. Dong, X. J. Huang, Y. J. Mo, L. Q. Chen, submitted to *Electrochem. Solid State Lett.* (2003).
[52] R. Yang, Y.Y. Wang, C.C.Wan, *J. Power Sources* 72, 66-70(1998).
[53] N. B. Colthup, L. H. Daly, S. E. Wiberley, *Introduction to Infrared and Raman Spectroscopy*, 3rd ed., Academic Press, San Diego 1990.
[54] Lin-Vien, N. B. Colthup, W. G. Fateley, J. G. Grasselli, *Library of Congress Cataloging in Publication Data. The Handbook of Infrared and Raman Characteristic Frequencies of Organic Molecules*. Academic Press, Inc., Boston 1991
[55] K. Nakanishi, P. H. Solomon, *Infrared Absorption Spectroscopy*, 2nd Edition, Holden-Day, Inc. San Francisco, 1977.
[56] Q. Xue, *Spectroscopic Methods in Structural Studies of Macromolecules*, High Education Press, Beijing, 1995.
[57] Z. Ogumi, A. Sano, M. Inaba, T. Abe, *J. Power Sources* 97/98, 156(2001).
[58] R. Yazami, *Electrochim. Acta* 45, 87(1999).
[59] X. R. Zhang, R. Kostecki, T. J. Richardson, J. K. Pugh, P. N. Ross, Jr, *J. Electrochem. Soc.*. 148, A1341 (2001).
[60] V. W. Behrendt, G.. Gattow, and M. Drager, *Z. Arorg. Allg. Chem.* 397, 237 (1973).
[61] J. Cho, G.Kim, H.S.Lim, *This Journal* 146, 3571(1999).
[62] Lee Smith, Applied Infrared Spectroscopy: fundamentals, techniques, and analytical *problem-solving*, John Wiley & Sons (1979).
[63] V. C. Farmer, *The Infrared Spectra of Minerals, Mineralogical Society*, London, 1974.
[64] Aurbach, M. L. Daroux, P. W. Faguy and E. Yeager, *J. Electrochem. Soc.* 134, 1611 (1987).
[65] W. Huang and R. Frech, *Solid State Ionics* 86-88, 395(1996).
[66] Aurbach, Y. Gofer, M. Ben-Zion and P. Aped, *J. Electroanal. Chem.* 339, 451(1992).
[67] H. Buqa, P. Golob, M.Winter, J.O.Besenhard, *J. Power Sources* 97/98, 122(2001)

[68] G. Amatucci, A. Blyr, C. Sigala, P. Alfonse, J. M. Tarascon, *Solid State Ionics* 104, 13(1997).

[69] J. Cho, Y. J. Kim, T. J. Kim, and B. Park, *J. Electrochem. Soc.*, 149, A127(2002).

[70] Z. X. Wang, Y. C. Sun, X. J. Huang, L. Q. Chen, *J. Electrochem. Soc.* 151, A914(2004).

[71] Y. Hase, V. Inez, P. Yoshida, *Chem.Phys.Lett.* 65, 46(1979).

[72] M. St. C. Flett, *Characteristic frequencies of chemical groups in the infrared*, Elsevier Publishing Company (1963).

[73] D.Aurbach, Y. Ein-Eli, O. Chusid, Y. Cameli, M. Babai and H. Yamin, *J. Electrochem. Soc.* 141 603 (1994).

[74] D. Aurbach, M. L. Daroux, P. W. Faguy, E. Yeager, *J. Electrochem. Soc.,* 135, 1863(1988).

[75] D. Aurbach, Y. Ein-Eli and A. Zaban, *J. Electrochem. Soc.* 141, L1(1994).

[76] D. Aurbach, *J. Electrochem. Soc.* 136, 906 (1989).

[77] Z. X. Wang, X. J. Huang and L. Q. Chen, *J. Electrochem. Soc.* 150, A199(2003).

[78] L.J. Liu, L.Q. Chen, X.J. Huang, X.Q. Yang, W.-S.Yoon, H.S. Lee, J. McBreen, *J. Electrochem. Soc.*151, A1344(2004).

[79] M G S R Thomas, P G Bruce, and J B Goodenough, *This Journal* 132, 1521(1985).

[80] L. A. Montoro, M. Abbate and J. M. Rosolen, *Electrochem Solid State Lett.* 3, 410(2000)).

[81] L. Kavan, P. Krtil, M. Gratzel, *J. Electroanal. Chem.*, 373, 123(1994).

[82] *Handbook of X-ray Photoelectron Spectroscopy*, Perkin-Elmer Corporation, Physical Electronics Division

[83] Y. Iriyama, H. Kurita, I. Yamada, T. Abe, Z. Ogumi , *J. Power Sources* 137, 111 (2004)

[84] S. Oh, J.K. Lee , D. Byun , W.I. Cho , B.W. Cho , *J. Power Sources* 132, 249 (2004)

[85] J.S. Gnanaraj, V.G. Pol, A. Gedanken, D. Aurbach, *Electrochem. Commun.* 5, 940 (2003)

[86] W. Hong, M.C. Chen, *Electrochem. Solid State Lett*, 9, A82 (2006)

[87] G.T.K. Fey, P. Muralidharan, C.Z. Lu, *Solid State Ionics* 176, 2759 (2005)

[88] J.S. Kim, C.S. Johnson, J.T. Vaughey, S.A. Hackney, K.A. Walz, W.A. Zeltner, M.A. Anderson, M.M. Thackeray, *J. Electrochem. Soc,* 151, A1755 (2004)

[89] A.M. Kannan, L. Rabenberg , A. Manthiram, *Electrochem. Solid State Lett,* 6,A16 (2003)

[90] N. van Landschoot, E.M. Kelder, P.J. Kooyman, C. Kwakernaak, J. Schoonman, *J. Power Sources* 138, 262 (2004).

[91] M. S. Whittingham, *Chem. Rev.* 104, 4271(2004).

[92] Z.X. Wang, X.J. Huang, L.Q. Chen, *J. Electrochem.Soc.* 151, A1641 (2004)

[93] Z.X. Wang, H. Dong, X.J. Huang, Y.J. Mo, L.Q, Chen, *Electrochem. Solid-State Lett,* 7, A353 (2004)

[94] Boese, W.E.S. Unger, E. Kemnitz, S.L.M. Schroeder, *Phys. Chem. Chem. Phys,* 4, 2824 (2002)

[95] L. Christopher, L. Campion, *J. Electrochem. Soc,* 152, A2327 (2005)

[96] J.S.T Myung, K. Izumi, S. Komaba, Y.K. Sun, H. Yashiro, N. Kumagai, *Chem. Mater.* 17, 3695 (2005)

[97] X. C Wang, J. C. Yu, Y. D. Hou, X. Z. Fu, *Adv. Mater.* 17, 99 (2005)

[98] K.K. Laali, *Chem. Rev.* 96, 1873 (1996)

[99] J.R. Sohn, *J. Ind. Eng. Chem.* 1, 15(2004)

[100] D. Aurbach, I. Weissman, A. Zaban, *Electrochimica Acta* 45, 1135 (1999)
[101] Z.H Chen, *Improved positive electrode materials for lithium ion batteries*, PhD Thesis, Dalhousie University, Canada, p43 (2003)
[102] J.Y. Xi, X.Z.Tang, *Chem. Phys. Lett.* 393, 271(2004)
[103] Y. Yang, L. Wang, N.E. Drysdale, M. Doyle, *Macromolecules,* 36, 8205(2003).
[104] Y. Bai, N. Liu, J.Y. Liu, Z.X. Wang, L.Q. Chen, *Electrochem. Solid-State Lett.* 9, A552(2006).

In: Diffusion and Reactivity of Solids
Editor: James Y. Murdoch, pp. 69-113
ISBN: 978-1-60021-890-3

Chapter 2

CATALYSTS DESIGN FOR HYDROGEN PRODUCTION: EMBEDDED RHODIUM NANOPARTICLES

Paolo Fornasiero**, *Tiziano Montini and Loredana De Rogatis
Chemistry Department, Center of Excellence for Nanostructured Materials (CENMAT) and INSTM-Trieste Research Unit, University of Trieste, Via L. Giorgieri 1, 34127, Trieste, Italy

Abstract

In a sustainable energy and mobility development, hydrogen will become very important as it is considered one of the key energy carriers in terms of energy source, as fuel for transportation and intermediate in the conversion of renewable energy sources. In addition, hydrogen is also of relevance as a clean fuel for fuel cells. Catalytic technologies will play a major role in the transformation towards hydrogen economy. Here, examples of the development of new catalysts for hydrogen production from fossil fuels (methane) and from bio masses (ethanol / water solution) are presented.

In particular, the partial oxidation of methane over Rh-based catalysts is discussed as an attractive process for the production of syngas. Despite its high cost, rhodium is widely investigated since it shows high yields, good selectivity and good resistance towards the deactivating effects of coke deposition. Nevertheless, the extreme working conditions encountered in syngas production, such as the high temperature and high space velocity, combined with the necessity of long lifetime for commercialisation of such catalysts, require the development of new catalytic materials with superior thermal stability than those currently available. It is showed that the controlled synthesis of Rh nanoparticles embedded in porous oxides results in catalysts which exhibit high hydrogen yield for partial oxidation of methane. Moreover, the process of encapsulation of the Rh nanoparticles during the synthesis stage largely prevents Rh sintering. Furthermore, the undesirable incorporation of Rh into the Al_2O_3 lattice, during high temperature oxidation treatments, can also be minimised. Consistently, under the working conditions employed, the embedded Rh nanoparticles present high thermal stability. Small and slow deactivation is observed due to coke formation and sintering of the support. The adoption of appropriate strategies, such as nature and texture modulation of the support or inclusion of extra components in the catalyst formulation, can be employed to minimise these drawbacks. In situ regeneration treatments have proved to strongly extend the

* E-mail address: pfornasiero@units.it. Fax + 39040 5583903

embedded catalyst life. The use of preformed metal nanoparticles, protected by a porous layer of nanocomposite oxides, is a successful strategy also for ethanol steam reforming. Notably, the rather low Rh loading, presently used, opens perspectives for technological transfer to industrial applications.

1. Introduction

Global energy consumption is expected to increase dramatically in the next decades, driven by rising standards of living and a growing population worldwide. The increased need for more energy will require enormous growth in energy generation capacity, more secure and diversified energy sources, and a successful strategy to tame greenhouse gas emissions. Among the various alternative energy strategies, building an energy infrastructure that uses hydrogen as the primary carrier that connects a host of energy sources to diverse end uses may enable a secure and clean energy future [1].

Before hydrogen-fueled future can become a reality, however, many complex challenges must be overcome.

Hydrogen production is the first step towards the transition to the hydrogen economy. A fundamental question in any strategy regarding how to establish it as a viable energy source economy entails how to produce hydrogen in sufficient quantities to meet the current energy demand. Hydrogen can be produced from a variety of feedstocks: from fossil resources such as natural gas and coal, from renewable resources such as biomass and from water. A variety of process technologies can be used, including chemical, biological, electrolytic, photolytic and thermo-chemical. Each technology is in a different stage of development, and each offers unique opportunities, benefits and challenges. Local availability of feedstock, maturity of technology, market applications and demand, policy issues and costs will influence the choice and timing of the various options for hydrogen production.

In the near- and mid-term, hydrogen production from hydrocarbons seems to be the best choice to achieve a gradual transition, given that the present infrastructure can be used and a certain reduction degree of greenhouse gas emissions can be achieved.

The commercially usable hydrogen currently being produced is extracted mostly from natural gas. Nearly 90 % of hydrogen is obtained by steam reforming of methane (SMR), the main component of natural gas. The process involves conversion of methane and steam to hydrogen and carbon monoxide (syngas). CH_4 is the most ideal source of hydrogen since it is the most abundant hydrocarbon resources and has the highest H/C ratio among all the hydrocarbons.

The process is highly endothermic, as shown in Eq.1, and very expensive, due to the high heat demand [2,3]. The typical temperatures and pressures for the process are 700 to 850 °C and 3 to 25 bar. The reaction produces the highest theoretical CO/H_2 ratio, but the ratio may be tailored by changing the reaction conditions [4,5]. CO can be further converted to CO_2 and H_2 through the water gas shift reaction (Eq.2). Finally the hydrogen produced has to be purified.

$$CH_4 + H_2O \longrightarrow CO + 3H_2 \qquad \Delta H^{O}_{298K} = 206.2\,\text{kJ mol}^{-1} \quad (1)$$

$$CO + H_2O \rightleftarrows CO_2 + H_2 \qquad \Delta H^{O}_{298K} = -41.1\,\text{kJ mol}^{-1} \quad (2)$$

Many late transition metals such as Pd, Pt, Ru, Rh and Ir can be used as catalysts for steam reforming, but nickel-based catalysts are the most feasible because of their acceptably high activity and significantly lower cost. More reactive metals such as iron and cobalt are in principle active but they oxidize easily under process conditions. Ruthenium, rhodium and other noble metals are more active than nickel, but are less attractive due their costs. A typical catalyst consists of relatively large Ni particles dispersed on α-Al_2O_3, MgO or $MgAl_2O_4$ spinel [5]. However, nickel-based catalysts are easily susceptible to deactivation from the deposition of carbon [6] and/or sintering of the metallic and support phases [7,8]. Both reforming and coke deposition are believed to be initiated by the same elementary hydrocarbon activation step [9]. Under steam-reforming conditions, metal surfaces are covered with various CH_x intermediates. Without a fast steam gasification step to convert these intermediates to CO and H_2, these CH_x species adsorbed on Ni surface can undergo further dehydrogenation, polymerization, and rearrangement into highly stable carbon species [10]. These stable carbon species not only show low reactivity towards the gasification reaction, but also may dissolve into or encapsulate the Ni particles. In some cases, the dissolution leads to whisker carbon growth that eventually destroys the catalyst and plugs the reactor. The catalyst poisoning can be suppressed by increasing the feed steam concentration, which results in higher rates of carbon oxidation and improved stability. Moreover high metal loadings are commonly used to overcome sintering induced deactivation. Commercial Ni-based systems may contain up to 15-25 wt % of metal. In order to enhance the overall performance characteristics of the basic formulation a number of additives can be included (e.g. Ca, noble metals, rare earths) [11-14].

A promising alternative reaction route, which has recently received the attention of many research groups, is the Catalytic Partial Oxidation of Methane (CPOM), in which the quantity of oxidizer is less than that stoichiometric required for the complete combustion of the hydrocarbon fuel. The process offers many advantages over the conventional steam reforming: it is a weakly exothermic reaction as expressed in Eq.3, so it requires lower energy costs than SMR; it can operate at low contact times (10^{-2} to 10^{-4} s), allowing the use of small reactors, thus lowering the cost of plant.

$$CH_4 + \frac{1}{2}O_2 \longrightarrow CO + 2H_2 \qquad \Delta H^O_{298K} = -35.7\ \text{kJ mol}^{-1} \quad (3)$$

Concerning the reaction pathway, two mechanistic schemes have been proposed: (a) an indirect scheme, which can be designated as the *combustion and reforming reaction* (CRR) *mechanism* and (b) a direct scheme, which can be designated as *direct partial oxidation* (DPO) *mechanism*. According to the CRR scheme, initial combustion of the hydrocarbon (Eq.4) is followed by the reforming reactions of the unconverted methane with H_2O (Eq.5) and CO_2 (Eq.6) produced in the first step. The water gas shift reaction is also involved.

Methane Complete Combustion (MCC)

$$CH_4 + 2O_2 \longrightarrow CO_2 + 2H_2O \qquad \Delta H^O_{298K} = -802.3\ \text{kJ mol}^{-1} \quad (4)$$

Methane Steam Reforming (MSR)

$$CH_4 + H_2O \longrightarrow CO + 3H_2 \qquad \Delta H^O_{298K} = 206.2\ \text{kJ mol}^{-1} \quad (5)$$

Methane Dry Reforming (MDR)

$$CH_4 + CO_2 \longrightarrow 2\,CO + 2\,H_2 \qquad \Delta H^{O}_{298K} = 247.3\,\text{kJ mol}^{-1} \quad (6)$$

A major problem encountered in this process is the highly uneven temperature profile along the catalyst bed, due to the huge amounts of heat produced by combustion reaction at the entrance of the catalyst bed, which can bring the local temperature well above 1000 °C. This results in several undesirable consequences such as catalyst deactivation due to sintering or danger of explosion [15-19].

On the other hand, in the DPO mechanism CH_4 and O_2 decompose on the surface of the catalyst and surface species recombine to produce CO and H_2 as primary products. CO_2 and H_2O are secondary products, obtained by successive reactions of the primary products with O species in the surface [17-19].

Supported noble metal catalysts are mainly used for the CPOM reaction, including Rh, Ru , Pd and Pt [19-37] and supported Ni catalysts [12,14,38-51]. The type of catalyst employed (e.g. metal, support type, etc) may strongly influence the reaction steps of the process. Despite intensive research efforts, this technology has encountered significant problem in the industrial scaling – up [36].

Another alternative approach to CPOM and SMR is the AutoThermal Reforming process (ATR), which results from a combination of both technologies. The ATR process uses methane or liquid hydrocarbons as fuel that undergoes a reaction with steam and air in a single reactor. Since the ATR process consists of the combination of CPOM and SMR, the balance of the specific heat for each reaction becomes a very distinctive characteristic of this process. This makes the whole process relatively more energy efficient since the heat produced from CPOM can transfer directly to be used by SMR. Under an ideal operating condition with the precise amount of air, fuel and steam, the reaction's theoretical efficiency can even be higher than in conventional SMR process. Potentially, ATR technologies show superior performances with respect to conventional SMR plants in terms of reduced size and weight, lower costs, faster starting time and improved transient-time [52].

Due to the fact that ATR is the least developed process with respect to CPOM and SMR processes, numerous scientist and engineers are putting great efforts in its development. One of the more promising result is that by reducing allowable oxygen to carbon ratios and using innovative catalysts, the operating temperature of the process is greatly reduced from the traditional operating temperature of 1200 °C to 650-900 °C [53-55]. However, although ATR has a higher theoretical efficiency than the SMR, preliminary experiments on pilot plants indicate a great need of process optimization.

The development of more efficient catalysts for steam reforming, catalytic partial oxidation and autothermal reforming processes would lead to the substitution of those currently employed in the industry with consequent environmental and economic benefits. Developing a comprehensive efficiency strategy is the fastest and cheapest way to reduce environmental impact of fossil fuels. Improving the energy efficiency of industrial processes offers impressive savings. However, any plan to substantially reduce greenhouse gas emissions can not succeed through increase in energy efficiency alone. Notably, due to the economic growth, the demand for energy will keep climbing the carbon emissions despite the introduction of more energy-efficient vehicles, buildings and appliances. Therefore, production of hydrogen by exploiting alternative / renewable sources seems imperative in this perspective.

In the long term, a hydrogen-based energy system would have to use renewable sources such as biomass for meeting sustainability goals. Renewables are essentially inexhaustible and their use usually entails much lower emissions of greenhouse gases or other pollutants, and fewer health hazards. In addition to their environmental benefits, renewable sources promise to enhance energy security by reducing the country's reliance on fossil fuels from other nation. Unfortunately the current costs of renewable energy sources are in many cases higher than those of conventional sources, and this has until recently retarded their deployment. Of special interests are methods in which biomass is converted to intermediate liquid bio-fuels. The main advantage of liquid bio-fuels is their high energy density and ease of handling, and the fact that they can be used for the on-demand production of hydrogen for example for fuel cells, with applications in mobile and stationary grid-independent power systems [56,57]. Among various liquid alcoholic fuels, ethanol is gaining increasing attention in recent years for several reasons [58-64]. In addition to advantages related to natural availability, storage and handling safety, it can be produced renewably from several biomass sources, including energy plants, waste materials from agroindustries or forestry residue materials, organic fraction of municipal solid waste, etc. In contrast to other fossil-fuel-based systems which have been proposed for fuel cells applications, namely methanol and gasoline, the bioethanol-to-hydrogen system has the significant advantage of reduced CO_2 emissions, since a significant fraction of the produced carbon dioxide is consumed for biomass growth. Notably, the carbon cycle is not a perfectly closed loop, due to the energy requirements for biomass cultivation, transformation and residue treatments. The conversion of ethanol to H_2 by steam reforming reaction has been widely investigated [64,65]. Most of the catalysts used are Al_2O_3-supported base metals [66-79], noble metals [65,70,72,74,80-85] or alloys [86,87]. The process operates at relatively higher temperatures (around 700 °C). To effectively integrate the ethanol reformer with fuel cells, the reformer should work at temperatures as low as possible. This process presents some disadvantages, such as formation of by-products and catalytic deactivation. Hence, the economic viability for the use of ethanol for this application depends on the development of new catalysts and determination of the appropriate reaction conditions.

To make the reactions described above economically viable and competitive in industrial market, it is necessary to design catalysts with high activity, selectivity and, more important, stability and log lifetime. The severe working conditions encountered in H_2-production processes lead to easy catalyst deactivation. Therefore it is important to develop new catalytic materials with superior thermal stability than those currently available and consequently the development of new appropriate synthetic procedure able to stabilize the catalyst nanostructure. In the case of base-metal catalyst, high metal loadings can be used to overcome deactivation induced by sintering. The same approach cannot be adopted for precious metals. A possible option in this case is the design of a catalyst with a very high metal dispersion, which must be stable under reaction conditions. This would allow metal loadings to be lowered to acceptable values.

Recently, great attention has been dedicated to the development of synthetic methods for the preparation of nanostructured catalysts, stable at high temperature. The solid phase crystallization (SPC) technique or the microemulsion method are two of these proposed approaches. The SPC strategy is based on the preparation of a crystalline oxide precursor (generally perovskite and hydrotalcite compounds) by sol-gel or co-precipitation method in the presence of ions of the active metal. After calcination, the material will homogeneously contains species of the active metal dispersed inside the bulk. The reduction at high temperature leads to the migration of the metal atoms to the surface to form small metallic particles which are homogeneously dispersed [88]. The metal–support interaction in this case is expected to be stronger than that obtainable by the usual impregnation or deposition methods. Using SPC active and thermally stable catalysts have been produced for reforming reaction involving methane [30,31,89-98] but the noble metal loading is too high for large scale exploitation. Concerning the preparation of catalytic materials, the microemulsion technique shows interesting advantages related to the possibility of controlling properties such as particle size, morphology and size distribution [99]. Nanosized particles with a narrow size distribution can often be achieved and, therefore, catalytic reactions can benefit from this preparation procedure as well as support materials where a high surface area and thermal stability are required. The synthetic strategy is quite successful in producing active and stable catalysts [100] but it usually requires expensive reagents and large quantities of them which have to be then removed during post synthesis treatments.

Our research focuses on development of novel Rh based catalysts with high surface area and high metal dispersion suitable for Catalytic Partial Oxidation of Methane (CPOM) and Steam Reforming of Ethanol (SRE) [37, 79, 85]. Rh is chosen as the active phase because evidence indicates that this noble metal is the best choice for C-H and C-C bonds activation [101-103] which is involved in the reforming reactions. Despite its high cost, Rh is widely investigated since it shows good yields, good selectivity and good resistance towards deactivation due to coke deposition [104]. Moreover, the high catalytic activity of Rh can consent the use of low metal loadings, which is a significant economic advantage for commercialization of such type of catalysts. The challenging approach for the catalyst design [37, 79, 85], that we believe can ensure the realization of innovative systems, is based on covering pre-formed metal nanoparticles with a protective layer of oxide support in order to limit metal sintering. The porous nature of the inorganic matrix prevents the total occlusion of the particles, and consequently favours accessibility of the catalytic sites to reactants (Scheme 1).

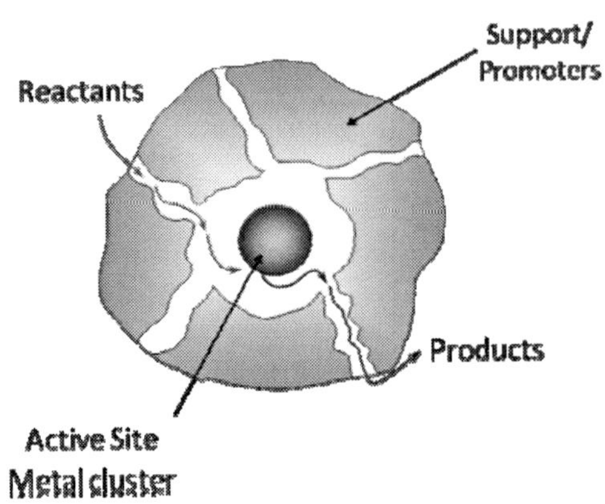

Scheme 1. Ideal structure of an embedded catalyst composed of metal nanoparticles protected by a porous layer of an oxide.

The validity of this approach has been recently presented also by Corma and co-workers in a very elegant work [105]. The authors presented a modified sol-gel procedure for the preparation of Au nanoparticles embedded in silica which is based on the incorporation of the metal nanoparticles into an open shell of silica. The role of the encapsulation is to limit the sintering of the metal particles for high temperature applications. In particularly the synthesis involves the formation of a three-component metal-organic structure composed of Au nanoparticles that are capped with alkanethiols and partially functionalized groups, and polymerized with tetraethyl orthosilicate. Alkanethiols reduce the tendency of Au particles to aggregate [106]. Moreover the particles reveals a narrow size distribution centered at approximately 2 nm. The reduced dimensions of Au particles together with interaction strength with the support is a important factor to convert the inert gold into highly active catalyst [107-110]. The material obtained in this way shows high activity in the oxidation of CO and in the Water Gas Shift Reaction (WGSR).

Here we present and discuss our simple and low cost synthetic strategy for the development of efficient and stable embedded Rh based catalysts for CPOM and SRE. The catalysts are obtained by co-precipitation method. The synthesis can be divided in two steps. The first step sees the preparation of a stable suspension of protected metal nanoparticles according to the method reported in the literature by Schulz and coworkers [111-113]. The metal colloid suspension is prepared using a highly water-soluble ionic surfactant as protective agent. Surfactants are usually organic compounds that are amphipathic, meaning they contain both hydrophobic groups (their tails) and hydrophilic groups (their heads). Due to their molecular structure they are able to assemble in solution into aggregates that are known as *micelles*. In water their tails form a core that is like an oil droplet, while their heads form an outer shell which maintains favorable contact with water. The charged heads of ionic surfactant prevent micelle aggregation. Moreover modifying parameters like pH, temperature, surfactant concentration, it is possible to tune the size of micelles and consequently metal particle sizes. The role of surfactant is to modulate the particle size and to prevent their aggregation and finally to control their encapsulation which represents the second step of the synthesis. During this step the growth of the porous oxide layers around metal nanoparticles also takes place. Post-synthesis treatments in order to remove organic materials lead to final catalyst.

The method allows the modulation of the nature of the support and its texture, the inclusion of extra components (promoters) in the catalyst formulation, thereby leading a strong flexibility to the approach.

2. Experimental

2.1. Starting Materials

$Rh(NO_3)_3 \cdot 2H_2O$ (Johnson Matthey), $NaBH_4$ (98+%, ACROS), 1-bromohexadecane (99%, ACROS) and N,N-dimethylethanolamine (99%, ACROS), and absolute ethanol (Carlo Erba) were used for the synthesis of the metal nanoparticles suspension.

$Al(NO_3)_3 \cdot 9H_2O$ (≥ 98.0%, Fluka), $Ce(NO_3)_3 \cdot 6H_2O$ (99.99%, Aldrich), $ZrO(NO_3)_2 \cdot xH_2O$ (44.85% ZrO_2, Aldrich), H_2O_2 (30%, Fluka), aqueous NH_4OH solution (30%, Carlo

Erba), NH_4NO_3 (> 99.0%, Aldrich) and 2-propanol (electronic grade, Carlo Erba) were used for the synthesis of the porous oxidic materials / supports.

2.2. Synthesis of the Catalysts

The cationic surfactant N-hexadecyl-N-(2-hydroxyethyl)-N,N-dimethyl ammonium bromide (HEAC16Br) was synthesised by reaction of 1-bromohexadecane with a 30% excess of N,N - dimethylethanolamine in absolute EtOH under reflux for 1 h [114], and purified by crystallisation in absolute EtOH [115] (Yield 70 – 80%, m.p. 200 – 210 °C, ^{1}H-NMR and ^{13}C-NMR spectra as reported in Ref. [113]).

Typically, 5g of the different catalysts were prepared. All samples contain a Rh nominal loading of 1 wt %.

Stabilized Rh nanoparticles were prepared under Ar at 20 °C, as reported by Schultz *et al.* [111]. An aqueous solution (122 mL) containing $NaBH_4$ (46 mg, 1.2 mmol) and the cationic surfactant HEAC16Br (0.364 g, 0.92 mmol) was quickly added under vigorous stirring to an aqueous solution of $Rh(NO_3)_3 \bullet 2H_2O$ (6.4 mL, 0.158 g, 0.49 mmol). The obtained suspension was then stirred for 2 hrs to decompose the excess of reductant. The HEAC16Br/Rh ratio was varied in order to study the effect of the surfactant amount on the dimensions of the obtained nanoparticles.

The preparation of the final catalysts was performed by embedding the protected Rh nanoparticles into nanostructured oxidic matrixes. The growth of the porous oxide around the Rh nanoparticles was obtained by the precipitation of the corresponding metal hydroxides in the presence of the colloidal suspension of Rh nanoparticles.

The *first generation* of catalysts was prepared by depositing the hydroxides in a single step. Following this preparation, 36.42 g of $Al(NO_3)_3 \bullet 9H_2O$ were added to the suspension of the Rh nanoparticles and stirred until the complete dissolution of the salt was achieved. Notably, the pH of the resultant solution was extremely low (< 1). The precipitation of the hydroxide was performed dropping the Al^{3+} solution, containing the Rh nanoparticles, into 300 mL of 10 wt% NH_4OH solution. After the complete precipitation of the aluminium hydroxide, the product was aged for 2 hrs at room temperature and filtered. The Br^- ions were removed by washing the precipitate with a NH_4OH / NH_4NO_3 buffer solution (pH = 10) until no Br^- ions were detected in the mother liquor (typically 4 washing cycles). The product was suspended into 500 mL of 2-propanol, stirred at room temperature overnight and then for 5 hrs under reflux, to stabilize the textural network of the precipitate [116,117]. The solid was filtered, dried at 120 °C overnight and finally calcined in a static furnace, firstly at 500 °C for 5 hrs and subsequently at 900 °C for 5 hrs (heating rate 3 °C min^{-1}).

To avoid some unwanted effects observed during the single step precipitation of the hydroxides (see §3.1 for details), a *second generation* of catalysts was developed.

Specifically, 36.42 g of $Al(NO_3)_3 \bullet 9H_2O$ were dissolved with water and diluted up to 250 mL for the preparation of 5 g of Rh(1 wt%)@Al_2O_3. Similarly 21.70 g of $Al(NO_3)_3 \bullet 9H_2O$, 4.28 g of $Ce(NO_3)_3 \bullet 6H_2O$ and 0.68 g of $ZrO(NO_3)_2 \bullet x\ H_2O$ were used for the preparation of 5 g Rh(1 wt%)@$Ce_{0.8}Zr_{0.2}O_2$(40%)-Al_2O_3. 25 mL (1/10) of this solution were added to the Rh nanoparticles. The pH of the final suspension was ~ 3.5, considerably higher with respect to the previous case. The first layer of mixed hydroxide was precipitated dropping the obtained suspension into 60 mL of 10 % NH_4OH solution. After aging for 2 hrs, the precipitate was washed 4 times with NH_4OH / NH_4NO_3 buffer solution (pH = 10) to remove the Br^- ions. After this, the precipitate was suspended by prolonged sonication into 300 mL of NH_4OH / NH_4NO_3 buffer solution (pH = 10) and the porous nanocomposite layer was expanded by adding dropwise the remaining 90% of the metal precursor solution. The final precipitate was aged overnight at room temperature, filtered, suspended into 500 mL of 2-propanol and refluxed for 5 hrs. After filtration, the solid was dried at 120 °C overnight and calcined at 500 °C for 5 hrs, (heating rate 0.5 °C min^{-1}). The catalysts were finally calcined at 900 °C for 5 hrs, (heating rate 0.5 °C min^{-1}). Hereinafter, these materials are indicated as calcined Rh(1wt%)@Al_2O_3 or Rh(1wt%)@$Ce_{0.8}Zr_{0.2}O_2$(40wt%)-Al_2O_3, where @ denote the fact that these are embedded catalyst.

As reference, a standard impregnated Rh(1wt%)/Al_2O_3 catalyst was prepared. The Al_2O_3 support was synthesised using the first preparation route. After calcination at 900 °C, the metal was deposited using a standard "incipient wetness impregnation" using a $Rh(NO_3)_3$ solution. After drying at 120 °C overnight, the material was calcined at 900 °C, (heating rate 3 °C min^{-1}).

2.3. Catalyst Characterization

H_2 chemisorption at -90°C and BET surface area measurements were conducted using a Micromeritics ASAP 2020C analyzer. H_2 adsorbed volumes were determined by extrapolation to zero pressure of the linear part of the adsorption isotherm (10 – 30 torr) after elimination of the so-called reversible hydrogen adsorption. A chemisorption stoichiometry H : Rh = 1 : 1 and a spherical geometry were assumed for metal particles estimation. N_2 physisorption isotherms were collected at -196°C on 0.1 g of sample, after overnight in vacuum at 350 °C.

Powder X-ray diffraction patterns were collected on a Siemens Kristalloflex Philips instrument using a Ni-filtered Cu-Kα radiation. The crystallites size were calculated using the Scherrer equation.

Temperature Programmed Reduction (TPR) experiments were performed according to procedures previously described [118]. Typically, 0.1 g of the calcined materials were pre-treated at 500 °C for 1 h in flowing O_2 (5%) / Ar, then purged with Ar at 500 °C for 15 min and cooled to room temperature. H_2 (5%) / Ar was admitted to the reactor and the flow allowed to stabilise for 30 min before increasing the temperature to 1000 °C (10 °C min^{-1}). H_2 consumption in the TPR experiment was estimated from the integrated peak areas by comparison with those obtained by using CuO as standard. In the case of the $Ce_{0.8}Zr_{0.2}O_2$-containing catalyst, after TPR the samples were out-gassed under Ar flow at 1000 °C for 15 min and cooled to 423 °C, at which temperature full oxidation of the cerium ions was carried out with pulses of O_2. Temperature Programmed Oxidation (TPO) experiments were carried out using a HPR200 Hiden Mass Spectrometer as detector, under O_2 (5%) / Ar (40 mL min^{-1}, 10 °C min^{-1}).

Transmission electron microscopy analysis was performed with a Philips CM200 UT microscope with a point resolution of 0.19 nm, operating at 200 kV. The images were recorded with photographic film or with a TV camera attached to the microscope. The particles were identified from the lattice or atomic column imaging using the digital Fast Fourier Transform (FFT). To prepare specimens, hexane suspensions of the powder were treated in an ultrasonic bath and then placed drop-wise on the ultrathin or holey carbon substrate.

Catalytic experiments were conducted in a U-shaped quartz microreactor. On-line mass spectrometry (MS) and gas chromatography (GC) activity tests were performed, respectively, using a Hiden HPR20 and a Hewlett Packard 5890 Series II gas chromatograph. For GC, a Molsieve 5A column (25 m x 0.53 mm ID) with Ar as carrier and a thermal conductivity detector (TCD) was used to analyse H_2, O_2, N_2, CH_4 and CO. A Select Permanent Gases/CO_2 PLOT column (parallel PoraPLOT 50m x 0.53mm ID and Molsieve 5A 10 m x 0.53 mm ID columns) with He as carrier and connected in series to a methanator and flame ionisation detector (FID) was used to analyse the C-containing compounds from CO to ethyl ether. C balance was always within ± 2%.

Before testing, the calcined materials were treated under O_2 (5%) / Ar at 500 °C for 1 h (40 mL min^{-1}, 10 °C min^{-1}) and activated by reduction in H_2 (5%) / Ar at 750 °C for 2 hrs (40 mL min^{-1}, 5 °C min^{-1}).

CPOM experiments were conducting on 8 – 10 mg of palletised catalyst (~ 2 tonn cm^{-2}, 250 - 300 μm), diluted in a 1:5 ratio with high purity Al_2O_3 (Grace Davison, calcined at 1100 °C for 5 hrs). A layer of quartz granules was placed on top to ensure thermal homogeneity of the reactive mixture passing over the catalyst bed. The temperature of the catalyst was measured with a K-type thermocouple. Gas flow rates were chosen in the range 60-100 mL min^{-1} to ensure GHSV values of 700000 mL g^{-1} h^{-1}.

SRE experiments were conducted on 10-12 mg of pelletized catalyst, diluted in a 1:2 weight ratio with high purity Al_2O_3 (Grace Davison, calcined at 1300 °C for 5 hrs) . EtOH/H_2O 1:5 mixture were injected into an Ar flow using a KDS Model 101 syringe pump. All the transfer lines between syringe, reactor and GC were kept at 120 °C. Gas flow rates were ~ 30 mL min^{-1}. On-line GC analysis was performed using a Hewlett Packard 5890 Series II gas chromatograph with TCD and FID detectors. Before testing the activity, the catalysts materials were pretreated under diluted O_2 at 500 °C for 1 h and activated by reduction in diluted H_2 at 750 °C for 2 hrs.

3. Results and Discussions

3.1. Synthesis

A stable suspension of Rh nanoparticles was prepared by the reduction of a metal salt ($Rh(NO_3)_3$) with $NaBH_4$ in the presence of HEAC16Br as protective agent. The procedure as proposed by Shultz *et al.* [111-113] is shown in Scheme 2.

In the absence of the stabilizer, aqueous particle dispersion is inherently unstable because of the formation of aggregates. The strong tendency to form aggregates is caused by attractive van der Waals forces, which basically always act at short separation distances between particles in water. These forces vary inversely as the sixth power of the distance between their surfaces. Consequently, the use of a stabilizing agent able to induce a repulsive force opposed to the van der Waals forces is necessary to provide stable nanoparticles in solution [119]. The aggregation leads to the loss of reactivity. In order to maximize the catalytic activity, it is important to optimize the dispersion of the metallic phase producing small particles. In fact, as a general rule, the efficiency of a catalyst is directly proportional to its surface area and particularly to the amount of exposed active sites. This is due to the fact that the reactions take place on the surface of the catalyst. Therefore, the active atoms of the catalyst are those of the surface layers only. To reduce the dimensions it means to increase the number of the active sites.

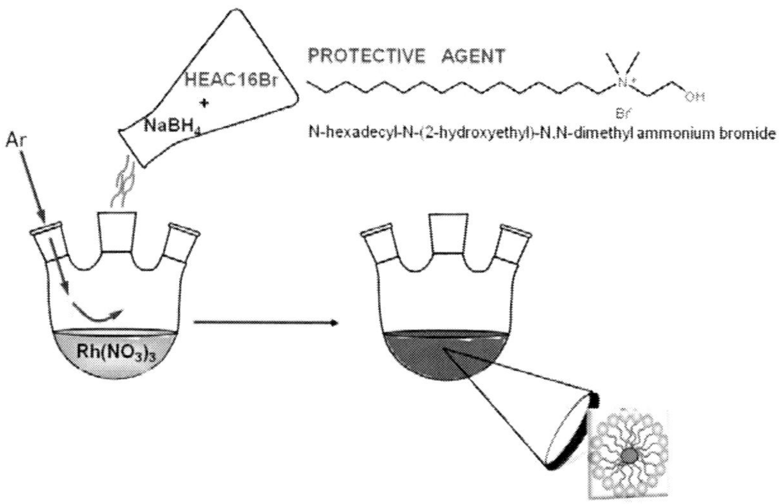

Scheme 2. Schematic representation of the synthesis of Rh nanoparticle colloidal suspension.

A metal/surfactant ratio of 0.53 led to the formation of metal particles with a size distribution centered at 2.1 nm. By increasing this ratio, the size distribution moved to greater values as shown in Figure 1 (3.1 and 3.8 nm using 0.26 and 0.11 metal/surfactant ratio respectively).

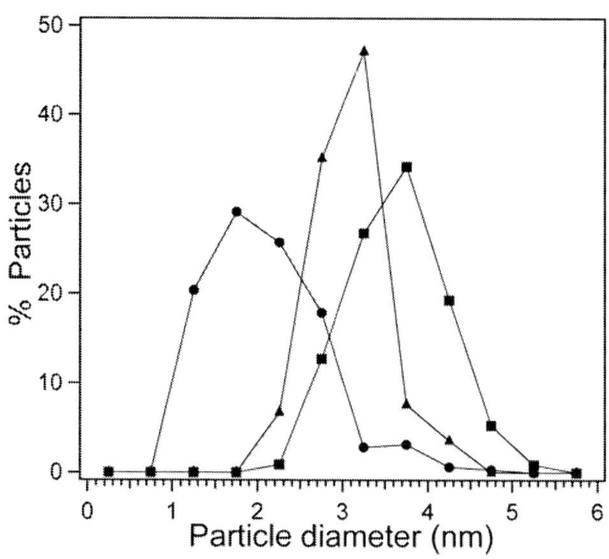

Figure 1. Rh particle size distribution obtained varying the metal/surfactant ratio: (•) 0.53; (▲) 0.26; (■) 0.11 (adapted from Ref. [79]).

The observed increase of Rh particle size with surfactant concentration was explained considering the geometrical and packing factors of surfactant molecules. It is important to emphasize that the particular molecular architecture of the surfactant molecule establishes the type of aggregate (e.g. spherical micelle, lamellar sheets, worm-like micelles, etc). Moreover, they are dynamic structures, able to rearrange in response to changing environmental conditions. Therefore, the increased surfactant addition to the solution could lead to an increased number of micelles of a given size or a modification of the micellar structure to allow a large number of molecules to be incorporated into each micelle [119]. The latter case can be achieved by a change of micelle shape from spherical micelles to a disc-like or cylindrical micelles for example. This is possible because the repulsion between head groups can be decreased by this change in structure. The morphology changes of surfactant may explain our observed trend.

The Rh particles with average diameter 2.1 nm were used in the preparation of the embedded catalysts. The complete catalyst preparation procedure is reported in Scheme 3 (see also § 2.2). The precipitation of support hydroxide takes place once all.

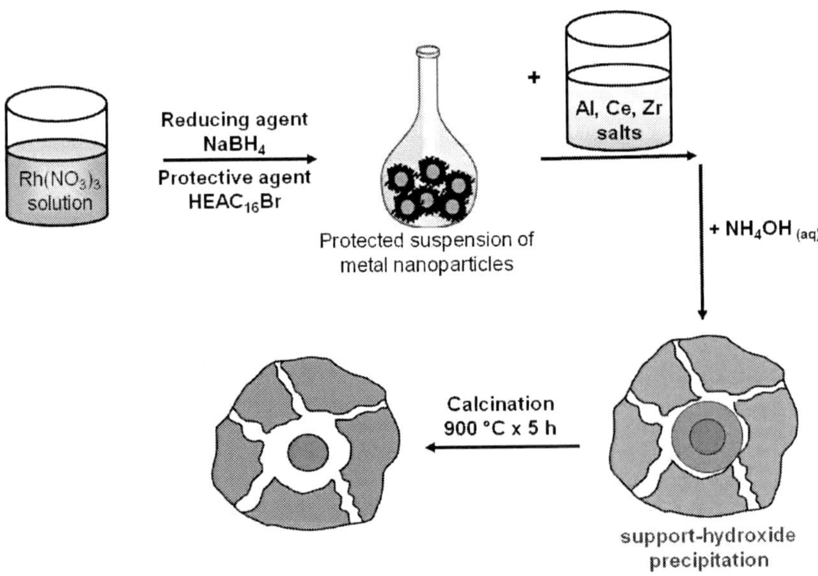

Scheme 3. Schematic representation of the synthesis of Rh embedded catalyst, single shell material.

A crucial aspect during the synthesis was the addition of the Al salt to the Rh nanoparticle suspension. In this step, particular attention must be dedicated to the stability of micelles which protect the metal particles. Depending on the nature of the stabilizing forces, aggregation may be triggered by an increase in the ionic strength or changes in the solution pH. A dispersion of charged colloidal particles, for instance, is usually stable at low salt concentrations, but can be destabilized by mixing with a salt solution of higher concentration. At a high concentration of salt ions, the particle charge is screened, and the repulsive electrostatic force is overcome by the attractive van der Waals force. Another example are particles with carboxylic groups on their surface. At high pH, the carboxylic groups are dissociated and repulsive electrostatic forces stabilize the dispersion, but lowering the pH discharges the particles and induces aggregation. In the present work the addition of Al salt, during the production of the Rh(1wt%)@Al_2O_3 sample, led to a drastic decrease of the pH and a strong increase of the ionic strength of the solution. As the ionic strength was increased, the electrostatic repulsive forces were progressively reduced allowing micelles collisions. Consequently the stability of the micelles decreased under these conditions, leading to a partial coalescence of the metal nanoparticles. The short inter-particle distances determined the aggregation behaviour as suggested by our HRTEM characterization [37]. To avoid this inconvenient, the synthetic procedure was modified for the production of a second generation of catalysts, as reported in Scheme 4.

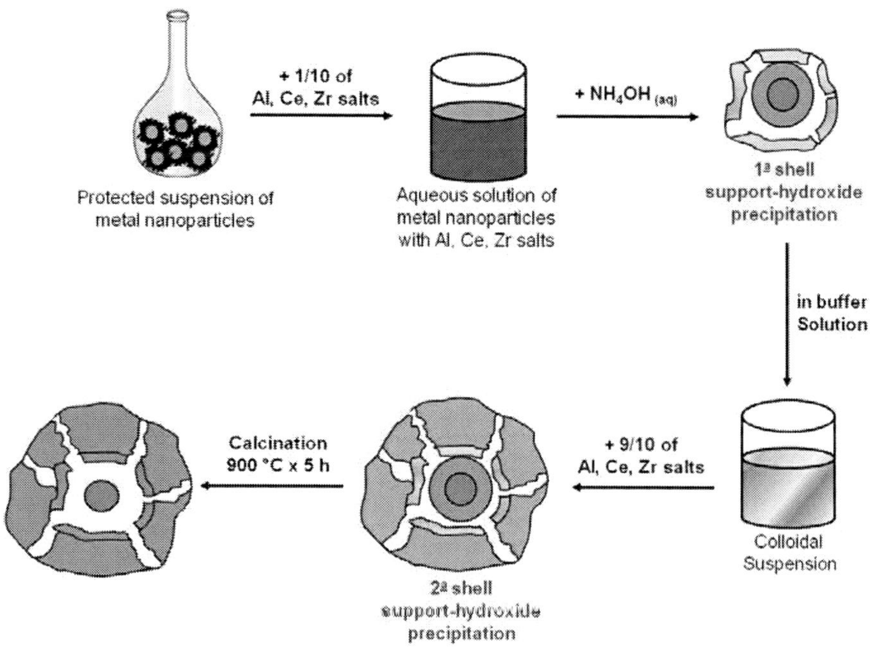

Scheme 4. Schematic representation of the synthesis of second generation of Rh embedded catalyst, 2-shell material.

In this case, the growth of the oxidic matrix occurred in a two step process, which involved the addition of only 10% of the required aluminium to the Rh nanoparticles solution. After the precipitation of a first protective layer under mild conditions of pH and ionic strength, the precipitate was washed and used to growth the second protective layer with the remaining 90% of the required aluminium. The same approach was used also for the preparation of a more complex nanocomposite matrix, $Ce_{0.8}Zr_{0.2}O_2$(40wt%)-Al_2O_3.

3.2. Characterization

Figure 2 shows the N_2 physisorption isotherms of the investigated embedded and impregnated Rh based catalysts, after standard activation procedure (reduction at 750 °C). All samples presented Type IV isotherms with hysteresis loops, typical of mesoporous materials [120]. The t-plot analysis indicated that the microporous volume is always negligible. The BJH analysis revealed that the materials have a bimodal pore distribution (Figure 3). The surface area of all materials was high despite the high temperature calcinations and the severe pre-reduced treatment at 750 °C. This confirms the importance of the alcohol thermal treatment to enhance and stabilize the porous structure [116]. The embedded Rh(1wt%)@Al_2O_3-1-shell showed a BET area slightly higher with respect to the corresponding 2-shell sample. The two samples had a very similar total pore volume (0.79 and 0.78 mLg^{-1}) (Figure 3). These small differences were mainly associated with the different heating rate adopted for the calcinations, 3 and 0.5 °C min^{-1} for first and second generation catalyst respectively. In fact, a slow heating rate can favor the low temperature crystallization and partially prevent a collapse of the porous structure of the hydroxide during their transformation to oxide in the calcinations step. This had certainly positive effects on the long term stability of the final materials. However, the total time of calcinations was higher at lower heating rate, justifying a slightly lower BET surface area of the 2-shell materials with respect to the 1-shell ones.

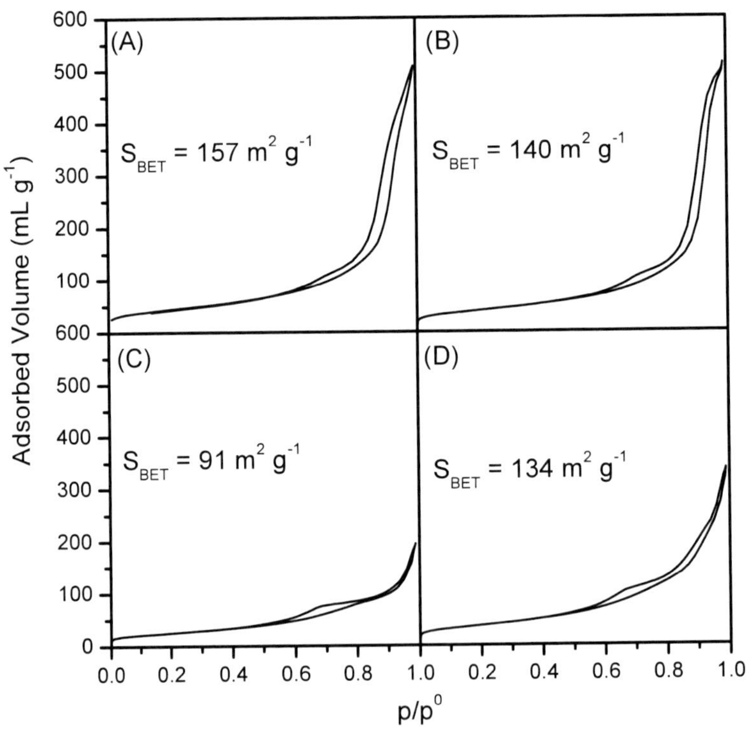

Figure 2. N_2 physisorption isotherms of (A) embedded Rh(1wt%)@Al_2O_3-1-shell, (B) embedded Rh(1wt%)@Al_2O_3-2-shell, (C) embedded Rh(1wt%)@$Ce_{0.8}Zr_{0.2}O_2$(40wt%)-Al_2O_3-2-shell and (D) impregnated Rh(1wt%)/Al_2O_3. The samples were subjected to standard activation procedure (reduction in H_2 (5%) / Ar at 750 °C for 2 hrs).

The introduction of ceria-zirconia mixed oxides into the Rh(1wt%)@Al_2O_3-2-shell reduced by ~ 35% the BET surface area and by ~ 62% the total pore volume. While the mesopores with average diameter around 6 nm (~ 0.1 mL g^{-1}) were marginally influenced by the addition of the rare earth components, a significant increase of the average pore diameter of the bigger mesopores and a decrease in the corresponding pore volume were also observed. These results are consistent with textural properties of $Ce_xZr_{1-x}O_2$-Al_2O_3 nanocomposites previously reported [121,122]. The impregnated sample had a surface area (134 m^2 g^{-1})

identical to that of the pure Al_2O_3 (140 m^2 g^{-1}), while the total pore volume was reduced from 0.77 mL g^{-1} (pure Al_2O_3) to 0.58 mL g^{-1}.

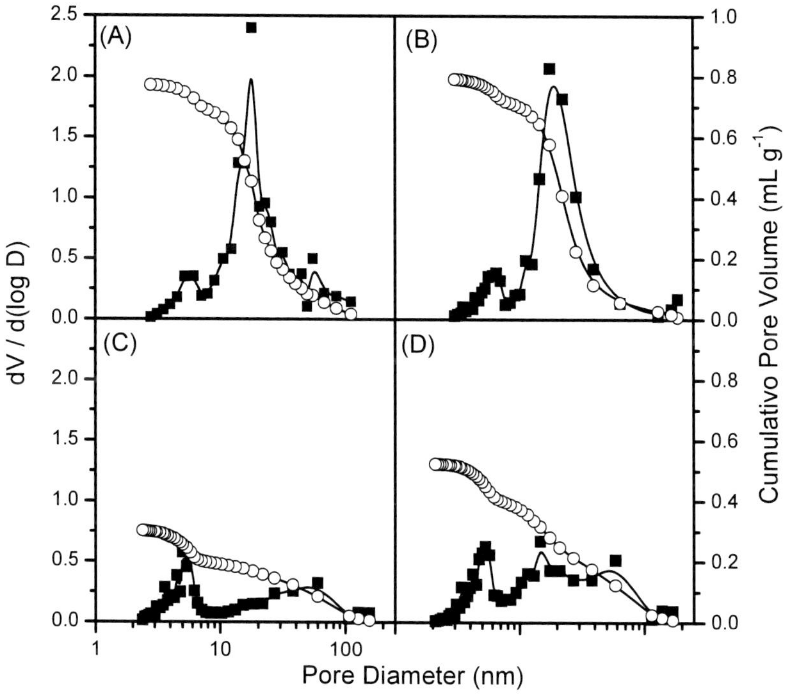

Figure 3. BJH pore distribution (filled symbols) and cumulative pore volume (open symbols) of (A) embedded Rh(1wt%)@Al_2O_3-1-shell, (B) embedded Rh(1wt%)@Al_2O_3-2-shell, (C) embedded Rh(1wt%)@$Ce_{0.8}Zr_{0.2}O_2$(40wt%)-Al_2O_3-2-shell and (D) impregnated Rh(1wt%)/Al_2O_3. The samples were subjected to standard activation procedure (reduction in H_2 (5%) / Ar at 750 °C for 2 hrs).

Powder XRD patterns of the samples after calcination are presented in Figure 4. Overlapping of the XRD peaks of the transition aluminas complicated phase attribution, as did their low crystallinity. However, analysis indicated compositions of approximately 50% γ-Al_2O_3 and 50% θ-Al_2O_3 for embedded Rh(1wt%)@Al_2O_3-1-shell and Rh(1 wt%)@Al_2O_3-2-shell and 40% γ-Al_2O_3 and 60% θ-Al_2O_3 for impregnated Rh(1wt%)/Al_2O_3. The slightly higher θ/γ-Al_2O_3 ratio observed in the impregnated sample with respect to the embedded ones was explained in terms of the additional calcination treatment applied to the impregnated material.The mean crystallite size, determined following the Scherrer equation, was ~ 6 nm for γ-Al_2O_3 and ~ 7 nm for θ-Al_2O_3, independently of the material.

In the case of the Rh(1wt%)@$Ce_{0.8}Zr_{0.2}O_2$(40wt%)-Al_2O_3-2-shell, the XRD pattern was dominated by the $Ce_{0.8}Zr_{0.2}O_2$ reflexes, due of the higher scattering factor of Ce and Zr with respect to Al. The $Ce_{0.8}Zr_{0.2}O_2$ phase was present as a solid solution with a mean crystallite size of 11 nm. A reasonable quantification of the amounts of the transitional forms of Al_2O_3 present in this sample was not possible, due to the low intensity of their reflections. In all the samples, phases related to the presence of Rh species could not be detected due to the low metal loading.

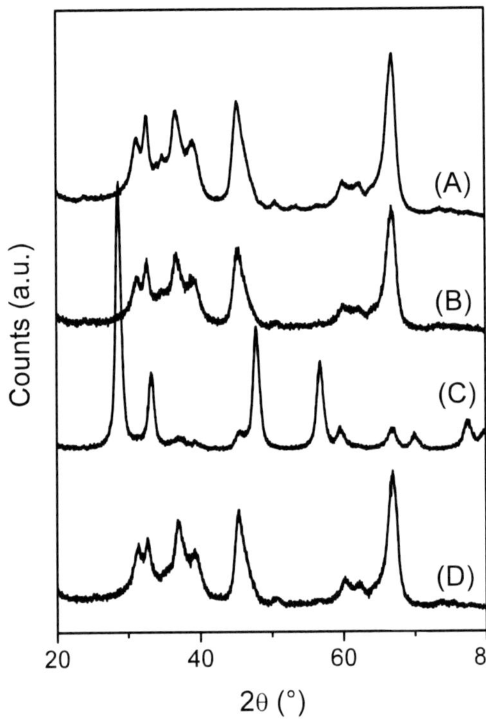

Figure 4. Powder XRD patterns of (A) embedded Rh(1 wt%)@Al_2O_3-1-shell, (B) embedded Rh(1wt%)@Al_2O_3-2-shell, (C) embedded Rh(1wt%)@$Ce_{0.8}Zr_{0.2}O_2$(40wt%)-Al_2O_3-2-shell and (D) impregnated Rh(1wt%)/Al_2O_3 (adapted from Refs. [37,85]).

The results obtained from the analysis of the powder XRD profiles agreed very well with their textural properties. Notably, the presence of α-Al_2O_3 was not observed (typical reflections around 2θ = 25°), in accordance with the high surface area of the samples. It is known that the transformation of the transitional aluminas into the thermodynamically stable α-Al_2O_3 is accompanied with a strong sinterization of the material [123]. In the case of the $Ce_xZr_{1-x}O_2$-containig material, it was reported that the addition of the ceria – zirconia solid solution to Al_2O_3 mutually stabilized the surface area of both components, preventing Al_2O_3 sinterization and its transformation into α-Al_2O_3 [122,124].

Figure 5 shows the Temperature Programmed Reduction (TPR) profiles of different samples. The presence of significant consumption of H_2 during the TPR experiments confirmed that the Rh nanoparticles were converted into RhO_x species during the calcination treatment at 900 °C. As a consequence the data indicated the need of an appropriate activation procedure to obtain active metallic Rh nanoparticles. The different types of RhO_x formed from the oxidation of metallic clusters dispersed on the surface of Al_2O_3 were extensively studied by Hwang *et al.* using the TPR technique [125]. It was shown that, increasing the oxidation temperature, the reduction peaks move to higher temperatures. The different reduction temperature of RhO_x species was related to a different interaction with the support: the stronger the interaction, the higher the reduction temperature. It was indicated that the nature of the RhO_x can range from O atoms adsorbed on the surface of metallic Rh (that reduces at sub-ambient temperature) to $Rh(AlO_2)_y$ species (that reduces at very high temperature) obtained from the diffusion of RhO_x into the atomic layers of Al_2O_3 during high temperature oxidation. It is worth noting that Wang *et al.* [126] reported that Rh nanoparticles supported on γ-Al_2O_3 are oxidized at temperature higher than 430 °C to form different species depending on their diameter, RhO_2 for Rh nanoparticles bigger than 1.5 nm or Rh_2O_3 for the smallest nanoparticles.

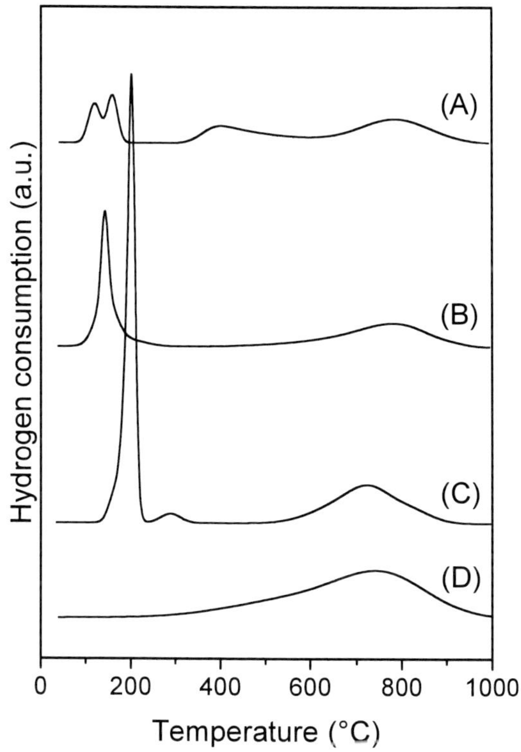

Figure 5. Temperature Programmed Reduction profiles of the samples after calcination and standard cleaning in O_2 (5%) /Ar at 500 °C for 1 h: (A) embedded Rh(1wt%)@Al_2O_3-1-shell, (B) embedded Rh(1wt%)@Al_2O_3-2-shell, (C) embedded Rh(1wt%)@$Ce_{0.8}Zr_{0.2}O_2$(40wt%)-Al_2O_3-2-shell and (D) impregnated Rh(1wt%)/Al_2O_3 (adapted from Refs. [37,85]).

The Rh(1wt%)@Al_2O_3-1-shell sample presented a structured TPR profile, characterized by different reduction peaks centered at 120, 160, 400 and 790 °C. The reduction peaks at 120 and 160 °C were associated to RhO_x species exposed on the surface of Al_2O_3. The different reduction temperature could be ascribed to a different interaction with the support [125] and/or with a different dimension of the RhO_x crystallites [127]. A reduction peak around 400 °C was not reported by Hwang *et al.* [125] but it was present in the TPR profiles of an hydrotalcitic precursor containing Rh prepared by Basile *et al.* [31]. Since in that case Rh^{3+} ions were homogeneously dispersed into the hydrotalcitic matrix, it was reasonable to ascribe the reduction peak at 400 °C in the TPR of the present Rh(1wt%)@Al_2O_3-1-shell to RhO_x species occluded into the Al_2O_3 matrix. Finally, the reduction peak around 790 °C was indicative of the presence of stable Rh species, such as the $Rh(AlO_2)_y$ species reported by Hwang *et al.* [125]. The quantification of the H_2 consumption during the TPR confirmed the presence of different RhO_x species. In fact, 13.1 μmol H_2 g^{-1} were measured during the TPR.

Assuming the presence of Rh_2O_3, this H_2 consumption would correspond to a reduction of ~ 90% of the Rh atoms, while assuming the presence of RhO_2 a 67% Rh reduction would be expected. This result is an indication that the collected information are not sufficient to clearly define the nature of the RhO_x species formed after calcination at 900 °C and only some qualitative considerations on the RhO_x species can be done on the basis of their reduction temperatures.

The TPR experiments highlighted that the redox behaviour of the embedded catalyst was significantly different to that of the impregnated sample. In fact, the TPR profile of the Rh(1wt%)/Al_2O_3 material was characterized by a single extended reduction process centered around 750 °C, indicating a strong interaction between the Rh species and the support.

The Rh(1wt%)@Al_2O_3-2-shell sample presented two major reduction peaks centered at 140 and 780 °C. The presence of a single, symmetric reduction peak at 140 °C was indicative of a high homogeneity of the RhO_x species exposed on the surface of the Al_2O_3 matrix. Notably, the deposition of the oxidic matrix in 2 steps was developed to avoid the coalescence of the nanoparticles during the addition of the Al salt and to improve their homogeneity in the final materials. No reduction peaks were observed around 400 °C, indicating the absence of RhO_x species deeply occluded into the Al_2O_3 matrix. The reduction peak centered at 780 °C suggested the formation of some $Rh(AlO_2)_y$ species [125].

The introduction of $Ce_{0.8}Zr_{0.2}O_2$ into the oxidic matrix led to an intense reduction peak centered around 200 °C, with a minor one around 280 °C. A reduction peak at 730 °C was also present. In the case of the Rh(1wt%)@$Ce_{0.8}Zr_{0.2}O_2$(40wt%)-Al_2O_3-2-shell the H_2 consumption at low temperature (200 – 300 °C) was ascribed to the reduction of both the RhO_x species exposed on the surface of the material and of the small $Ce_{0.8}Zr_{0.2}O_2$ crystallites in contact with the Rh nanoparticles [128,129]. The reduction peak at high temperature could be related to $Ce_{0.8}Zr_{0.2}O_2$ with larger dimensions and/or not directly in contact with the Rh nanoparticles. Part of the H_2 consumption at high temperature could be ascribed also to the reduction of some $Rh(AlO_2)_y$ species formed by the Rh nanoparticles directly in contact with Al_2O_3. After TPR, the amount of reduced Ce was evaluated by O_2 pulsing at 427 °C: 7.9 mL g^{-1} of O_2 were consumed, corresponding to the reoxidation of ~ 54 % of the total Ce(III) species. This reduction degree was in good agreement with the typical Oxygen Storage Capacity (OSC) values reported for $Ce_{0.8}Zr_{0.2}O_2$ composition [130].

The TPR data indicated that an isothermal reduction at 750 °C was the best pretreatment to fully reduce all samples, preventing undesirable sintering. In fact, the standard activation treatment did not modify the surface area and the porosity of the materials.

It must be mentioned that the present catalysts were subjected to both severe reducing (standard activation) and severe oxidising treatments (regeneration, see § 3.3.1). Therefore it is important to recall the tendency of Rh/Al_2O_3 catalysts to deactivate under oxidising conditions. The phenomenon was attributed to sintering of the Rh particles, but also to the formation of difficult-to-reduce Rh species, either because of strong interaction with the support or the formation of a rhodium aluminium oxide (Scheme 5). Reduction treatments at least partially reverse this deactivation. Formation of the above-mentioned $Rh(AlO_2)_y$ species was reported by Hwang et. al. [125] under oxidising conditions comparable to those used in the present investigation, while reduction caused segregation of the Rh from the Al_2O_3, leading to low temperature reduction. The present Rh(1wt%)/Al_2O_3 and Rh(1wt%)@Al_2O_3-1-shell presented this behavior but with different extent [37]. The Rh incorporation into the alumina under oxidising conditions, followed by its segregation during reduction, occured more easily for the impregnated sample.

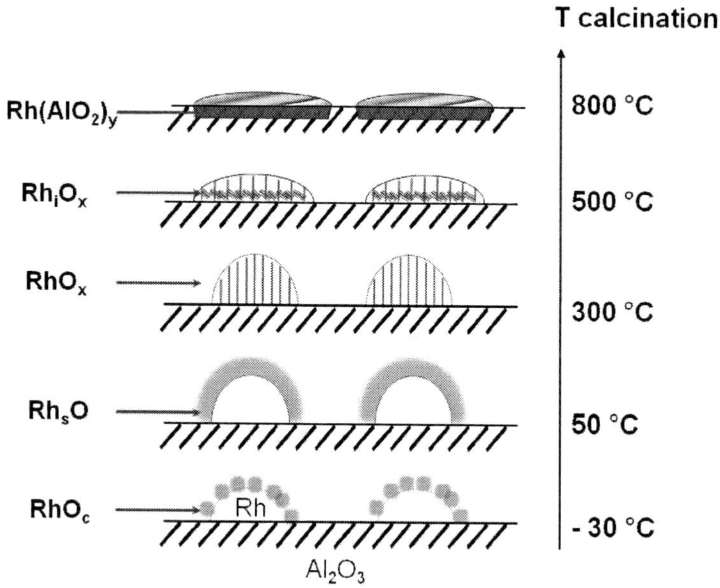

Scheme 5. Schematic representation of the Rh species proposed for Rh/Al_2O_3 system as a function of the calcination treatments. RhO_c = chemisorbed oxygen on the Rh surface; Rh_sO = Rh particles with oxidised surfaces; RhO_x = bulk Rh oxide species; Rh_iO_x = bulk Rh oxide species strongly interacting with the support; $Rh(AlO_2)_y$ = Rh species formed by the diffusion of Rh^{3+} into the outermost layers of Al_2O_3 (adapted from Ref. [125]).

The low metal content and the high surface area of the present samples were intrinsic limitations for evaluation of the metal dispersion by Transmission Electron Microscopy (TEM) analysis. In fact, TEM investigations are often carried out on model systems (low surface area (< 30 m^2 g^{-1}) and/or high metal loading (4-5 wt %)) [131,132]. Nevertheless, HRTEM analysis of the first generation of embedded Rh(1wt%)@Al_2O_3 was successfully conducted. Notably, the second generation of catalysts presented further significant difficulties in the identification of the Rh nanoparticles. In fact, the synthesis was designed to both prevent the metal particles coalescence and to maximise the incorporation of them into the porous matrix, reducing the presence of Rh nanoparticles on the external surface of the alumina or of the $Ce_{0.8}Zr_{0.2}O_2$(40wt%)-Al_2O_3. Moreover, the nanocomposite Rh(1wt%)@$Ce_{0.8}Zr_{0.2}O_2$(40wt%)-Al_2O_3-2-shell showed a relatively tiny contrast between the Rh phase and the dense ceria-zirconia phase. All these facts, precluded a statistically significant analysis of the Rh particle size distribution in the case of the second generation of embedded catalysts.

The TEM analysis of the embedded Rh(1wt%)@Al_2O_3-1-shell catalyst after calcination revealed the presence of a very small number of Rh particles. We associated this with the formation of RhO_x and $Rh(AlO_2)_y$ during the calcination treatment. It is very difficult to distinguish the former from the Al_2O_3 support, while the latter would not be visible as particles. Thus, a statistical analysis was not possible, nor would it be meaningful. Figure 6 shows one of the rare example of the identification of Rh particles.

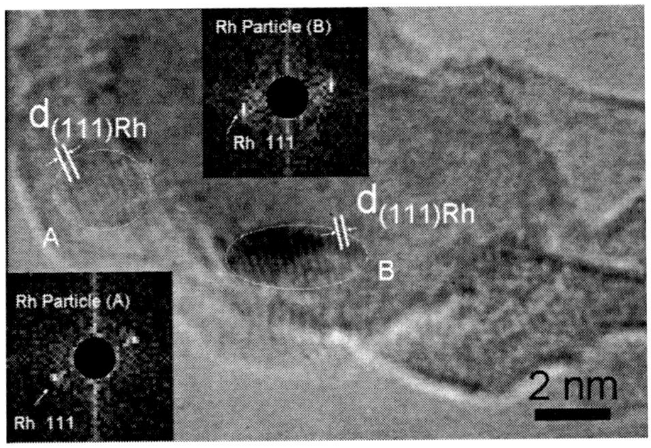

Figure 6. Example of HRTEM analysis and corresponding FFT pattern (insets) on the embedded Rh(1wt%)@Al_2O_3-1-shell after calcination at 900 °C for 5 hrs.

Figure 7 reports representative examples of the embedded Rh(1 wt%)@Al_2O_3-1-shell after the activation procedure. Nanoparticles situated on the surface of a support grain, with various coverages of Al_2O_3, are observed. On the other hand, Rh fully embedded in the Al_2O_3 matrix were also present. Notably, Rh particles on the external surface of the support have lower thermal stability with respect to those protected. The size distribution of the particles ranged from 1 to 8 nm, with a mean size of 2.3 nm (Figure 7).

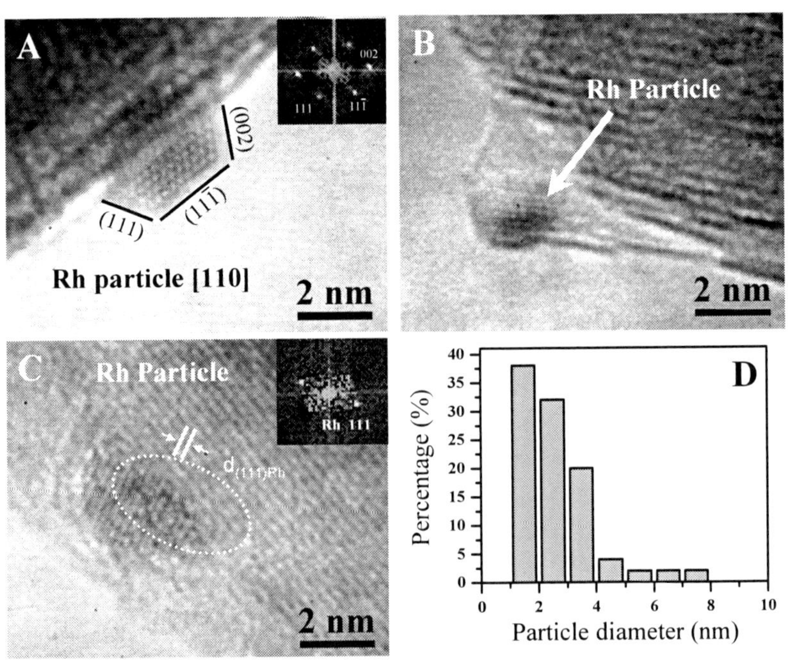

Figure 7. Representative HRTEM analysis on activated embedded Rh(1wt%)@Al_2O_3-1-shell (calcined at 900 °C for 5 hrs and reduced in H_2 (5%)/ Ar at 750 °C for 2 hrs). (A) Rh nanoparticle where the exposed surface shows facets parallel to the (111) and (002) crystallographic planes, (B) Rh nanoparticle (highlighted with the arrow) partially embedded and (C) Rh particle (encircled) which appears to be embedded in the Al_2O_3 matrix, (D) size distributions of the Rh nanoparticles.

Hydrogen chemisorption technique was used to characterize the accessible metallic phase and estimate the average Rh particle diameter of all samples, including those not suitable for HRTEM analysis. In the case of ceria containing catalyst, hydrogen chemisorption experiments must be carried out at low adsorption temperature and in the low pressure region, in order to prevent hydrogen spillover onto the ceria based support [131-133]. In fact, previous HRTEM investigation and chemisorption data clearly demonstrated that true metal dispersion and hence estimation of NM (Rh and Pt) particle size in NM/$Ce_{0.68}Zr_{0.32}O_2$ samples, can be achieved by means of volumetric H_2 chemisorption measurements at low temperature [131,132]. Furthermore, negligible hydrogen spillover at low temperature, was observed in Pt(1wt%)/$Ce_{0.6}Zr_{0.4}O_2$(13wt%)/Al_2O_3 nanocomposites [134]. Notably, it was showed that for Rh/$Ce_{0.68}Zr_{0.32}O_2$ reduced at 700 °C, the metal particles maintain their regular shape and there is still no evidence of decoration of the metal particle; nevertheless loss of chemisorptive capability due to chemical deactivation was observed. The chemisorptive capability could be however recovered by a relatively mild oxidation followed by low temperature re-reduction [132]. For this reason the apparent H/Rh ratios of the present Rh(1wt%)@$Ce_{0.8}Zr_{0.2}O_2$(40wt%)-Al_2O_3-2-shell catalysts were determined both after high temperature reduction - standard activation procedure and after the same treatment followed by the redox cycle of mild re-oxidation / mild reduction. Parallel investigation on Rh(1 wt%)@Al_2O_3 indicated that the adopted mild re-oxidation did not induce metal re-dispersion [85]. The chemisorption results on Rh(1wt%)@$Ce_{0.8}Zr_{0.2}O_2$(40wt%)-Al_2O_3-2-shell showed that minor chemisorption deactivation was observed upon reduction at 750 °C. This fact was interpreted in terms of the low ceria-zirconia loading. Moreover, it must be noted that the standard reduction temperature (750 °C) was only slightly above the temperature at which Rh starts to be chemically deactivated by strongly reduced ceria-based materials [132]. For a correct comparison, all catalyst were subjected to the pretreatments required for the ceria containing systems.

The results of hydrogen chemisorption experiments are detailed in Table 1. The impregnated Rh(1wt%)/Al_2O_3 and the embedded Rh(1wt%)@Al_2O_3-1-shell, after standard activation procedure, showed comparable metal surface area. Since the impregnated Rh(1wt%)/Al_2O_3 was calcined at rather high temperature (900 °C), significant occurrence of Rh incorporation into the support can be expected, as also confirmed by TPR evidences. In addition, high temperature sintering of the Rh phase could certainly justify the modest apparent metal dispersion (around 35 %). Consistently, the same impregnated Rh(1 wt%)/Al_2O_3 but calcined only at 500 °C and reduced at 150 °C showed, as expected, a high metal dispersion (77%). Considering the embedded samples, it is more correct to discuss the chemisorption data in terms of metal accessibility. In fact, as showed by HRTEM of the first generation of catalysts (Figure 7), there were various type of Rh nanoparticles: (i) on the external surface of the alumina support, (ii) partially embedded and (iii) deeply embedded. Different hydrogen diffusion rate and different extent of interaction can be expected with the different Rh particles. In this respect it must be highlighted that the chemisorption for the embedded Rh(1 wt%)@Al_2O_3-1-shell overestimated the apparent dimensions respect to those obtained by HRTEM. This could be partially explained considering the fact that the nanoparticles were partially surrounded by the porous matrix and therefore the Rh atoms at the surface of contact between Rh and the oxide could not contribute to hydrogen chemisorption. However, some of the Rh nanoparticles were also fully buried into the support.

The embedded Rh(1wt%)@Al_2O_3-2-shell showed a lower exposed metal surface area and higher apparent metal particles diameter with respect to the corresponding 1-shell catalyst. This result was apparently in contradiction with the fact that the 2-shell synthesis was designed to optimise the growth of the protective porous oxide layer around the nanoparticles, preventing, at the same time, micelles instability during the synthesis. Notably, the heating rate during the calcination treatments of the 2 shell material was optimised for enhancing the thermal stability of the material. Therefore, we can not exclude some influence of the heating rate on the final Rh particle size. However we must note that the chemisorption data do not necessarily indicate that the true metal nanoparticles were bigger, but showed a lower metal accessibility. Recalling the fact that HRTEM was not able to observe a statistically significant number of Rh particles, we can assume that most of the nanoparticles are truly embedded (much more than in the case of the 1-shell synthesis) and therefore a higher fraction can be deeply buried and not accessible to hydrogen. Even if same Rh was lost into the support, it is reasonable to expect that the accessible nanoparticles were smaller than those of the 1 shell synthesis.

From the data reported in Table 1, it appears clear that the addition of ceria – zirconia favoured the apparent metal dispersion. This behaviour was previously observed for Rh/Al_2O_3 and Rh/$Ce_xZr_{1-x}O_2$ impregnated catalysts [135] due to the different interaction of the metal precursor with the support surface. In the present case, other factors were predominant. The ionic strength during the first step of the growth of the protective $Ce_xZr_{1-x}O_2$-Al_2O_3 porous layer played the major role in determining the differences in the final dimensions of the metal particles. The higher was the ionic strength, the lower was the protecting effect of the surfactant around the Rh nanoparticles in the suspension and more likely agglomeration occured. Consistently with this hypothesis, the trend in the apparent particle dimensions (Table 1) followed the ionic strength.

Table 1. Hydrogen chemisorption on embedded and impregnated Rh based catalysts pre-reduced at 750 °C (adapted from Ref. [85]).

Sample[a]	H/M	Rh surface (m^2/g)	Φ [b] (nm)
Impregnated Rh(1wt%)/Al_2O_3	0.345	1.34	3.2
Embedded Rh(1wt%)@Al_2O_3-1-shell	0.376	1.46	2.9
Embedded Rh(1wt%)@Al_2O_3-2-shell	0.270	1.19	4.1
Rh(1 wt%)@$Ce_{0.8}Zr_{0.2}O_2$(40wt%)-Al_2O_3-2-shell	0.641	2.82	1.7

[a] Samples calcined at 900 °C for 5 hrs in air, followed by *in-situ* standard cleaning procedure (O_2 (5%) / Ar from rt to 500 °C, 1 h), standard activation (H_2 (5%) / Ar from rt to 750 °C, 2 hrs), mild re-oxidation (O_2 (5%) / Ar from rt to 427 °C, 1 h), and re-reduction in a flow of H_2 (35 ml min^{-1}) at a heating rate of 10 °C min^{-1} up to 150 °C and evacuated at 400 °C for 5 hrs before chemisorption experiments at -90 °C.

[b] Rhodium particle diameter determined from hydrogen chemisorption assuming spherical geometry.

3.3. Catalytic Partial Oxidation of Methane (CPOM)

3.3.1. Embedded Rh(1wt%)@Al_2O_3-1-shell vs Impregnated Rh(1wt%)/Al_2O_3 catalyst

The catalytic performances of our embedded Rh(1wt%)@Al_2O_3-1-shell catalyst were tested in the CPOM and compared with those of a standard impregnated Rh(1wt%)/Al_2O_3 catalyst. In order to control the temperature gradients in the catalyst we operated with high dilution of both the catalyst bed and the reactants. Figure 8 shows the difference between the temperature of the catalyst (measured by the thermocouple inserted in the catalytic bed) and the temperature of the furnace at various dilutions of the reactants during a typical reaction test under reaction conditions. Notably that at low CH_4 conversion (below 300 °C), the temperatures of catalyst were only marginally lower than that of the furnace. As the CH_4 conversion increased, the temperature of the catalyst raised over the temperature of the furnace. As expected, with high reactants concentrations a significant difference in temperature was observed. Due to the exothermicity of the combustion reaction, an efficient control of the temperature of the catalyst was possible only with very diluted reaction mixtures [34,136].

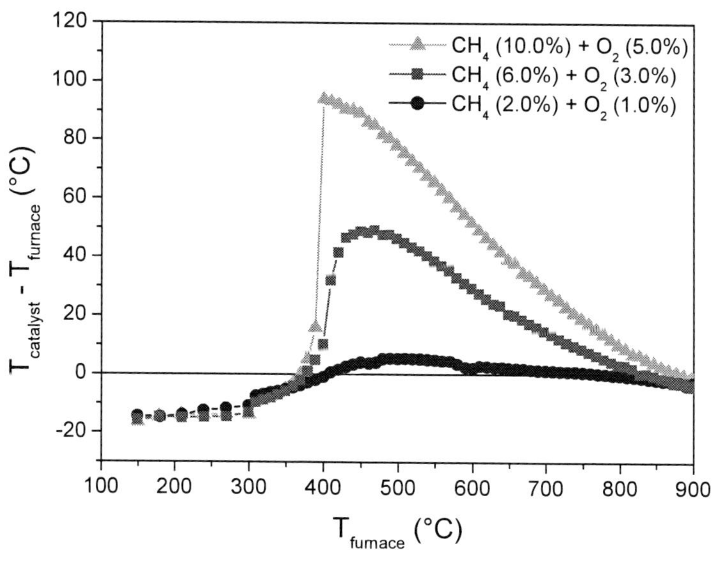

Figure 8. Difference between the temperature of the catalyst and the temperature of the furnace varying the concentration of the reactants on the Rh(1wt%)@Al_2O_3-1-shell catalyst. Condition: GHSV = 700000 mL g^{-1} h^{-1}.

Figure 9 presents the results of CPOM in a run-up experiment on the embedded Rh(1wt%)@Al_2O_3-1-shell and impregnated Rh(1wt%)/Al_2O_3. On the Rh(1 wt%)@Al_2O_3-1 shell catalyst, CH_4 and O_2 conversions started around 300 °C. O_2 was completed converted around 400 °C while CH_4 conversion reached 25%. Up to this temperature, the only detectable products were CO_2 and H_2O, with the exact stoichiometry of the combustion reaction. CO and H_2 were observed only when O_2 was fully converted. Above 400 °C, CH_4

conversion increased continuously up to 750 °C and remained constant (99 %) up to 900 °C. While CH_4 conversion increased, CO and H_2 yields increased and H_2O yield decreased. Meanwhile CO_2 yield remained almost constant (around 25%) up to 480 °C and started to decrease above this temperature. Consequently, the CO and H_2 selectivity increased above 400 °C while CO_2 one decreased. At the same time, the experimental H_2 / CO ratio (calculated on the basis of the amounts of CO and H_2) decreased monotonically from 3.37 at 400 °C to 1.95 at 750 °C, in good agreement with the theoretical value of 2 expected for the CPOM reaction.

The products distribution was consistent with a reaction pathway involving the combustion – reforming reactions, as previously reported for other Rh catalysts supported on Al_2O_3-based oxides [18,30,31,95]. This hypothesis was further supported by the high thermal difference between the furnace and the catalyst observed at increasing reagents concentration (see Figure 8). Basile *et al.* [30,31,95] reported that strong temperature gradient was observed along the catalytic bed, suggesting that a strong exothermic reaction (complete methane combustion) takes place at the entrance of the reactor and endothermic reactions (steam and dry reforming) occur in the remaining part. Notably, on our Rh(1wt%)@Al_2O_3 catalyst, increasing the furnace temperature, the thermal difference between catalyst and furnace decreased, suggesting that the heat produced by the combustion of part of the methane was consumed by the endothermic reforming reactions. Using *in-situ* IR spectroscopy, Weng *et al.* [18] demonstrated that over Rh/Al_2O_3 the primary products of the reaction under CPOM conditions were CO_2 and H_2O, while CO and H_2 were formed by the subsequent reforming with CH_4. Moreover, an isotopic study using CH_4/CD_4 suggested that the reaction pathway could change with the catalyst temperature [137]. In particular, using Rh/γ-Al_2O_3 an isotopic effect was observed in the conversion of methane and the formation of CO and CO_2 at low temperature, while at high temperature this effect was lowered. This suggested that the combustion – reforming pathway was active at low temperature while the direct partial oxidation (known also as pyrolysis – oxidation mechanism) was operative at high temperature. The reaction pathway is influenced by the metal loading: with a very low metal loading (0.05%) the combustion – reforming mechanism is preferred, while increasing the metal loading (1.0%) the pyrolysis – oxidation seems to be predominant [137]. In the case of our Rh(1wt%)@Al_2O_3-1-shell catalyst, despite the "high" metal loading, the products distribution suggested that the combustion – reforming pathway is predominant at least up to 600 – 650 °C. Recently, Weng *et al.* [138] reported that the calcination at 900 °C of Rh(1 wt%)/Al_2O_3 results in the formation of Rh species in which the metal may partially substitute Al in the support oxide, irreducible at the temperature below 600 °C (in a similar manner of the Rh$(AlO_2)_y$ species reported by Hwang *et al.* [125]). These species possess a strong Rh – O and a high oxygen affinity, resulting in a higher oxygen concentration on the surface of the catalyst. This condition was recognized as the major factor that promotes the CRR pathway with respect to the DPO one [19,138]. In the case of our Rh(1wt%)@Al_2O_3-1-shell catalyst, significant amounts of hardly reducible species were present (see § 3.2). The presence of high temperature reduction species further supported the predominance of the CRR mechanism.

The reference standard impregnated Rh(1wt%)/Al_2O_3 (Figure 9) showed a reaction profile similar to that of the embedded catalyst, with the only difference of a slight shift of the reactivity to higher temperature of 10 – 20 °C. This result is in good agreement with the similar exposed metallic surface area measured by H_2 chemisorption (see § 3.2).

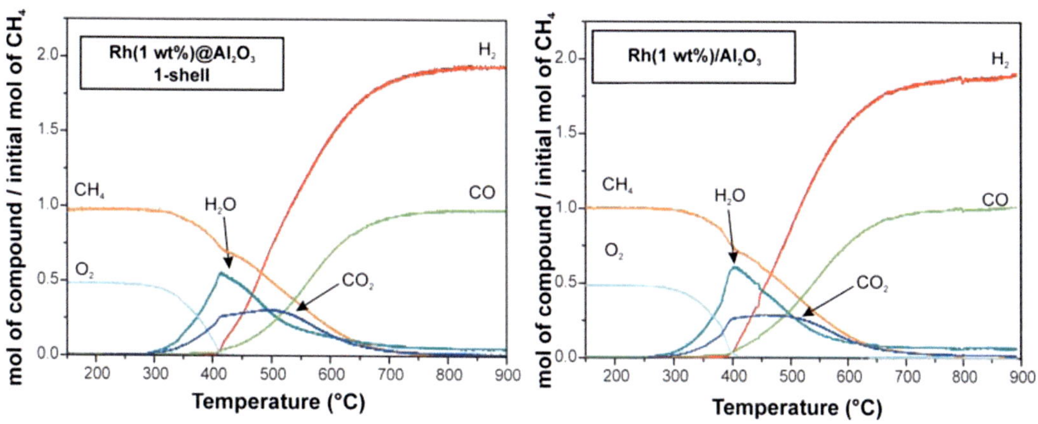

Figure 9. Catalytic activity of the embedded Rh(1wt%)@Al_2O_3-1-shell and the impregnated Rh(1wt%)/Al_2O_3 catalysts under CPOM conditions (CH_4 (2.0%) + O_2 (1.0%) in Ar, GHSV = 700000 mL g^{-1} h^{-1}) (adapted from Ref. [37]).

To further test the potentiality of the Rh(1wt%)@Al_2O_3-1-shell catalyst, it was tested also under different reforming conditions. In particular, its activity was tested with respect to the Methane Dry Reforming (MDR) and the Methane Steam Reforming (MSR) reactions (Figure 10). Besides the importance of these processes in the hydrogen production, these experiments were able to provide further information regarding the reaction pathway. Figure 10 shows the catalytic performances of the Rh(1wt%)@Al_2O_3-1-shell under MDR and MSR conditions.

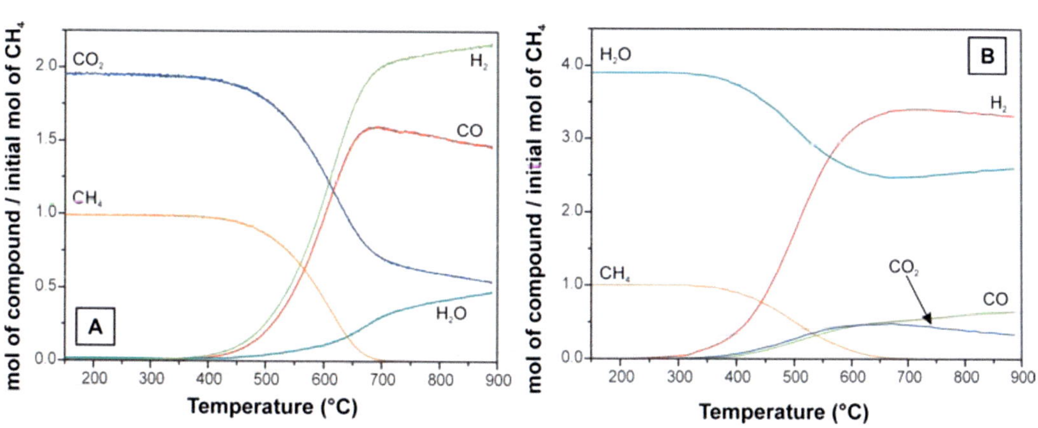

Figure 10. Catalytic activity of the embedded Rh(1wt%)@Al_2O_3-1-shell under (A) MDR conditions (CH_4 (1.0%) + CO_2 (2.0%) in Ar, GHSV = 700000 mL g^{-1} h^{-1}) and (B) MSR conditions (CH_4 (1.5%) + H_2O (6.0%) in Ar, GHSV = 700000 mL g^{-1} h^{-1}) (adapted from Ref. [37]).

Regarding the MDR process (Figure 10, part A), CH_4 and CO_2 conversions started around 400 °C, with concomitant H_2 and CO production. CH_4 consumption was completed above 700 °C and H_2 and CO yields raised continuously up to this temperature. The H_2O formation above 450 °C confirmed the occurrence of the Reverse Water Gas Shift Reaction (R – WGSR). The H_2O amount raised continuously also in the temperature range investigated. At the same time, above 700 °C the CO_2 and H_2 amounts decreased, forming CO and H_2O as suggested by the thermodynamic behaviour of the R – WGSR reaction (endothermic process).

In the case of the MSR process, a large excess of H_2O is used (H_2O / C ratio = 4) in order to simulate the reaction conditions employed in the industrial process. CH_4 and H_2O conversions started above 300 °C, with production of H_2, CO and CO_2. In fact, the excess of water shifts the WGSR equilibrium to the formation of H_2 and CO_2 and the occurrence of this latter product was an indication that the Rh(1wt%)@Al_2O_3-1-shell catalyst is able to favour the equilibration of the gaseous mixture [37]. Increasing the temperature, CH_4 was completed converted above 650 °C. At this temperature, CO_2 and H_2 production reached their maximum values. Notably, the occurrence of the WGSR reaction had a positive effect on the process, due to the H_2 production above the limit of the pure MSR process (3 mol of H_2 per mol of CH_4) and reducing the CO production. Above 650 °C, the H_2 and CO_2 production lowered while the CO and H_2O amount raised, in agreement with the thermodynamic behaviour of the WGSR (exothermic reaction).

Despite of their similar reactivity during the first run-up experiments, the impregnated and embedded Rh catalysts presented significant differences in their stability under CPOM conditions, either under consecutive run-up experiments or during isothermal reaction at high temperature for a long time. The embedded Rh(1wt%)@Al_2O_3-1-shell presented a constant reactivity profile during at least 6 consecutive run-up experiments. On the contrary, the impregnated Rh(1wt%)/Al_2O_3 catalyst showed a progressive worsening of its catalytic performance, with a significant decrease of the CH_4 conversion above 550 °C [37].

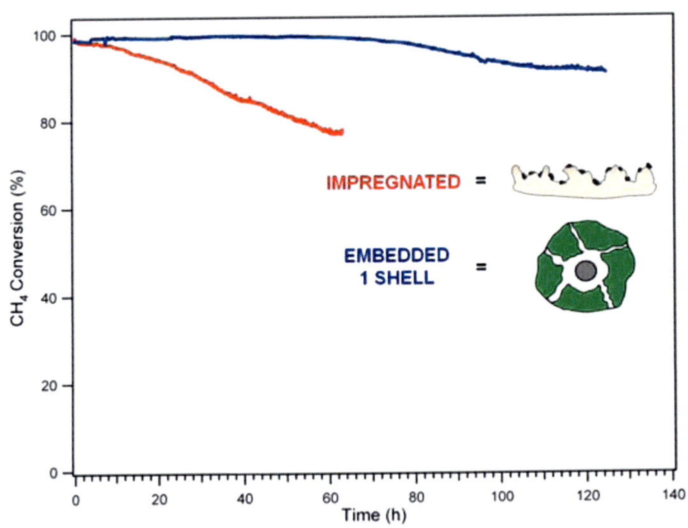

Figure 11. Stability of the catalytic activity of the embedded Rh(1wt%)@Al_2O_3-1-shell and the impregnated Rh(1wt%)/Al_2O_3 catalysts under CPOM conditions at 750 °C (CH_4 (2.0%) + O_2 (1.0%) in Ar, GHSV = 700000 mL g^{-1} h^{-1}).

The stability of the catalytic activity under isothermal conditions was tested at 750 °C. The comparison of the embedded and impregnated samples is reported in Figure 11. Under CPOM conditions, the catalytic activity of Rh(1wt%)@Al_2O_3-1-shell catalyst was constant for at least 60 hrs. After this period of time, the CH_4 conversion decreased slowly reaching the value of 88 % after 125 hrs. CO and H_2 selectivity followed the same trend, decreasing from 99.5 and 95 % to 93 and 93 %, respectively. At the same time, CO_2 selectivity increased form 0.5 % to 7 %. This could be due to the fact that, as a result of the catalyst deactivation, the steam and dry reforming reactions were not completed. Therefore it was possible to detect the presence of unreacted CH_4, CO_2 and H_2O, with a consequent decrease in the CO and H_2 selectivity and an increase in the CO_2 selectivity. In the case of the reference impregnated Rh(1 wt%)/Al_2O_3 catalyst, CH_4 conversion remained above 99 % for a significantly shorter period (~ 10 hrs) and then decreased rapidly, reaching a conversion of 78 % after 60 hrs. (Figure 11).

Reactivation of the partially deactivated embedded Rh(1wt%)@Al_2O_3-1-shell catalyst was attempted with oxidative treatments under diluted oxygen at different temperatures. After each oxidative treatment, the catalytic activity at 750 °C was monitored and the stability compared with that of the fresh system. Partial recovery of the initial activity was obtained by exposing the aged sample to diluted O_2 at 750 °C for 5 minutes. Notably, that the CH_4 conversion started to decrease almost immediately after this relatively mild reactivation process. After a prolonged oxidative treatment (750 °C, 75 minutes), the catalytic activity was completely recovered and it remained constant for 24 hrs before the beginning of a decrease, reaching a conversion of 96% after 48 hrs. An almost complete reactivation of the catalyst was obtained only after exposure of the aged sample to diluted O_2 at 850 °C for 75 minutes: the CH_4 conversion remained stable above 99 % for at least 50 hrs (Figure 12). During the oxidative treatments, CO_2 evolution was observed, suggesting the removal of coke deposited on the catalyst during the aging under reaction conditions. It is worth noting that the identification of the exact conditions of catalyst regeneration was out of the scope of the present research activity. In fact, the prolonged oxidation at 850 °C almost certainly exceeded the needed thermal treatment. Furthermore the conditions required for full reactivation strongly depends on the aging conditions. In fact, during the aging the coke was both progressively built up (increased coverage) and transformed into a graphitic deposit.

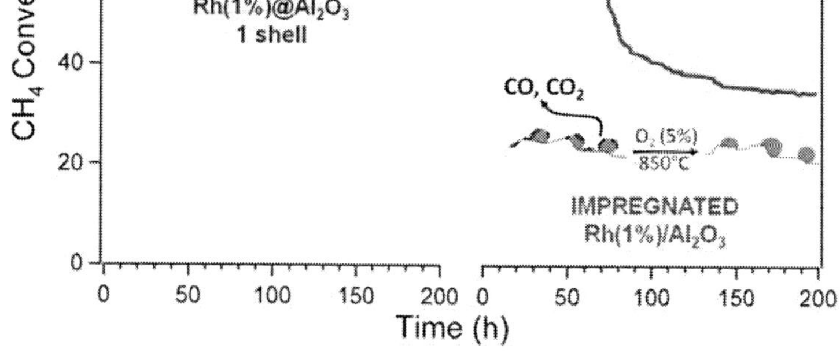

Figure 12. Stability of the catalytic activity of the embedded Rh(1wt%)@Al_2O_3-1-shell (left) and Rh(1wt%)/Al_2O_3 (right) under CPOM conditions at 750 °C (CH_4 (2.0%) + O_2 (1.0%) in Ar, GHSV = 700000 mL g^{-1} h^{-1}) and schematic representation of the reactivation processes (O_2 (5.0%) in Ar, 40 mL min^{-1}, 75 min).

The stability of the catalytic activity of the Rh(1wt%)@Al_2O_3 catalyst was studied also under different experimental conditions. At 850 °C, under diluted reaction mixture, minor deactivation was observed after 120 hrs of reaction: CH_4 conversion decreases only to ~ 95%. A similar behaviour (CH_4 conversion of ~ 96%) was observed using more concentrated reaction mixture (CH_4 (10.0%) + O_2 (5.0%) in Ar). Notably, the last behavior could be related to the higher effective catalyst temperature realized under concentrated conditions.

In the case of impregnated Rh(1wt%)/Al_2O_3 catalyst, a transient restoration of the catalytic activity was obtained after analogous oxidative treatment at 850 °C. In fact, just after a few hrs of reaction at 750 °C the reactivated catalyst showed a methane conversion below 40% (Figure 12).

The full characterization of the aged samples revealed the origin of the different stability of the two catalysts. N_2 physisorption data indicated that, after aging under CPOM conditions at 750 °C for 125 hrs, both the catalysts underwent a partial sintering of the alumina with the loss of surface area and pore volume (surface area from 157 and 134 m^2 g^{-1} to 98 and 103 m^2 g^{-1} and pore volume from 0.79 and 0.58 to 0.41 and 0.42 mL g^{-1}, respectively for the embedded and the impregnated samples). After oxidative treatment used for the reactivation (O_2 (5%) / Ar at 750 °C for 75 minutes), a partial recovery of the pore volume was observed for both the catalysts (0.53 and 0.49 mL g^{-1} for the embedded and the impregnated catalysts, respectively). At the same time, a significant crystallization of the θ-Al_2O_3 was observed. The fitting of the XRD spectrum of the Rh(1wt%)@Al_2O_3-1-shell sample subjected to aging, under CPOM reaction at 750 °C for 125 hrs, revealed that the oxidic matrix is constituted by ~ 27% of γ-Al_2O_3 (7 nm), ~ 20% of δ-Al_2O_3 (7 nm) and ~ 53% of θ-Al_2O_3 (18 nm). Similar results were obtained for the aged Rh(1 wt%)/Al_2O_3: ~ 25% of γ-Al_2O_3 (7 nm), ~ 14% of δ-Al_2O_3 (7 nm) and ~ 61 wt% of θ-Al_2O_3 (18 nm). These data indicated that more likely the differences in the stability of the two catalysts were linked to the metal phase. Therefore H_2 chemisorption characterization was conducted on the aged samples (Table 2)

After prolonged aging under CPOM conditions, the Rh(1wt%)@Al_2O_3 catalyst presented a appreciable reduction (20%) of the exposed metal surface (Table 2). In the case of the impregnated Rh(1wt%)/Al_2O_3, the aging treatment led to an almost complete deactivation of the capacity of H_2 chemisorption. Although the chemisorption data are unable to discriminate between metal sintering and poisoning effects, it was reasonable to expect comparable coking effects. The reoxidation treatment followed by reduction at 750 °C, restored almost completely the initial hydrogen chemisorption capability of both samples. It was demonstrated that Rh/Al_2O_3 can present Rh incorporation phenomena during high temperature oxidation and Rh redispersion during high temperature reduction [139]. Thus, the recovery of chemisorption was attributed to a combination of coke removal and overall metal redispersion. The samples were affected to different extents when the samples were reduced at low temperature after reactivation. We explained this behaviour on the bases of differences in the extent Rh incorporation during oxidation of the aged samples, combined with the reduction temperature which is to low to reverse the process. Notably, in the activity investigation, reactivation was conducted by oxidation only. For the embedded Rh(1wt%)@Al_2O_3-1-shell, the overall catalytic behaviour was restored. This was consistent with a predominance of the process of coke removal, with no measurable effect of the reactivation procedure on Rh dispersion. It also suggested that the Rh exposed surface area was not strongly influenced by the reaction conditions.

Table 2. H_2 chemisorption results on the Rh(1wt%)@Al_2O_3-1-shell and the Rh(1wt%)/Al_2O_3 samples, after various pretraments.

Sample	Pre-reduction Temperature (°C)	H/M	Rh surface (m^2/g)	Φ [a] (nm)
Rh(1wt%)@Al_2O_3-1-shell				
Activated [b]	750	0.376	1.46	2.9
Aged [c]	150	0.274	1.06	4.0
Reactivated [d]	150	0.128	0.56	8.5
Reactivated [d]	750	0.344	1.33	3.2
Impregnated Rh(1wt%)/Al_2O_3				
Activated [b]	750	0.345	1.34	3.2
Aged [c]	150	0.034	0.15	32.2
Reactivated [d]	150	0.044	0.19	25.0
Reactivated [d]	750	0.301	1.30	3.6

[a] Diameter of the metal particles by H_2 chemisorption assuming a spherical geometry.

[b] Samples calcined at 900 °C for 5 hrs in air, followed by *in-situ* standard cleaning procedure (O_2 (5%) / Ar from rt to 500 °C, 1 h), standard activation (H_2 (5%) / Ar from room temperature to 750 °C, 2 hrs), mild re-oxidation (O_2 (5%) / Ar from rt to 427 °C, 1 h), re-reduction at 150 °C and evacuation at 400 °C for 5 hrs before chemisorption experiments at -90 °C.

[c] After aging pretreatment under CPOM conditions at 750 °C for 125 hrs. No *in-situ* standard cleaning procedure to avoid carbon removal.

[d] Aged samples reactivated under oxidative treatment (O_2 (5%) / Ar, 750 °C, 75 min).

The deactivation/reactivation behaviour of the impregnated Rh(1wt%)/Al_2O_3 is strikingly different. First, a progressive deactivation is observed during the isothermal test (Figure 11), which is more rapid with respect to the embedded sample. Second, the oxidation procedure results in a short transient reactivation, followed by accelerated deactivation (Figure 12). The transient reactivation after oxidation was consistent with the initial increase of accessible catalytic sites (Table 2) due to coke removal. Given that the oxidation removes coke, the phenomenon could simply be related to a more rapid coking of Rh. This phenomenon was previously observed for Ni-based catalysts used as reforming catalyst [46,140,141]. In particular, this effect was reported in the case of the steam reforming [141] and the partial oxidation [46] of methane. In both cases, the strong interaction between the Ni nanoparticles and the support (as a results of tailored synthetic strategies) reduced the metal sintering and the coke deposition. Moreover, Zhang *et al.* [140] reported the influence of the Ni particle size on the coke deposition during partial oxidation and steam reforming of higher hydrocarbons, indicating that the rate of deactivation is higher for large particles. However, the impregnated Rh(1wt%)/Al_2O_3 exhibited a stronger tendency to incorporate Rh than the embedded Rh(1wt%)@Al_2O_3 under oxidising conditions [37], again leaving a smaller number of sites to poison. In both cases, the reduced number of sites could have been sufficient to justify the high initial activity of the regenerated Rh(1wt%)/Al_2O_3. Clearly, the opposing tendencies of sintering during reaction and incorporation during oxidation complicate matters.

Occurrence of sintering phenomena during aging at high temperature was confirmed by HRTEM investigation carried out on Rh(1wt%)@Al_2O_3-1-shell catalyst. Figure 13 shows representative results. The aged Rh(1wt%)@Al_2O_3-1-shell catalyst showed a slight growth of the particle dimensions. The mean diameter of the particles was 3.9 nm. Some big metal nanoparticles were observed (diameter up to 30 nm). Some Rh nanoparticles were covered by one layer of the graphitic structure (0.335 nm). Notably, no spacing near this value is attributable to the transitional aluminas present in the oxidic matrix. IR measurements confirmed the formation of this graphitic coke deposits.

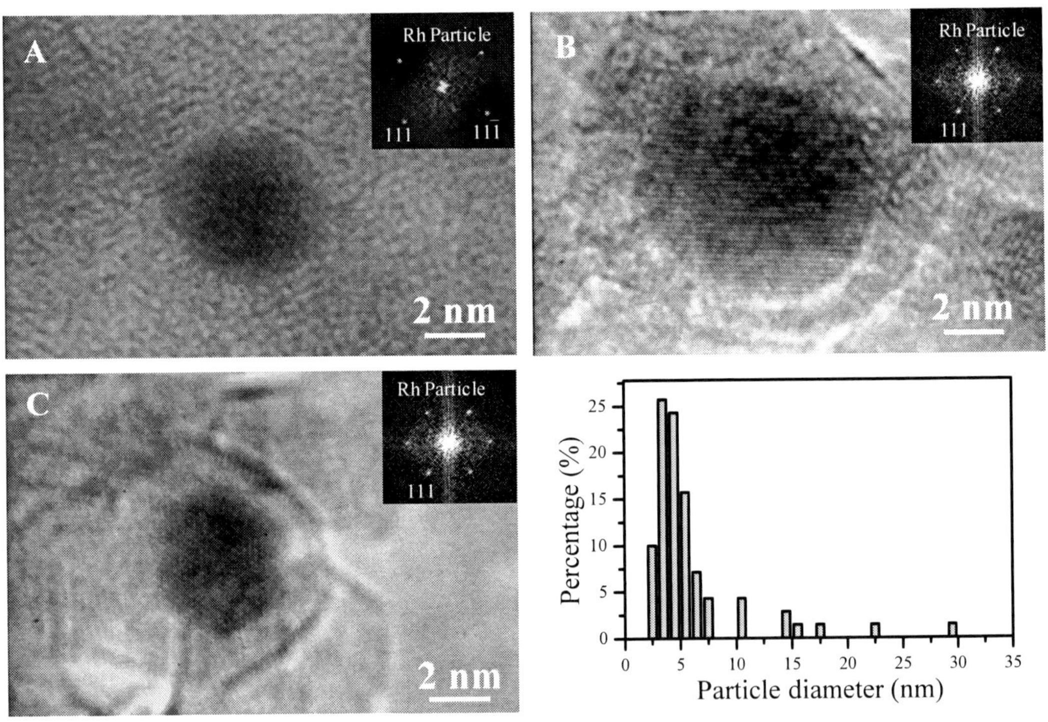

Figure 13. Representative HRTEM analysis on aged embedded Rh(1wt%)@Al_2O_3-1-shell under CPOM conditions (CH_4 (2.0%) + O_2 (1.0%) in Ar, GHSV = 700000 mL g^{-1} h^{-1}). (A – C) Rh nanoparticles which appear to be embedded in the Al_2O_3 matrix, (D) size distribution of the Rh nanoparticles (adapted form Ref. [37]).

The *ex situ* FT-IR characterization of the aged Rh(1wt%)@Al_2O_3-1-shell catalyst evidenced the presence of significant absorption bands in the 1450 – 1700 cm^{-1} range, ascribable to the presence of polynuclear aromatic hydrocarbons [142]. The more intense band, centred at 1606 cm^{-1}, was associated to pseudo-graphitic or polycyclic aromatic structures [142]. The absorption bands at 1515, 1529, 1561 and 1580 cm^{-1} could be assigned to alkyl-substituted naphthalenes or to polyphenylenic structures, although the latter band could be associated with the asymmetric stretching of carboxylate groups [142]. The band at 1638 cm^{-1} was related to the stretching of C = O bonds of acetilyc groups. The bands centred at 1652 and 1677 cm^{-1} were indicative of the presence of double C = C bonds (twisting and stretching vibrations, respectively). Moreover, an absorption band at 1488 cm^{-1} was observed, ascribable to the asymmetric stretching of the C – H bonds of methyl groups [142].

3.3.2. Increasing the Performance of the Embedded Catalyst

Various strategies were adopted to enhance the stability of the Rh(1 wt%)@Al_2O_3 under CPOM, and in particular to reduce the formation of coke deposits: (i) the addition of small quantities of Pt as an oxidation co-catalyst, (ii) the use of *in situ* short regeneration steps, (iii) the adoption of the improved 2-shell synthesis strategy and (iv) the addition of ceria-zirconia to the formulation of the catalyst.

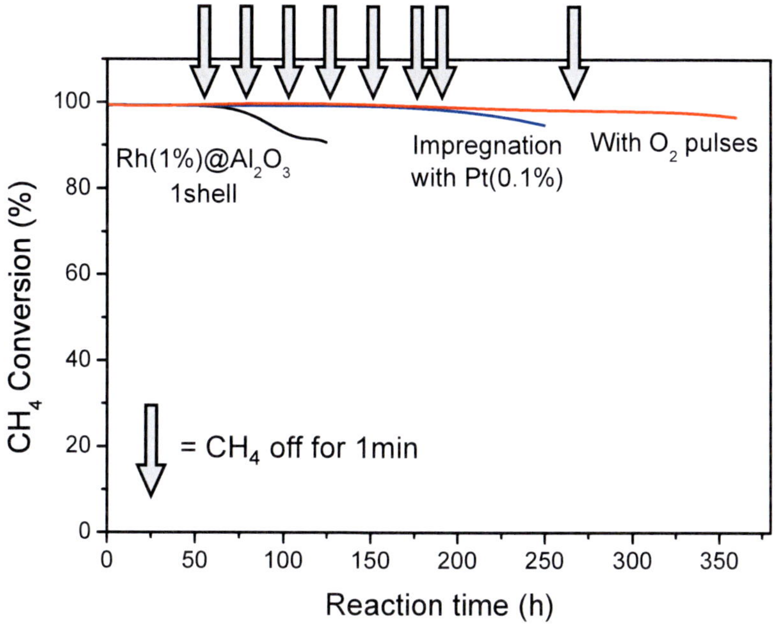

Figure 14. Improvement of the stability of the catalytic activity of the Rh(1wt%)@Al_2O_3 catalyst preventing coke deposition under CPOM conditions (CH_4 (2.0%) + O_2 (1.0%) in Ar, GHSV = 700000 mL g^{-1} h^{-1}): effect of impregnation with Pt (0.1 wt%) and of periodic coke removal by oxidative treatment (adapted from Ref. [37]).

The stability of the embedded Rh(1wt%)@Al_2O_3-1-shell was significantly improved by the presence of 0.1 wt % of Pt, added by standard impregnation method on the embedded catalyst (Figure 14). An *in situ* regeneration treatment involving brief switching to an oxidising atmosphere during reaction by stopping the CH_4 flow for 1 min was also very effective. In the case of the embedded Rh(1wt%)@Al_2O_3-1-shell, the first of these switches was made after 60 hrs on-stream, i.e., before the on-set of deactivation, and repeated at maximum every 24 hrs. Using this procedure, negligible deactivation was observed up to 360 hrs on-stream (Figure 14). The sample investigated was previously aged during run-up experiments and steady state experiments, after which it was regenerated under oxidizing conditions to remove any carbonaceous species. Notably, the aged impregnated Rh(1wt%)/Al_2O_3 sample presented severe deactivation even in the presence of the oxidising switches and only transient reactivation was possible (0.5 h). The results obtained using the O_2 switching procedure showed that it may be used to prolong the embedded Rh(1wt%)@Al_2O_3 lifetime, ideally forever. The data indicated that this may be attributed to removal of carbonaceous species from the surface. Notably, timely removal of deposited species prevents build-up of the deposits to the point where catalyst deactivation begins. Furthermore, the nature of the coke changes with aging time: the hydrogen content progressively decreases to form more stable, graphitic deposits. The necessity of severe reactivation procedures (high temperature – prolonged time) to remove the coke formed after prolonged aging is therefore evident. Increasing the frequency of reactivation (oxygen treatment before severe deactivation), on the other hand, facilitates easier removal of the graphitic coke precursors. Notably, the reactivation process has two effect: (i) the positive burning of coke; and (ii) the diffusion of part of Rh into the alumina support. The latter aspect is more evident at high temperature and for long treatment.

The effects of the improved 2-shell synthesis strategy (see § 2.2) and of the addition of the $Ce_{0.8}Zr_{0.2}O_2$ promoter on the CPOM reaction are reported in Figure 15. The catalytic activity of the Rh(1wt%)@Al_2O_3-2-shell catalyst was very similar to that of the 1-shell analogous (compare Figure 15 and Figure 9). The only difference observed was a slight shift to high temperature (~ 30 °C) of the complete conversion of O_2 and of onset of H_2 and CO production.

The introduction of the $Ce_{0.8}Zr_{0.2}O_2$ solid solution into the formulation of the catalyst, significantly changed the reactivity (Figure 15). The conversion of CH_4 and O_2 increased slowly with temperature and a instantaneous kick-off of the reforming reactions was observed at 470 °C. Above this temperature the catalytic behaviour was similar to that of the Rh(1wt%)@Al_2O_3-2-shell, although the complete CH_4 conversion was obtained only above 850 °C. The activity remained unperturbed during subsequent run up experiments, suggesting that the fast low temperature onset of the reforming reactions was a characteristic of the catalyst and not a transient behaviour.

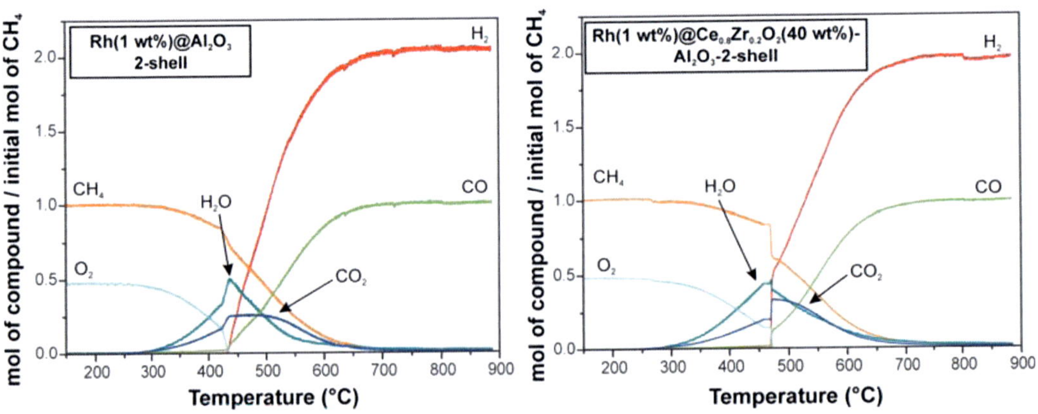

Figure 15. Catalytic activity of the embedded Rh(1 wt%)@Al_2O_3-2-shell and the Rh(1wt%)@$Ce_{0.8}Zr_{0.2}O_2$(40wt%)-Al_2O_3-2-shell catalysts under CPOM conditions (CH_4 (2.0%) + O_2 (1.0%) in Ar, GHSV = 700000 mL g^{-1} h^{-1}).

The catalytic behaviour of the Rh(1wt%)@Al_2O_3-2-shell could be interpreted in term of a partial inhibition of the complete combustion of methane, which represents the initial step of the CRR pathway. In fact, it is well known that $Ce_xZr_{1-x}O_2$ solid solutions could increase the concentration of oxygen species on the surface of the metal particles through the back-spillover phenomenon [143-148]. Although an appropriate amount of oxygen species bonded to the Rh surface represents a necessary condition to favour the CRR mechanism with respect to the DPO one [138], a too high concentration could inhibit the combustion reaction. In fact, the strong tendency of Rh surfaces to accommodate O species [149,150] could result in the passivation of the surface and in a decrease of the number of sites able to activate CH_4 for its combustion.

The synthesis modification (2-shell vs 1-shell) induced also a significant improvement in the catalyst stability under reaction conditions (Figure 16).

In the case of the Rh(1wt%)@Al_2O_3-2-shell catalyst, the conversion of CH_4 at 750 °C was complete and constant for at least 150 hrs, more then 2 time longer with respect to the performance of the Rh(1wt%)@Al_2O_3-1-shell catalyst (Figure 16). This improvement could be attributed to a lower rate of deposition of coke on 2-shell system, as a result of the reduced nanoparticles coalescence during the preparation, of the more uniform distribution (See TPR, § 3.1) and of the improved thermal stabilization against sintering offered by the protective layer. Notably, it was reported that Ni based reforming catalysts showed reduced coking effects on small particles with respect to big Ni particles [46,141,151].

The addition of $Ce_{0.8}Zr_{0.2}O_2$ into the formulation of the catalyst produced a further reduction of the coke deposition rate, although a slight reduction of CH_4 conversion was reached. This was in agreement with the run up experiments (Figure 15). The lower rate of coke deposition could be related to the ability of $Ce_xZr_{1-x}O_2$ to transfer oxygen species to the metal nanoparticles, preventing the polymerization of the CH_X species and the formation of graphitic coke. Moreover, doped-CeO_2 has widely studied for its catalytic activity in soot oxidation [152-155]. Similarly, in the studied Rh(1wt%)@$Ce_{0.8}Zr_{0.2}O_2$(40wt%)Al_2O_3-2-shell catalyst, the promoter could burn the carbon deposited directly on the oxide surface, reducing the pore blocking effect.

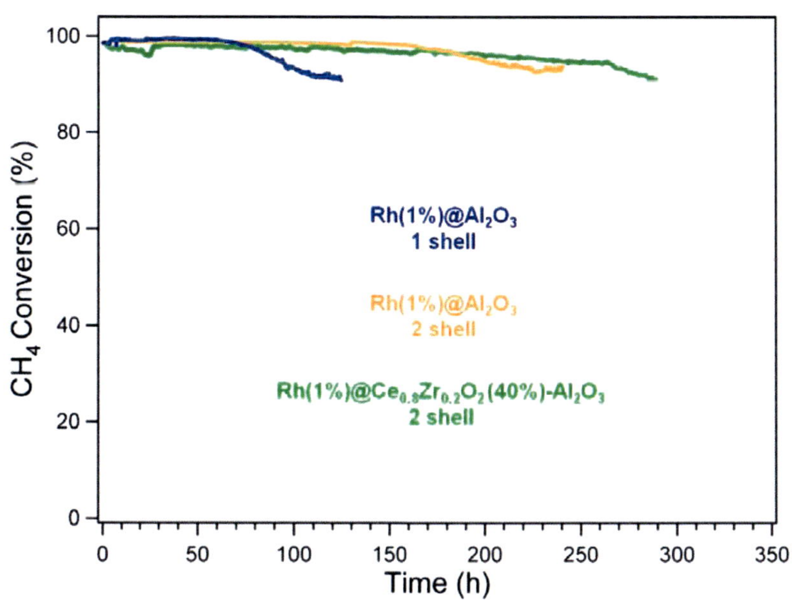

Figure 16. Stability of the catalytic activity of the embedded Rh(1 wt%)@Al_2O_3-2-shell and the Rh(1wt%)@$Ce_{0.8}Zr_{0.2}O_2$(40 wt%)-Al_2O_3-2-shell catalysts under CPOM conditions at 750 °C (CH_4 (2.0%) + O_2 (1.0%) in Ar, GHSV = 700000 mL g^{-1} h^{-1}).

3.4. Steam Reforming of Ethanol (SRE)

The reaction network involved in ethanol steam reforming is very complex. Many pathways are possible as shown in Scheme 6 [65,74,75,156-162]. The process includes several steps that require catalytic sites able to dehydrogenate ethanol, to break the C – C bonds of surface intermediates producing CO and CH_4 and to promote the steam reforming of CH_4. Moreover, also the Water Gas Shift Reaction (WGSR) is involved, contributing to reduce the CO concentration and increasing the H_2 production. Some of these steps can be favoured depending on the catalyst used [73,160,163-167]. However, other secondary reactions can be involved. Dehydration reaction leads to the formation of ethylene, especially when acid supports are used (such as Al_2O_3 [165]). Ethylene can be easily transformed into carbon that is deposited on the active phase producing the deactivation of the catalyst. Also ethane, formed by methane coupling, acts as very strong promoters for carbon formation. Acetone could be produced from acetaldehyde through a series of reactions involving aldol condensation, oxidation and decarboxylation. Significant formation of this product is observed when the support is able to provide structural oxygen for the oxidation step (such as CeO_2 [161,162] or $Ce_xZr_{1-x}O_2$ [85]). It should be borne in mind that methane, ethane, acetaldehyde, acetic acid and acetone are undesirable products because they decrease the hydrogen production efficiency and can reduce the operational time of the catalyst. Therefore, it is necessary to design the formulation of the catalyst to obtain the highest ethanol conversion with the highest H_2 production.

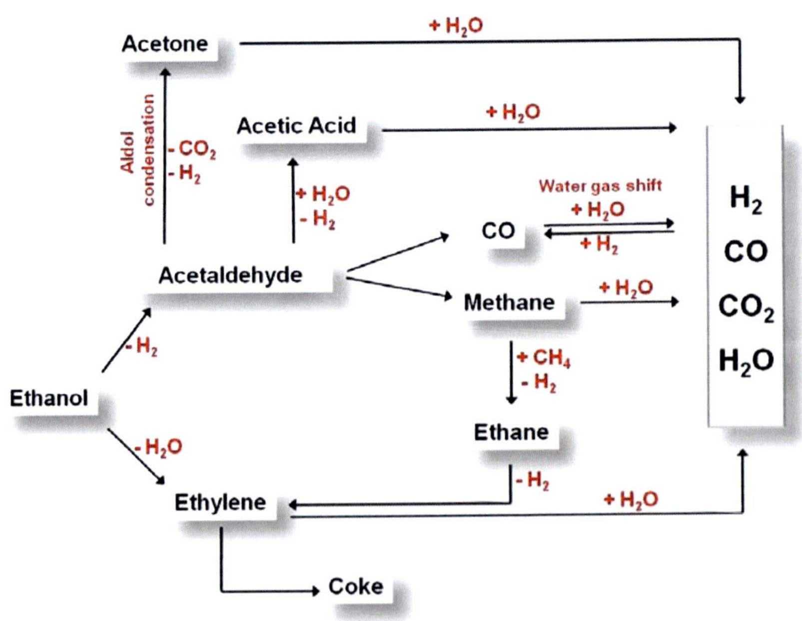

Scheme 6. Reaction network involved in the Steam Reforming of Ethanol (SRE) over metal catalysts (adapted from Refs. [65,74,75,156-160]).

The SRE results obtained on pure alumina, embedded Rh(1wt%)@Al_2O_3-2-shell and embedded Rh(1wt%)@$Ce_{0.8}Zr_{0.2}O_2$(40wt%)-Al_2O_3-2-shell are presented in Figure 17, in which the moles of the different compounds detected at the outlet of the reactor are divided by the moles of ethanol in the feed of the reactor.

On bare Al_2O_3, ethanol was converted above 250 °C, producing essentially ethylene through the dehydration reaction (Figure 17). Only above 600 °C, some reforming products was detected, obtained mainly from the cracking of the carbonaceous compounds.

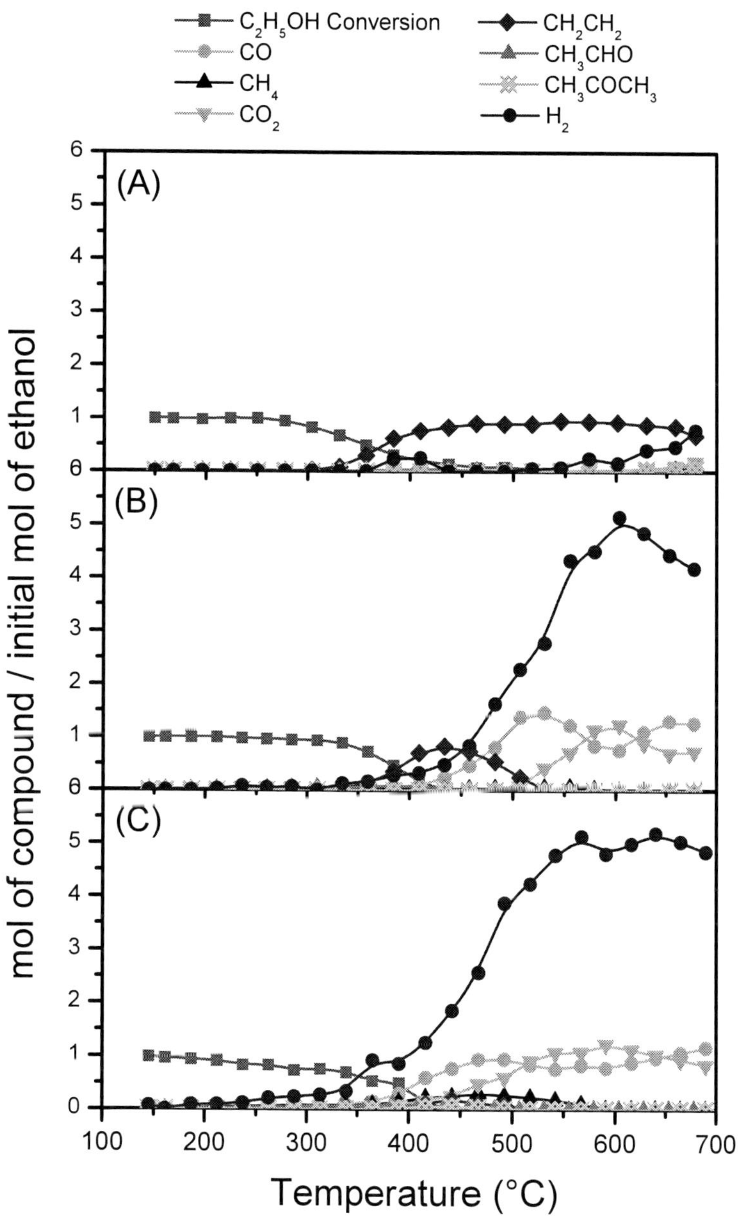

Figure 17. Catalytic activity of (A) Al_2O_3, (B) embedded Rh(1wt%)@Al_2O_3-2-shell and (C) embedded Rh(1wt%)@$Ce_{0.8}Zr_{0.2}O_2$(40wt%)-Al_2O_3-2-shell under SRE conditions (C_2H_5OH (1.0%) + H_2O (5.0%) in Ar, GHSV = 150000 mL g^{-1} h^{-1}) (adapted from Ref. [85]).

For Rh(1wt%)@Al_2O_3-2-shell, significant steam reforming of ethanol occurred above 230 °C (Figure 17). Above this temperature and up to 450 °C, the dehydration of alcohol was the predominant reaction. Small quantities of acetaldehyde, produced by dehydrogenation process, were also observed. Increasing reaction temperature resulted in a progressive decrease of selectivity towards ethylene, which dropped to zero at temperatures above 520 °C. Above 450 °C ethanol conversion was completed and CO, CO_2 and H_2 produced by reforming reactions became the main products. Above 600 °C, the changes of H_2, CO and CO_2 amounts were mainly due to the influence of mild exothermic WGSR, which reached the thermodynamic equilibrium in this temperature range.

It is known [74,165,168] that the dehydration of ethanol to ethylene is strongly active on Al_2O_3 supported catalysts. This is due to the fact that Al_2O_3 possesses acid sites, which promote the dehydration route.

In the case of the Rh(1wt%)@Al_2O_3-2-shell sample, ethanol was converted to ethylene on the acid site of the protective oxide. Ethylene selectivity was almost complete around 430 °C. Above this temperature ethylene can undergo reforming reaction which is catalyzed by the metallic phase [82]. Indeed Rh(1wt%)@Al_2O_3-2-shell catalyst was able to promote the steam reforming of ethylene (Figure 18).

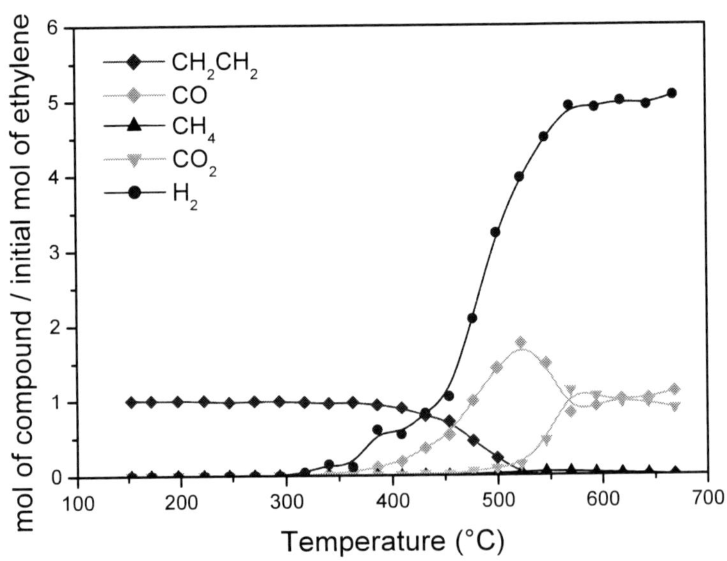

Figure 18. Catalytic activity of embedded Rh(1wt%)@Al_2O_3-2-shell under Steam Reforming of Ethylene conditions (C_2H_4 (1.0%) + H_2O (6.0%) in Ar, GHSV = 150000 mL g^{-1} h^{-1}).

The SRE behaviour of the Rh(1wt%)@$Ce_{0.8}Zr_{0.2}O_2$(40wt%)-Al_2O_3-2-shell was significantly different. Ethanol conversion started at considerably lower temperature (around 200 °C). The introduction of $Ce_{0.8}Zr_{0.2}O_2$ mixed oxide in the catalyst formulation resulted in a change in the reaction selectivity at intermediate temperature. In fact, differently from the Rh(1wt%)@Al_2O_3-2-shell catalyst, very low amounts of ethylene were observed and considerable amounts of H_2, CO, CO_2 and CH_4 were produced above 350 °C. This was an indication that the preferred pathway for the SRE reaction was the dehydrogenation to acetaldehyde, followed by its decomposition to CO and CH_4. Moreover, the higher amount of

CO with respect to that of CH_4 suggested that part of methane was directly reformed into H_2 and CO. Above 600 °C, ethanol was completely converted into a mixture of H_2, CO and CO_2, reaching the thermodynamic equilibrium with respect to the WGSR. The observed promotion of the reforming reaction agrees with previously reported works on $Rh/Ce_XZr_{1-X}O_2$ catalysts [82]. Moreover, the presence of O at low coverage in model Rh surfaces favours ethanol decomposition, via C – C activation, and H_2 production [169]. The presence of $Ce_xZr_{1-x}O_2$ in contact with the metal nanoparticles could increase the O coverage via back-spillover [143-148] and it favours the ethanol decomposition and reforming.

The strong reduction of ethylene production was correlated to multiple factors, such as the reduction of the acidic sites of the Al_2O_3 due to the partial coverage with ceria-zirconia mixed oxide and the lower surface area of the rare earths doped sample. Furthermore the mixed oxide influenced the stability of the transitional aluminas.

Traces of acetone were also observed at intermediate temperatures. Its formation on CeO_2-based catalysts was reported by Elliott *et al.* [161] and recently by Nishiguci *et al.* [162] for CuO/CeO_2 ethanol steam reforming catalyst. The formation of acetone is the result of a network of reactions involving aldol condensation of acetaldehyde, oxidation of the intermediate aldehyde and finally decarboxylation and dehydrogenation to acetone, H_2 and CO_2. The presence of ceria-zirconia mixed oxide clearly influenced the oxidation step of the reaction. Notably, acetone was not observed over Rh(1wt%)@Al_2O_3-2-shell catalyst.

Stability tests on the present catalysts were conducted under SRE conditions at 600 °C, the lower temperature at which only the reforming products (H_2, CO and CO_2) were observed over both the embedded Rh catalysts. The stability tests were conducted for 160 hrs. Although during this period of time no appreciable deactivation was observed, the post-reaction Temperature Programmed Oxidation (TPO) experiments showed CO_2 evolution, suggesting the presence of carbonaceous deposits.

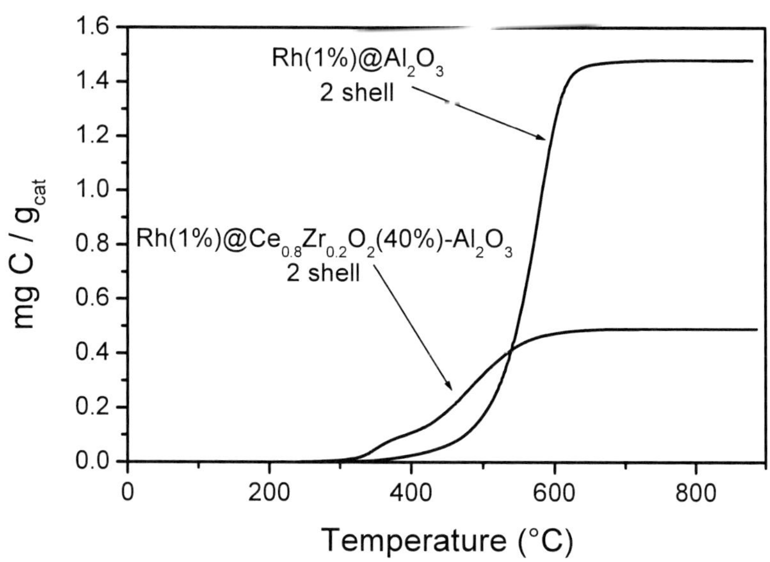

Figure 19. Carbon removal during TPO on the embedded Rh(1wt%)@Al_2O_3-2-shell and embedded Rh(1wt%)@$Ce_{0.8}Zr_{0.2}O_2$(40wt%)-Al_2O_3-2 shell after aging under SRE conditions (C_2H_5OH (1.0%) + H_2O (5.0%) in Ar, GHSV = 150000 mL g^{-1} h^{-1}) for 160 hrs.

Figure 19 shows the cumulative amount of carbon, normalized with respect to the mass of the catalyst, recorded from the evolved CO_2 during the TPO experiments on the aged samples. A high amount of carbon deposits was estimated for the aged Rh(1 wt%)@Al_2O_3-2-shell catalyst. The introduction of $Ce_{0.8}Zr_{0.2}O_2$ strongly reduced the carbon amount. Moreover, the carbon removal was easier in the case of the $Ce_{0.8}Zr_{0.2}O_2$-containing catalyst, as evidenced by the evolution of CO_2 at lower temperature.

In order to investigate the nature of the coke deposits formed after prolonged aging at high temperature, IR investigations were conducted on aged samples. The spectrum of Rh(1wt%)@Al_2O_3-2-shell showed very weak absorption band at 1606 cm^{-1} that were associated with pseudo-graphite or polyaromatics structures [45]. This is consistent with the CO_2 evolution peak observed at high temperature during the TPO experiment. A number of less intense bands, at 1486, 1518, 1523, 1564, 1582, 1636, 1655 and 1674 cm^{-1}, which may all be assigned to carbon-containing species, were also present [142]. Very weak bands at 2918 and 2836 cm^{-1} indicate the presence of traces of aliphatic residues. Significantly less evident are the IR bands associated with coke deposits on Rh(1wt%)@$Ce_{0.8}Zr_{0.2}O_2$(40wt%)-Al_2O_3-2-shell.

4. Conclusion

Active catalysts, resistant to the sinterization under severe reaction conditions, were developed through a simple and low cost synthetic route, combining the high reactivity of nanosized noble metal particles with the excellent high temperature stability of Al_2O_3-based nanocomposites. The catalyst design was based on the encapsulation of pre-formed Rh nanoparticles into a porous oxide matrix, which limits the mobility of the metal particles at high temperature. As a major effect, the encapsulation of the Rh nanoparticles inhibits their sinterization and prevents their total occlusion, favouring the accessibility of the catalytic sites to the reactants. As suggested by H_2 chemisorption data, catalytic activity tests and TEM characterization, the samples maintained their nanostructured design. Moreover, this challenging approach allowed the modulation of the texture of the support and its nature by the introduction of extra components like ceria-based mixed oxides as promoters.

Two generations of catalysts were prepared. In the first generation the deposition and the growth of protective oxide layers around Rh nanoparticles were realized in a single step. This strategy led to a material with some particles surrounded by porous oxide, while some of them were situated at the surface of the support and, therefore, only partially embedded. Instead, in the second generation of catalysts we obtained a better incorporation and preservation of particle sizes due to the milder conditions of pH and ionic strength during the synthetic procedure. In this case, the oxidic matrix was realized in a two step process.

The catalysts were tested for the Catalytic Partial Oxidation of Methane and the Steam Reforming of Ethanol.

The embedded Rh(1wt%)@Al_2O_3-1-shell presented higher thermal stability under CPOM conditions with respect to the traditional catalyst obtained by incipient wetness impregnation. This higher stability was correlated to the protection offered by the surrounding layer of porous oxide which prevented extensive sintering of the active metal phase. However, some coke deposition was observed and resulted in a partial deactivation of the catalytic activity. This partial deactivation was essentially reversible in the case of the Rh(1wt%)@Al_2O_3-1-

shell catalyst. Notably, in the case of CPOM, an *in situ* treatment with O_2 at high temperature was able to restore the properties of the catalyst while brief switches to O_2 during the reaction prevented deactivation. Furthermore, the embedded catalyst life time was improved by the addition of Pt through standard impregnation method or by the introduction of ceria-zirconia mixed oxides in the support formulation. Both components played a key role in preventing coke deposition and in favoring its removal. On the contrary, the deactivation observed on the impregnated Rh(1 wt%)/Al_2O_3 catalyst was mainly irreversible. The major factor inducing the deactivation was the sinterization of the metal phase and/or the incorporation of Rh into the Al_2O_3 lattice during high temperature treatments.

In the SRE process, the formulation of the nanocomposite oxide covering the Rh nanoparticles strongly influenced the product distribution. In the case of the Rh(1wt%)@Al_2O_3-2-shell catalyst, the reaction proceeded through the dehydration of ethanol on the acid site of the Al_2O_3 forming ethylene, which was reformed on the metal nanoparticles at higher temperature. Introducing the $Ce_{0.8}Zr_{0.2}O_2$ promoter, the main reaction pathway followed the dehydrogenation of ethanol to acetaldehyde, its decomposition to CO and CH_4 and the steam reforming of the latter product. Traces of ethylene (produced on the acid sites of Al_2O_3) and acetone (obtained from acetaldehyde) was also observed. The different reaction pathway active on the un-promoted and $Ce_{0.8}Zr_{0.2}O_2$-containing catalysts resulted in a different tendency of deactivation by coke deposition. A high amount of carbonaceous deposits were evidenced by Temperature Programmed Oxidation on the Rh(1wt%)@Al_2O_3-2-shell catalyst aged under SRE conditions for 160 hrs. On the contrary, the reduction of ethylene production induced by the introduction of $Ce_{0.8}Zr_{0.2}O_2$ into the protective matrix led to a less coke deposition. Moreover, the $Ce_{0.8}Zr_{0.2}O_2$ mixed oxide favoured the removal of the carbonaceous deposits during the oxidative treatment.

Acknowledgements

Prof. M. Graziani (University of Trieste, Italy), Prof. F.C. Lovey and Dr A.M. Condò, (Centro Atómico Bariloce, Argentina), Prof. S. Polizzi (University of Venice, Italy), Dr. M.F. Casula (University of Cagliari, Italy) are acknowledged for helpful discussions. Dr. F. Colombo, Mr. R. Crevatin and Mr. E. Merlach (University of Trieste) are acknowledged for qualified technical assistance. University of Trieste, Centre of Excellence for Nanostructured Materials (CENMAT), INSTM, FISR2002 "Nanosistemi inorganici ed ibridi per lo sviluppo e l'innovazione di celle a combustibile", FIRB2001 contract n° RBNE0155X7, are gratefully acknowledged for financial support.

References

[1] Graziani, M.; Fornasiero, P., Editors *Renewable Resourches and Renewable Energy: a Global Challenge;* Taylor & Francis: New York, 2006.

[2] Shishido, T.; Sukenobu, M.; Morioka, H.; Kondo, M.; Wang, Y.; Takaki, K.; Takehira, K. *Appl. Catal. A: Gen.* 2002, 223 (1-2), 35-42.

[3] Albertazzi, S.; Arpentinier, P.; Basile, F.; Del Gallo, P.; Fornasari, G.; Gary, D.; Vaccari, A. *Appl. Catal. A: Gen.* 2003, 247 (1), 1-7.

[4] Pena, M. A.; Gomez, J. P.; Fierro, J. L. G. *Appl. Catal. A: Gen.* 1996, 144 (1-2), 7-57.

[5] Freni, S.; Calogero, G.; Cavallaro, S. *J. Power Sources* 2000, 87 (1-2), 28-38.

[6] Sehested, J. *Catal. Today* 2006, 111 (1-2), 103-110.

[7] Rasmussen, F. B.; Sehested, J.; Teunissen, H. T.; Molenbroek, A. M.; Clausen, B. S. *Appl. Catal. A: Gen.* 2004, 267 (1-2), 165-173.

[8] Sehested, J.; Gelten, J. A. P.; Helveg, S. *Appl. Catal. A: Gen.* 2006, 309 (2), 237-246.

[9] Bengaard, H. S.; Norskov, J. K.; Sehested, J.; Clausen, B. S.; Nielsen, L. P.; Molenbroek, A. M.; Rostrup-Nielsen, J. R. *J. Catal.* 2002, 209 (2), 365-384.

[10] Trimm, D. L. *Catal. Today* 1997, 37 (3), 233-238.

[11] Choudhary, V. R.; Mamman, A. S.; Sansare, S. D. *Angew. Chem. Int. Ed.* 1992, 31 (9), 1189-1190.

[12] Choudhary, V. R.; Rane, V. H.; Rajput, A. M. *Catal. Lett.* 1993, 22 (4), 289-297.

[13] Choudhary, V. R.; Prabhakar, B.; Rajput, A. M. *J. Catal.* 1995, 157, 752-754.

[14] Choudhary, V. R.; Uphade, B. S.; Mamman, A. S. *Catal. Lett.* 1995, 32 (3-4), 387-390.

[15] Guerrero-Ruiz, A.; Ferreira-Aparicio, P.; Bachiller-Baeza, M. B.; Rodriguez-Ramos, I. *Catal. Today* 1998, 46 (2-3), 99-105.

[16] Avci, A. K.; Trimm, D. L.; Onsan, Z. I. *Chem. Eng. J.* 2002, 90 (1-2), 77-87.

[17] Weng, W. Z.; Chen, M. S.; Yan, Q. G.; Wu, T. H.; Chao, Z. S.; Liao, Y. Y.; Wan, H. L. *Catal. Today* 2000, 63 (2-4), 317-326.

[18] Weng, W. Z.; Luo, C. R.; Huang, J. J.; Liao, Y. Y.; Wan, H. L. *Top. Catal.* 2003, 22 (1-2), 87-93.

[19] Weng, W. Z.; Yan, Q. G.; Luo, C. R.; Liao, Y. Y.; Wan, H. L. *Catal. Lett.* 2001, 74 (1-2), 37-43.

[20] Walter, K.; Buyevskaya, O. V.; Wolf, D.; Baerns, M. *Catal. Lett.* 1994, 29 (1-2), 261-270.

[21] Buyevskaya, O. V.; Wolf, D.; Baerns, M. *Catal. Lett.* 1994, 29 (1-2), 249-260.

[22] Huff, M.; Torniainen, P. M.; Schmidt, L. D. *Catal. Today* 1994, 21 (1), 113-128.

[23] Buyevskaya, O. V.; Walter, K.; Wolf, D.; Baerns, M. *Catal. Lett.* 1996, 38 (1-2), 81-88.

[24] Grunert, W.; Wolf, D.; Buyevskaya, O. V.; Walter, K.; Baerns, M. *Z.Phys.Chem.* 1996, 197, 49-65.

[25] Witt, P. M.; Schmidt, L. D. *J. Catal.* 1996, 163 (2), 465-475.

[26] Grunwaldt, J. D.; Basini, L.; Clausen, B. S. *J. Catal.* 2001, 200 (2), 321-329.

[27] Panuccio, G. J.; Schmidt, L. D. *Appl. Catal. A: Gen.* 2006, 313 (1), 63-73.

[28] Grunwaldt, J. D.; Baiker, A. *Catal. Lett.* 2005, 99 (1-2), 5-12.

[29] Perkas, N.; Zhong, Z. Y.; Chen, L. W.; Besson, M.; Gedanken, A. *Catal. Lett.* 2005, 103 (1-2), 9-14.

[30] Basile, F.; Fornasari, G.; Rosetti, V.; Trifiro, E.; Vaccari, A. *Catal. Today* 2004, 91-92, 293-297.

[31] Basile, F.; Fornasari, G.; Gazzano, M.; Kiennemann, A.; Vaccari, A. *J. Catal.* 2003, 217 (2), 245-252.

[32] Wang, D. Z.; Li, Z. L.; Luo, C. R.; Weng, W. Z.; Wan, H. L. *Chem. Eng. Sc.* 2003, 58 (3-6), 887-893.

[33] Pantu, P.; Kim, K.; Gavalas, G. R. *Appl. Catal. A: Gen.* 2000, 193 (1-2), 203-214.

[34] Boucouvalas, Y.; Zhang, Z. L.; Verykios, X. E. *Catal. Lett.* 1996, 40 (3-4), 189-195.

[35] Basini, L.; Guarinoni, A.; Aragno, A. *J. Catal.* 2000, 190 (2), 284-295.

[36] Basini, L.; Aasberg-Petersen, K.; Guarinoni, A.; Ostberg, M. *Catal. Today* 2001, 64 (1-2), 9-20.
[37] Montini, T.; Condò, A. M.; Hickey, N.; Lovey, F. C.; De Rogatis, L.; Fornasiero, P.; Graziani, M. *Appl. Catal. B: Environ.* 2007, 73 (1-2), 84-97.
[38] Vermeiren, W. J. M.; Blomsma, E.; Jacobs, P. A. *Catal. Today* 1992, 13 (2-3), 427-436.
[39] Miao, Q.; Xiong, G. X.; Sheng, S. S.; Cui, W.; Xu, L.; Guo, X. X. *Appl. Catal. A: Gen.* 1997, 154 (1-2), 17-27.
[40] Lu, Y.; Liu, Y.; Shen, S. K. *J. Catal.* 1998, 177 (2), 386-388.
[41] Liu, S. L.; Xiong, G. X.; Sheng, S. S.; Yang, W. S. *Appl. Catal. A: Gen.* 2000, 198 (1-2), 261-266.
[42] Chu, W. L.; Yang, W. S.; Lin, L. W. *Appl. Catal. A: Gen.*2002, 235 (1-2), 39-45.
[43] Jun, J. H.; Lee, T. J.; Lim, T. H.; Nam, S. W.; Hong, S. A.; Yoon, K. J. *J. Catal.* 2004, 221 (1), 178-190.
[44] Kim, P.; Kim, Y.; Kim, H.; Song, I. K.; Yi, J. *Appl. Catal. A: Gen.* 2004, 272 (1-2), 157-166.
[45] Leroi, P.; Madani, E.; Pham-Huu, C.; Ledoux, M. J.; Savin-Poncet, S.; Bousquet, J. L. *Catal. Today* 2004, 91-92, 53-58.
[46] Xu, S.; Zhao, R.; Wang, X. L. *Fuel Process. Techn.* 2004, 86 (2), 123-133.
[47] Xu, S.; Wang, X. L. *Fuel* 2005, 84 (5), 563-567.
[48] Ma, D.; Mei, D. J.; Li, X.; Gong, M. C.; Chen, Y. Q. *J. Rare Earths* 2006, 24 (4), 451-455.
[49] Dissanayake, D.; Rosynek, M. P.; Kharas, K. C. C.; Lunsford, J. H. *J. Catal.* 1991, 132, 117-132.
[50] Choudhary, V. R.; Rajput, A. M.; Prabhakar, B. *Catal. Lett.* 1992, 15 (4), 363-370.
[51] Choudhary, V. R.; Rajput, A. M.; Prabhakar, B. *Angew. Chem. Int. Ed.* 1994, 33 (20), 2104-2106.
[52] Hagh, B. F. *Int. J. Hydrogen Energy* 2003, 28 (12), 1369-1377.
[53] Choudhary, V. R.; Uphade, B. S.; Mamman, A. S. *Microp. Mesop. Mat.* 1998, 23 (1-2), 61-66.
[54] Ahmed, S.; Krumpelt, M. *Int. J. Hydrogen Energy* 2001, 26 (4), 291-301.
[55] Moon, D. J.; Sreekumar, K.; Lee, S. D.; Lee, B. G.; Kim, H. S. *Appl. Catal. A: Gen.* 2001, 215 (1-2), 1-9.
[56] Deluga, G. A.; Salge, J. R.; Schmidt, L. D.; Verykios, X. E. *Science* 2004, 303 (5660), 993-997.
[57] Liguras, D. K.; Kondarides, D. I.; Verykios, X. E. *Appl. Catal. B: Environ.* 2003, 43 (4), 345-354.
[58] Velu, S.; Satoh, N.; Gopinath, C. S.; Suzuki, K. *Catal. Lett.* 2002, 82 (1-2), 145-152.
[59] Kugai, J.; Velu, S.; Song, C. S. *Catal. Lett.* 2005, 101 (3-4), 255-264.
[60] Fierro, V.; Klouz, V.; Akdim, O.; Mirodatos, C. *Catal. Today* 2002, 75 (1-4), 141-144.
[61] Klouz, V.; Fierro, V.; Denton, P.; Katz, H.; Lisse, J. P.; Bouvot-Mauduit, S.; Mirodatos, C. *J.Power Sources* 2002, 105 (1), 26-34.
[62] Benito, M.; Sanz, J. L.; Isabel, R.; Padilla, R.; Arjona, R.; Daza, L. *J. Power Sources* 2005, 151, 11-17.
[63] Morgenstern, D. A.; Fornango, J. P. *Energy & Fuels* 2005, 19 (4), 1708-1716.

[64] Haryanto, A.; Fernando, S.; Murali, N.; Adhikari, S. *Energy & Fuels* 2005, 19 (5), 2098-2106.
[65] Cavallaro, S.; Chiodo, V.; Freni, S.; Mondello, N.; Frusteri, F. *Appl. Catal. A: Gen.* 2003, 249 (1), 119-128.
[66] Cavallaro, S.; Freni, S. *Int. J. Hydrogen Energy* 1996, 21 (6), 465-469.
[67] Haga, F.; Nakajima, T.; Miya, H.; Mishima, S. *Catal. Lett.* 1997, 48 (3-4), 223-227.
[68] Marino, F. J.; Cerrella, E. G.; Duhalde, S.; Jobbagy, M.; Laborde, M. A. *Int. J. Hydrogen Energy* 1998, 23 (12), 1095-1101.
[69] Osaki, T.; Horiuchi, T.; Sugiyama, T.; Suzuki, K.; Mori, T. *J. Non-Cryst. Solid.* 1998, 225 (1), 111-114.
[70] Cavallaro, S.; Mondello, N.; Freni, S. *J. Power Sources* 2001, 102 (1-2), 198-204.
[71] Marino, F.; Boveri, M.; Baronetti, G.; Laborde, M. *Int. J. Hydrogen Energy* 2001, 26 (7), 665-668.
[72] Aupretre, F.; Descorme, C.; Duprez, D. *Catal. Commun.* 2002, 3 (6), 263-267.
[73] Batista, M. S.; Santos, R. K. S.; Assaf, E. M.; Assaf, J. M.; Ticianelli, E. A. *J. Power Sources* 2003, 124 (1), 99-103.
[74] Fatsikostas, A. N.; Verykios, X. E. *J. Catal.* 2004, 225 (2), 439-452.
[75] Sun, J.; Qiu, X. P.; Wu, F.; Zhu, W. T. *Int. J. Hydrogen Energy* 2005, 30 (4), 437-445.
[76] Fajardo, H. V.; Probst, L. F. D. *Appl. Catal. A: Gen.* 2006, 306, 134-141.
[77] Sahoo, D. R.; Vajpai, S.; Patel, S.; Pant, K. K. *Chem. Eng. J.* 2007, 125 (3), 139-147.
[78] Song, S. M.; Akande, A. J.; Idem, R. O.; Mahinpey, N. *Eng. Appl. Artif. Intell.* 2007, 20 (2), 261-271.
[79] De Rogatis, L.; Montini, T.; Casula, M. F.; Fornasiero, P. *J. Alloy. Compd.*, doi:10.1016/j.jallcom.2007.04.231.
[80] Cavallaro, S.; Chiodo, V.; Vita, A.; Freni, S. *J.Power Sources* 2003, 123 (1), 10-16.
[81] Navarro, R. M.; Alvarez-Galvan, M. C.; Sanchez-Sanchez, M. C.; Rosa, F.; Fierro, J. L. G. *Appl. Catal. B: Environ.* 2005, 55 (4), 229-241.
[82] Breen, J. P.; Burch, R.; Coleman, H. M. *Appl. Catal. B: Environ.* 2002, 39 (1), 65-74.
[83] Erdohelyi, A.; Rasko, J.; Kecskes, T.; Toth, M.; Domok, M.; Baan, K. *Catal. Today* 2006, 116 (3), 367-376.
[84] Vaidya, P. D.; Rodrigues, A. E. *Ind. Eng. Chem. Res.* 2006, 45 (19), 6614-6618.
[85] Montini, T.; De Rogatis, L.; Gombac, V.; Fornasiero, P.; Graziani, M. *Appl. Catal. B: Environ.* 2007, 71 (3-4), 125-134.
[86] Casanovas, A.; Llorca, J.; Homs, N.; Fierro, J. L. G.; de la Piscina, P. R. *J. Mol. Catal. A: Chem.* 2006, 250 (1-2), 44-49.
[87] Homs, N.; Llorca, J.; de la Piscina, P. R. *Catal. Today* 2006, 116 (3), 361-366.
[88] Takehira, K. *Catal. Surv. Japan* 2002, 6 (1-2), 19-32.
[89] Morioka, H.; Shimizu, Y.; Sukenobu, M.; Ito, K.; Tanabe, E.; Shishido, T.; Takehira, K. *Appl. Catal. A: Gen.* 2001, 215 (1-2), 11-19.
[90] Shishido, T.; Wang, P.; Kosaka, T.; Takehira, K. *Chem. Lett.* 2002, (7), 752-753.
[91] Takehira, K.; Shishido, T.; Kondo, M. *J. Catal.* 2002, 207 (2), 307-316.
[92] Takehira, K.; Shishido, T.; Wang, P.; Kosaka, T.; Takaki, K. *Phys. Chem. Chem. Phys.* 2003, 5 (17), 3801-3810.
[93] Takehira, K.; Shishido, T.; Wang, P.; Kosaka, T.; Takaki, K. *J. Catal.* 2004, 221 (1), 43-54.

[94] Basile, F.; Basini, L.; Fornasari, G.; Gazzano, M.; Trifiro, F.; Vaccari, A. *Chem. Commun.* 1996, (21), 2435-2436.
[95] Basile, F.; Fornasari, G.; Trifiro, F.; Vaccari, A. *Catal. Today* 2001, 64 (1-2), 21-30.
[96] Basile, F.; Fornasari, G.; Trifiro, F.; Vaccari, A. *Catal. Today* 2002, 77 (3), 215-223.
[97] Arpentinier, P.; Basile, F.; Del Gallo, P.; Fornasari, G.; Gary, D.; Rosetti, V.; Vaccari, A. *Catal. Today* 2005, 99 (1-2), 99-104.
[98] Arpentinier, P.; Basile, F.; Del Gallo, P.; Fomasari, G.; Gary, D.; Rosetti, V.; Vaccari, A. *Catal. Today* 2006, 117 (4), 462-467.
[99] Eriksson, S.; Nylen, U.; Rojas, S.; Boutonnet, M. *Appl. Catal. A: Gen.* 2004, 265 (2), 207-219.
[100] Schicks, J.; Neumann, D.; Specht, U.; Veser, G. *Catal. Today* 2003, 81 (2), 287-296.
[101] Wittborn, A. M. C.; Costas, M.; Blomberg, M. R. A.; Siegbahn, P. E. M. *J. Chem. Phys.* 1997, 107 (11), 4318-4328.
[102] Siegbahn, P. E. M.; Blomberg, M. R. A.; Svensson, M. *J. Am. Chem. Soc.* 1993, 115 (5), 1952-1958.
[103] Blomberg, M. R. A.; Siegbahn, P. E. M.; Nagashima, U.; Wennerberg, J. *J. Am. Chem. Soc.* 1991, 113 (2), 424-433.
[104] Hickman, D. A.; Schmidt, L. D. *Science* 1993, 259 (5093), 343-346.
[105] Budroni, G.; Corma, A. *Angew. Chem. Int. Ed.* 2006, 45 (20), 3328-3331.
[106] Martin, J. E.; Wilcoxon, J. P.; Odinek, J.; Provencio, P. *J. Phys. Chem. B* 2000, 104 (40), 9475-9486.
[107] Haruta, M.; Tsubota, S.; Kobayashi, T.; Kageyama, H.; Genet, M. J.; Delmon, B. *J. Catal.* 1993, 144 (1), 175-192.
[108] Haruta, M. *Catal. Today* 1997, 36 (1), 153-166.
[109] Sakurai, H.; Ueda, A.; Kobayashi, T.; Haruta, M. *Chem. Commun.* 1997, (3), 271-272.
[110] Hickey, N.; Arneodo Larochette, P.; Gentilini, C.; Sordelli, L.; Olivi, L.; Polizzi, S.; Montini, T.; Fornasiero, P.; Pasquato, L.; Graziani, M. *Chem. Mater.* 2007, 19 (4), 650-651.
[111] Schulz, J.; Roucoux, A.; Patin, H. *Chem.-Eur. J.* 2000, 6 (4), 618-624.
[112] Roucoux, A.; Schulz, J.; Patin, H. *Chem. Rev.* 2002, 102 (10), 3757-3778.
[113] Roucoux, A.; Schulz, J.; Patin, H. *Adv. Syn. Catal.* 2003, 345 (1-2), 222-229.
[114] Scott, A. B.; Tartar, H. V. *J. Am. Chem. Soc.* 1943, 65, 692.
[115] Bunton, C. A.; Ionescu, L. G. *J. Am. Chem. Soc.* 2004, 95, 2912.
[116] Jones, S. L.; Norman, C. *J. Amer. Ceram. Soc.* 1988, 71 (4), C190-C191.
[117] Tadokoro, S. K.; Muccillo, E. N. S. *J. Eur. Ceram. Soc.* 2002, 22 (9-10), 1723-1728.
[118] Fornasiero, P.; Hickey, N.; Kaspar, J.; Dossi, C.; Gava, D.; Graziani, M. *J. Catal.* 2000, 189 (2), 326-338.
[119] Everett, D. H. *Basic Principles of Colloid Science;* Royal Society of Chemistry: London, 1988.
[120] Sing, K. S. W.; Everett, D. H.; Haul, R. A. W.; Moscou, L.; Pierotti, R. A.; Rouquérol, J.; Siemieniewska, T. *Pure Appl. Chem.* 1985, 57 (4), 603-619.
[121] Podzorova, L. I.; Il'icheva, A. A.; Mikhailina, N. A.; Shevchenko, V. Y.; Shvorneva, L. I. *Inorg. Mater.* 2002, 38 (12), 1235-1240.
[122] Di Monte, R.; Fornasiero, P.; Desinan, S.; Kaspar, J. *Chem. Mater.* 2004, 16 (22), 4273-4285.
[123] Zhou, R. S.; Snyder, R. L. *Acta Cryst. B: Struct. Sc.* 1991, 47, 617-630.

[124] Di Monte, R.; Fornasiero, P.; Kaspar, J.; Graziani, M.; Gatica, J. M.; Bernal, S.; Gomez-Herrero, A. *Chem. Commun.* 2000, (21), 2167-2168.
[125] Hwang, C. P.; Yeh, C. T.; Zhu, Q. M. *Catal. Today* 1999, 51 (1), 93-101.
[126] Wang, C. B.; Yeh, C. T. *J. Mol. Catal. A: Chem.* 1997, 120 (1-3), 179-184.
[127] Vis, J. C.; van't Blik, H. F. J.; Huizinga, T.; van Grondelle, J.; Prins, R. *J. Catal.* 1985, 95, 333-345.
[128] Fornasiero, P.; Kaspar, J.; Graziani, M. *J. Catal.* 1997, 167 (2), 576-580.
[129] Fornasiero, P.; Kaspar, J.; Sergo, V.; Graziani, M. *J. Catal.* 1999, 182 (1), 56-69.
[130] Kaspar, J.; Fornasiero, P.; Graziani, M. *Catal. Today* 1999, 50 (2), 285-298.
[131] Gatica, J. M.; Baker, R. T.; Fornasiero, P.; Bernal, S.; Blanco, G.; Kaspar, J. *J. Phys. Chem. B* 2000, 104 (19), 4667-4672.
[132] Gatica, J. M.; Baker, R. T.; Fornasiero, P.; Bernal, S.; Kaspar, J. *J. Phys. Chem. B* 2001, 105 (6), 1191-1199.
[133] Bernal, S.; Calvino, J. J.; Cauqui, M. A.; Gatica, J. M.; Larese, C.; Omil, J. A. P.; Pintado, J. M. *Catal. Today* 1999, 50 (2), 175-206.
[134] Fornasiero, P.; Kaspar, J.; Montini, T.; Graziani, M.; Dal Santo, V.; Psaro, R.; Recchia, S. *J. Mol. Catal. A: Chem.* 2003, 204, 683-691.
[135] Vlaic, G.; Fornasiero, P.; Martra, G.; Fonda, E.; Kaspar, J.; Marchese, L.; Tomat, E.; Coluccia, S.; Graziani, M. *J. Catal.* 2000, 190 (1), 182-190.
[136] Boucouvalas, Y.; Zhang, Z. L.; Verykios, X. E. *Catal. Letters* 1994, 27 (1-2), 131-142.
[137] Wang, H. Y.; Ruckenstein, E. *J. Phys. Chem. B* 1999, 103 (51), 11327-11331.
[138] Weng, W. Z.; Pei, X. Q.; Li, H. M.; Luo, C. R.; Liu, Y.; Lin, H. Q.; Huang, C. J.; Wan, H. L. *Catal. Today* 2006, 117 (1-3), 53-61.
[139] Kaspar, J.; de Leitenburg, C.; Fornasiero, P.; Trovarelli, A.; Graziani, M. *J. Catal.* 1994, 146 (1), 136-143.
[140] Zhang, J. C.; Wang, Y. H.; Ma, R. Y.; Wu, D. Y. *Appl. Catal. A: Gen.* 2003, 243 (2), 251-259.
[141] Christensen, K. O.; Chen, D.; Lodeng, R.; Holmen, A. *Appl. Catal. A: Gen.* 2006, 314 (1), 9-22.
[142] Sarbak, Z. *React. Kin. Catal. Lett.* 2000, 69 (1), 177-181.
[143] Haneda, M.; Mizushima, T.; Kakuta, N. *J. Phys. Chem. B* 1998, 102 (34), 6579-6587.
[144] Nicole, J.; Tsiplakides, D.; Pliangos, C.; Verykios, X. E.; Comninellis, C.; Vayenas, C. G. *J. Catal.* 2001, 204 (1), 23-34.
[145] Lambrou, P. S.; Costa, C. N.; Christou, S. Y.; Efstathiou, A. M. *Appl. Catal. B: Environ.* 2004, 54 (4), 237-250.
[146] Men, Y.; Gnaser, H.; Zapf, R.; Hessel, V.; Ziegler, C. *Catal. Commun.* 2004, 5 (11), 671-675.
[147] Men, Y.; Gnaser, H.; Ziegler, C.; Zapf, R.; Hessel, V.; Kolb, G. *Catal. Lett.* 2005, 105 (1-2), 35-40.
[148] Men, Y.; Gnaser, H.; Zapf, R.; Hessel, V.; Ziegler, C.; Kolb, G. *Appl. Catal. A: Gen.* 2004, 277 (1-2), 83-90.
[149] Baraldi, A.; Cerda, J.; Martin-Gago, J. A.; Comelli, G.; Lizzit, S.; Paolucci, G.; Rosei, R. *Phys. Rev. Lett.* 1999, 82 (24), 4874-4877.
[150] Comelli, G.; Dhanak, V. R.; Kiskinova, M.; Prince, K. C.; Rosei, R. *Surf. Sc. Rep.* 1998, 32 (5), 167-231.
[151] Zhang, J. C.; Wang, Y. H.; Wu, D. Y. *Canad. J. Chem. Eng.* 2003, 81 (2), 307-311.

[152] Bueno-Lopez, A.; Krishna, K.; Makkee, M.; Moulijn, J. A. *J. Catal.* 2005, 230 (1), 237-248.
[153] Lopez, A.; Krishna, K.; Makkee, M.; Moulijn, J. *Catal. Lett.* 2005, 99 (3-4), 203-205.
[154] Aneggi, E.; de Leitenburg, C.; Dolcetti, G.; Trovarelli, A. *Catal. Today* 2006, 114 (1), 40-47.
[155] Aneggi, E.; Boaro, M.; de Leitenburg, C.; Dolcetti, G.; Trovarelli, A. *J. Alloy. Compd.* 2006, 408, 1096-1102.
[156] Freni, S. *J. Power Sources* 2001, 94 (1), 14-19.
[157] Comas, J.; Marino, F.; Laborde, M.; Amadeo, N. *Chem. Eng. J.* 2004, 98 (1-2), 61-68.
[158] Goula, M. A.; Kontou, S. K.; Tsiakaras, P. E. *Appl. Catal. B: Environ.* 2004, 49 (2), 135-144.
[159] Sun, J.; Qiu, X. P.; Wu, F.; Zhu, W. T.; Wang, W. D.; Hao, S. J. *Int. J. Hydrogen Energy* 2004, 29 (10), 1075-1081.
[160] Marino, F.; Boveri, M.; Baronetti, G.; Laborde, M. *Int. J. Hydrogen Energy* 2004, 29 (1), 67-71.
[161] Elliott, D. J.; Pennella, F. *J. Catal.* 1989, 119, 359.
[162] Nishiguchi, T.; Matsumoto, T.; Kanai, H.; Utani, K.; Matsumura, Y.; Shen, W. J.; Imamura, S. *Appl. Catal. A: Gen.* 2005, 279 (1-2), 273-277.
[163] Batista, M. S.; Santos, R. K. S.; Assaf, E. M.; Assaf, J. M.; Ticianelli, E. A. *J. Power Sources* 2004, 134 (1), 27-32.
[164] Marino, F.; Baronetti, G.; Jobbagy, M.; Laborde, M. *Appl. Catal. A: Gen.* 2003, 238 (1), 41-54.
[165] Llorca, J.; de la Piscina, P. R.; Sales, J.; Homs, N. *Chem. Commun.* 2001, (7), 641-642.
[166] Kaddouri, A.; Mazzocchia, C. *Cataly. Commun.* 2004, 5 (6), 339-345.
[167] Frusteri, F.; Freni, S.; Chiodo, V.; Spadaro, L.; Di Blasi, O.; Bonura, G.; Cavallaro, S. *Appl. Catal. A: Gen.* 2004, 270 (1-2), 1-7.
[168] Cavallaro, S. *Energy & Fuels* 2000, 14 (6), 1195-1199.
[169] Vesselli, E.; Comelli, G.; Rosei, R.; Freni, S.; Frusteri, F.; Cavallaro, S. *Appl. Catal. A: Gen.* 2005, 281 (1-2), 139-147.

In: New Developments in Soliton Research
Editor: L.V. Chen, pp. 115-161

ISBN 1-59554-561-8

Chapter 3

AC MEASUREMENTS OF HIGH IONIC CONDUCTIVITY DUE TO OXYGEN MIGRATIONS IN DOPED LANTHANUM GALLATES

***E. Iguchi*[1], *D. I. Savytskii*[2] and *M. Kurumada*[3*]**

[1]Honorary Professor, Yokohama National University, Tokiwadai, Hodogaya-Ku, Yokohama, 240-8501 Japan,
AGC Seimi Chemical Co., Ltd. Chigasaki 3-2-10, Chigasaki, 253-8585 Japan

[2]Department of Semiconductor Electronics, Institute of Telecommunication, Radioelectronics and Electronic Technique, Lviv Polytechnic National University, 12 Bandera Street, 79013 Lviv, Ukraine

[3]New Business Development Division, AGC Seimi Chemical Co., Ltd. Chigasaki 3-2-10, Chigasaki, 253-8585 Japan

Abstract

Oxygen ionic conduction in oxides results from the self diffusion of O^{2-} ions, the elementary process of which constitutes the migration of an O^{2-} ion from a lattice site to the next vacant lattice site across a saddle point in a diffusion path. The ac experimental method provides important knowledge about the dynamics of O^{2-} migrations in ceramic oxides because there are two effective techniques in the ac method, i.e., impedance analysis and measurements of the dielectric properties. In the impedance analysis, intra-grain conduction can be distinguished from inter-grain conduction and the parameter that represents the degree of the distribution of the relaxation times involved in O^{2-} migrations is provided directly. In the measurements of the dielectric properties, the relaxation processes due to O^{2-} migrations in different zones in a ceramic oxide can be recognized separately and the energy required for O^{2-} migration in each relaxation process is obtained directly. If both the impedance analysis and measurements of the dielectric relaxation processes are conducted together, ionic transport properties in oxides can be elucidated more directly with high-precision. The ionic conductivity of doped lanthanum gallates is very high as compared to that of most other oxides. In order to investigate the reasons for the high ionic conductivity

*E-mail address: msmkurumada@seimichemical.co.jp. Tel: +81 467 82 4131; Fax: +81 467 88 1778. (Corresponding author.)

of doped lanthanum gallates, ac measurements have been carried out using a single crystal of $La_{0.95}Sr_{0.05}Ga_{0.9}Mg_{0.1}O_{3-\delta}$ grown by the Czochralski method along with the dc measurements. This single crystal comprises a twin structure that consists of domains and the domain walls. These experiments have succeeded in revealing the dynamics of O^{2-} migrations in the domains and along the domain walls. These two different types of O^{2-} migrations constitute a parallel circuit of two independent R-C combinations. This parallel circuit corresponds to the conventional equivalent circuit of the twin structure modeled in the ac treatments. As a consequence of the parallel circuit, the resultant resistivity in the twin structure is considerably low. In order to examine whether this speculation holds in polycrystalline doped lanthanum gallates also, similar experiments have been carried out with $La_{1-x}Sr_xGa_{1.1-x}Zr_{x-0.1}O_{3-\delta}$ ceramics (x = 0.2-0.5). Subsequently, it is observed that the ionic conductive behaviors of these ceramics can be explained in terms of the twin structures in the bulks when the value of x is small.

Introduction

We have studied oxygen ionic conduction in oxides such as doped lanthanum gallates and yttria-stabilized zirconia (YSZ) [1-7]. As indicated by the Nernst-Einstein relation for ionic conductivity [8-10], oxygen ionic conduction results from the self diffusion of O^{2-} ions. The elementary process of oxygen diffusion constitutes the migration of an O^{2-} ion from a lattice site to the next vacant lattice site across a saddle point in a diffusion path. Therefore, the presence of oxygen vacancies is the minimum requirement for migrations of O^{2-} ions. When an O^{2-} ion arrives at a saddle point, the ions surrounding the O^{2-} ion are displaced from their lattice positions. After the O^{2-} ion has passed through the saddle point, these displaced ions return to their original positions. Consequently, the O^{2-} migration process involves the displacement of the ions that possess intrinsic electronic charges.

When the O^{2-} ion is at the original lattice site, the lattice energy is minimum (the ground state); the lattice energy becomes maximum when the O^{2-} ion is at the saddle point because local lattice distortion occurs around the saddle point due to the ionic displacement (the excited state). After the O^{2-} ion has passed through the saddle point, the lattice returns to the ground state. This implies that O^{2-} migration includes a relaxation process. Since the product of displacement and electronic charge results in a dipole moment, it is expected that such a relaxation process due to ionic displacement changes the dielectric properties to some extent. In fact, as revealed in the previous studies [1-7], alternating current (ac) measurements show that the dielectric properties of oxides vary due to O^{2-} migrations. Therefore, ac measurements provide very important knowledge of ionic conduction by oxygen diffusion in oxides.

Doped perovskite lanthanum gallates have been studied from various viewpoints because their ionic conductivity values are very high in comparison with those of most other oxides [1-7,11-21]. The chemical formula of perovskite oxide is ABO_3, and an illustration of the unit cell of the crystal structure is shown in Fig. 1, where A ions occupy the corners, B ions are located at the body-centered sites, and O^{2-} ions are located at the face-centered sites of the unit cells. Owing to the high ionic conductivity values of doped lanthanum gallates, they can be considered as suitable candidates for use as electrolytes in solid oxide fuel cells (SOFC); it might be possible to employ them in the manufacture of intermediate-temperature SOFC. In order to understand the reasons for high ionic conductivity and to further encourage ionic conduction in doped lanthanum gallates, it is essential to elucidate the dynamics of O^{2-} migrations. In particular, the investigation of the origin of high ionic conductivity in doped lanthanum gallates is essential. With this intention, several attempts have been made to vary the parameters of the Nernst-Einstein formula. If a change in some parameter of the Nernst-Einstein relation increases the ionic conductivity of doped lanthanum gallates, then this parameter must be the main factor causing high conductivity.

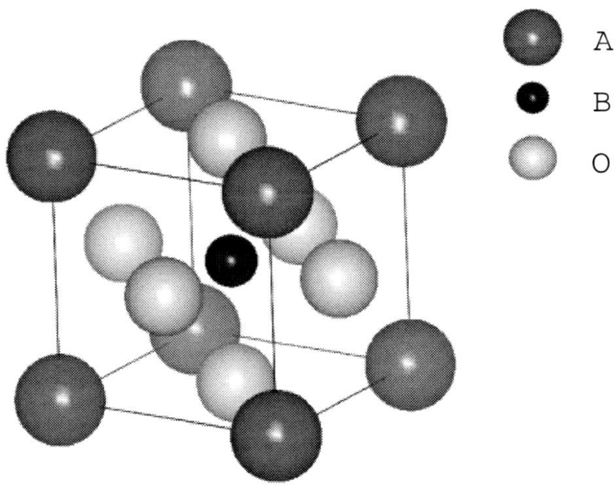

Figure 1. Perovskite structure unit cell of ABO_3.

The two most important parameters that dominate oxygen ionic conduction are the migration energy of O^{2-} ions required for ionic conduction E_M and the density of oxygen vacancies. Ionic conduction is further enhanced as the migration energy decreases. This phenomenon is witnessed more clearly at low temperatures. As described previously, O^{2-} migration requires the assist of oxygen vacancies and therefore the ionic conductivity due to O^{2-} migrations is proportional to the number of free mobile oxygen vacancies that assist O^{2-} migrations [1-4]. In order to decrease E_M by increasing the critical radius of the cation triangle of the perovskite lattice, r_{crit} [18,22,23], various doping ions and their combinations have been tested. Despite the marginal increase in the values of oxygen ionic conductivity by these attempts, it is still difficult to explain the reasons for high ionic conductivity in doped lanthanum gallates.

According to the Nernst-Einstein relation, the ionic conductivity should theoretically be proportional to the number of oxygen vacancies. However, when the density of oxygen vacancies is large, these vacancies tend to form stabilized clusters. In such cases, the process of thermal dissociation of the vacancies from stabilized clusters to free mobile states is indispensable for O^{2-} migrations; further, this process of dissociation from the stabilized clusters requires high energy. As a result, when the density of oxygen vacancies is high, oxygen ionic conduction requires high activation energy, resulting in low ionic conductivity as compared to the conductivity in the case where the density of oxygen vacancies is low. Therefore, it can be inferred that too many oxygen vacancies do not encourage oxygen ionic conduction in oxides. This implies that, in any oxide, there is intrinsically an optimum density of oxygen vacancies that is peculiar to it. In the other parameters of the Nernst-Einstein relation, between the doped lanthanum gallates and the other oxides, there are no large differences that can account for the high ionic conductivity in doped lanthanum gallates.

Therefore, we have to consider the migration dynamics of the O^{2-} ions in doped lanthanum gallates from a completely different viewpoint. In particular, the previous studies have not considered the relation between high ionic conductivities and crystallographic structures within the crystal grains in doped lanthanum gallates very carefully [1-7,11-21]. High-resolution synchrotron techniques have clarified the twinning that occurs in a Laue pattern as peak splitting in doped lanthanum gallate solid solutions such as $La_{0.95}Sr_{0.05}Ga_{0.9}Mg_{0.1}O_{3-\delta}$ produced using the Czochralski method [24,25]. In the ferroelastic noncubic phases such as those in this solid solution, the twin systems are composed of different domain states. The size of the domains in single crystals is less than 1 μm [24]. Furthermore, twinning forms chevron-like structures with the domain walls parallel to the crystallographic axis of a perovskite unit cell [24-28]. In doped lanthanum gallate ceramics, there is a high possibility that the crystal grains include the twin structure that plays an important role in the enhancement of ionic conduction. The legitimacy of this possibility will be revealed subsequently.

Although the details of the synchrotron study of the twin structures of doped lanthanum gallate solid solutions are described elsewhere [24,25], in the present review, we present an outline of this structure by considering $La_{0.95}Sr_{0.05}Ga_{0.9}Mg_{0.1}O_{3-\delta}$ as an example. In this oxide, there are three phase transitions: (i) orthorhombic ($Imma$) to monoclinic ($I2/a$) at approximately 550 K, (ii) monoclinic to trigonal ($R3c$) at approximately 690 K, and (iii) trigonal to trigonal ($R3\overline{c}$) at approximately 870 K. The twin structures of the orthorhombic and trigonal phases reveal that the system comprises four different domain states. In both the orthorhombic and trigonal phases that are ferroelastic, the domain walls form a chevron-like structure. The domains are related to each other by the symmetry elements that have been lost with respect to the prototype cubic symmetry. Figure 2 shows a simple schematic representation of the twin structure in an $La_{0.95}Sr_{0.05}Ga_{0.9}Mg_{0.1}O_{3-\delta}$ plate, where $[001]_p$ is the crystallographic direction of the perovskite unit cell, i.e., the c_p axis. The local ionic arrangement in the domain wall zones is different from that within the domains.

With regard to the crystal structures in single crystals doped with large amounts of impurity ions as compared to $La_{0.95}Sr_{0.05}Ga_{0.9}Mg_{0.1}O_{3-\delta}$, there are no experimental results of high-resolution synchrotron x-ray diffractions. This must be because synthesizing such single crystals is difficult.

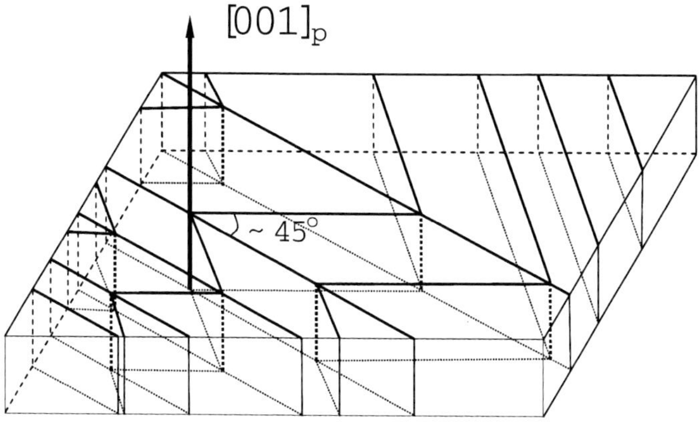

Figure 2. A simple schematic representation of the twin structure in the plate of the LSGM single crystal, where $[001]_p$ is the crystallographic direction of the perovskite unit cell, i.e., the c_p axis.

Experimentally, we have limited knowledge of the electric transport properties of the twin structures. Although it is difficult to produce single crystals, once they are obtained it would be possible to examine the reasons why doped lanthanum gallates have high ionic conductivity by measuring the electric transport properties of single crystals. It is expected that the oxygen ionic conduction along and/or across the domain walls in the twin structures differs from that within the domains [29-31]. The ac method is very effective to explain ionic conduction based on O^{2-} migrations in different zones in an oxide; in the ac method, the O^{2-} migrations in different zones can be distinguished because the relaxation processes of the O^{2-} migrations in individual zones are different from each other. From this viewpoint, in the present review, we have investigated the electric transport properties of a single crystal of the $La_{0.95}Sr_{0.05}Ga_{0.9}Mg_{0.1}O_{3-\delta}$ solid solution produced using the Czochralski method mainly by using the ac method. By examining these ac results, attempts have been made to pursue the reasons for the high ionic conductivity of doped lanthanum gallates.

Since most of the oxides actually used as electrolytes in SOFC are polycrystalline ceramics, the investigation of the characteristics of ceramic oxides is also important from the industrial point of view. Because the twin structure is intrinsic to doped lanthanum gallates, as described before, it has been experimentally clarified that the bulks in the ceramics of doped lanthanum gallates comprise the twin structures as well; further, it has been clarified that the sizes of the domains are in the range of 4 $\sim$ 20 nm [32,33]. Therefore, it is extremely important to elucidate the correlation between the twin structures in the bulks and the ionic conductivity of the doped lanthanum gallate ceramics. For $La_{1-x}Sr_xGa_{1-y}Mg_yO_{3-\delta}$, the optimum oxygen deficiency δ that yields the best electrical conductivity is approximately 0.185. When $(x + y)$ is greater than 0.37, however, short-range interactions among the oxygen vacancies interfere with O^{2-} migrations, thereby decreasing the electric conductivity [16]. From the crystallographic point of view, when $(x + y)$ is greater than 0.35, this oxide system is cubic at room temperature, whereas the crystal symmetry is low as compared to the cubic phase when the amount of impurity ions is small. Such low symmetric lanthanum gallates are ferroelastics, and their electric conductivity is dominated by the relative relation between the oxygen deficiency and the formation of the ferroelastic structure, i.e., the twin structure. $La_{1-x}Sr_xGa_{1-y}Mg_yO_{3-\delta}$ in the cubic phase at rather large concentrations of Sr^{2+} and Mg^{2+} ions does not involve the twin structure and consequently the conductivity is low.

Further, in these polycrystalline ceramics, the impurity phases are rather easily formed and they have adverse effects on ionic conduction. The easily-formed impurity phases are a disadvantage of doped lanthanum gallate ceramics, and if possible, they should be eliminated for practical applications of these oxides. Therefore, it is also necessary to investigate whether or not the impurity phases exist in the ceramics using x-ray diffraction (XRD) just after the polycrystalline ceramic oxides are synthesized.

Although we have studied the ionic conduction induced by O^{2-} migrations in polycrystalline ceramics of doped lanthanum gallates and yttria-stabilized zirconia thus far [1-7], one of the objectives of the present review is to clarify the relation between the twin structures within the bulks and the ionic conduction in polycrystalline ceramics by considering $La_{1-x}Sr_xGa_{1.1-x}Zr_{x-0.1}O_{3-\delta}$ (x = 0.2-0.5) as examples. For simplification, $La_{1-x}Sr_xGa_{1.1-x}Zr_{x-0.1}O_{3-\delta}$ is abbreviated in the present review as LSGZ(x). The study of LSGZ(x) is a continuation of the study of $La_{1-x}Sr_xGa_{1.1-x}Ti_{x-0.1}O_{3-\delta}$ [2,7]. In the previous report [2], we synthesized a solid solution of $(LaGaO_3)_{1-x}(SrTiO_3)_x$ in order to investigate the variations in the electronic and magnetic properties due to a partial substitution of $Ti^{4+}(3d^0)$ for $Ga^{3+}(3d^{10})$ at the B site in perovskite $LaGaO_3$.

The dc electric conductivity of $(LaGaO_3)_{0.5}(SrTiO_3)_{0.5}$ is higher than that of $LaGaO_3$ by two or three orders. Despite such a great increase in the electric conductivity by this substitution, there is no noticeable change in the activation energy of the dc conduction. These results indicate that this oxide system can potentially be used as an electrolyte in SOFC. Further, a study of the ionic conduction of oxides based on this system in greater detail would certainly be advantageous. As the first trail in this research field, $La_{1-x}Sr_xGa_{1.1-x}Ti_{x-0.1}O_{3-\delta}$, which is abbreviated as LSGT(x), was synthesized in order to increase the number of oxygen vacancies. This synthesis was done by taking into consideration that there is an optimum density of oxygen vacancies that is peculiar to each oxide [2,7]. Under the assumption that the ionic charge of Ti^{4+} remains unchanged even in the oxygen-deficient state, the electrical neutrality nominally yields δ in $La_{1-x}Sr_xGa_{1.1-x}Ti_{x-0.1}O_{3-\delta}$ = 0.05. This implies that LSGT(x) contains oxygen vacancies, although nominally there are no vacancies in $(LaGaO_3)_{1-x}(SrTiO_3)_x$. In fact, as expected, very high ionic conductivity has been observed in LSGT(x) in comparison with $(LaGaO_3)_{1-x}(SrTiO_3)_x$ [2,7]. Although the degree of the oxygen-deficiency in LSGT(x) is nominally fixed irrespective of x, the ionic conductivity of LSGT(x) varies with x [7]. There are chiefly two reasons for such an unexpected phenomenon. One is that the ionic charge of Ti^{4+} varies in the oxygen-deficient state. The other reason must be the crystallographic structure. The experimental results of LSGT(x) might indicate that the twin structure has to be considered during the analysis of the electric transport properties of LSGT(x) and the other doped lanthanum gallates.

LSGT(x) might also be used as an SOFC electrolyte in addition to the other doped lanthanum gallates. However, if LSGT(x) ceramics are actually used in SOFC, the strong reducing atmosphere at the anode side due to very low oxygen partial pressures might possibly cause problems with the Ti^{4+}/Ti^{3+} couples. This is because a $3d$ electron in Ti^{3+} ($3d^1$) in LSGT(x) causes high electronic conductivity due to polaronic conduction [34,35]. The Ti^{4+}/Ti^{3+} couples might make some contribution to the x dependence of the ionic conductivity of LSGT(x), as described above. Such high electronic conductivity due to polaronic conduction has adverse effects on SOFC functions. In order to eliminate this limitation, the Ti^{4+} ions have been replaced with non-magnetic Zr^{4+} ions in the present study and $La_{1-x}Sr_xGa_{1.1-x}Zr_{x-0.1}O_{3-\delta}$ ceramics denoted by LSGZ(x) have been synthesized. This oxide system is expected to be stable even in a strong reducing atmosphere because Zr ions are never multivalent and consequently Zr^{4+} cannot accommodate any electrons that cause electronic conduction.

Unfortunately, there are few studies on the electric conduction of the LSGZ(x) system. Therefore, several issues regarding this oxide system have to be examined experimentally. They are as follows:

(i). Since the impurity phases are easily formed in LSGT(x) and since they might have adverse effects on ionic conduction, it is important to investigate whether the impurity phases are formed in LSGZ(x) as well as in LSGT(x). When LSGZ(x) contains impurity phases, it is necessary to investigate the relation between the ionic conductivity and the volume ratio of the impurity phases.

(ii). If the impurity phases in LSGZ(x) have adverse effects on the ionic conduction that are comparable to those of the impurity phases in LSGT(x), it is very important to establish the way the growth of the impurity phases in LSGZ(x) can be suppressed. When the conductivity of LSGZ(x) is lower than that of LSGT(x), we have to determine the best way in which the ionic conductivity of LSGZ(x) can be increased.

(iii). If O^{2-} migrations in the twin structures within the bulks enhance ionic conductivity, an increase in ionic conductivity would be possible when x is reduced to 0.2 from 0.5. This is because at high doping levels of Sr^{2+} and Zr^{4+} ions, a cubic structure is expected in LSGZ(x) as well as in $La_{1-x}Sr_xGa_{1-y}Mg_yO_{3-\delta}$, which is cubic when $(x + y) > 0.35$. The cubic $La_{1-x}Sr_xGa_{1-y}Mg_yO_{3-\delta}$ does not contain the twin structure and the electric conductivity is low as compared to that of the orthorhombic one, for which the point symmetries described here are the ones at room temperature. Although the oxygen deficiency in LSGZ(x) is independent of x, i.e., $\delta = 0.05$ for every x, the formation of the twin structure is very sensitive to the value of x, which implies that electric conduction is also dependent on x. Therefore, the best conductivity of LSGZ(x) could be obtained by varying x.

In order to establish a technical means to suitably overcome these issues, the value of x in LSGZ(x) is changed from 0.5 to 0.2. Further, the sintering temperature during the synthesis of LSGZ(x) has been varied at each value of x so that the volume ratio of the impurity phases may become minimum. From this point of view, ac measurements of the LSGZ(x) samples that have the best ionic conductivity at each value of x have been carried out in order to investigate the correlation between the twin structure and the ionic conduction.

Fundamental Theories of Experimental Measurements

1. Dielectric relaxation process due to O^{2-} migration

In the Debye's theory [34,36-40], the complex dielectric constant due to a single relaxation process of dipole moments is defined as follows:

$$\epsilon(f) = \epsilon'(f) - j\epsilon''(f), \tag{1}$$

$$\epsilon'(f) = \epsilon_0 + \frac{\epsilon_s - \epsilon_0}{1 + (2\pi f \tau)^2}, \tag{2}$$

$$\epsilon''(f) = (\epsilon_s - \epsilon_0)\frac{2\pi f}{1 + (2\pi f \tau)^2}. \tag{3}$$

where ϵ_0, ϵ_s, f and τ are the high-frequency dielectric constant (or the optical dielectric constant), the static dielectric constant, the frequency of the applied ac field, and the relaxation time, respectively. With regard to the relaxation processes in solids including the electronic relaxation processes [41], the first approximation of $(\epsilon_s - \epsilon_0)$ has the following form,

$$(\epsilon_s - \epsilon_0) \cong \frac{4\pi}{3}\frac{N(qe)^2 a^2}{k_{\mathrm{B}}T}, \tag{4}$$

where N is the number of dipole moments per unit volume that cause the relaxation process in the ac electric field; qe is the electronic charge of an ion (or an electronic carrier), the displacement of which results in a dipole moment; a is the displacement distance; and k_{B} is Boltzmann's constant.

As described above, O^{2-} migration in oxides includes transfer of the energy between the ground state and the excited state when an O^{2-} ion moves from a lattice site to the next vacant lattice site through a saddle point. Therefore, O^{2-} migration involves a dielectric relaxation process owing to the displacement of the ions around the O^{2-} ion at the saddle point. This dielectric relaxation process requires the energy corresponding to the energy difference between the ground state and the excited state. In other words, this is the energy required when the O^{2-} ion passes through the saddle point. This is the migration energy of an O^{2-} ion that is denoted by E_M. The relaxation process of the O^{2-} migrations is characterized by the relaxation time given as follows:

$$\tau = \tau_0 \exp\left(\frac{E_M}{k_{\mathrm{B}}T}\right) = \frac{1}{2\pi f_0}\left(\frac{E_M}{k_{\mathrm{B}}T}\right), \tag{5}$$

where f_0 is the optical phonon frequency in an oxide.

The imaginary part of the complex dielectric constant ϵ'' is termed as the dielectric loss factor. As shown in Eq.(3), the dielectric loss factor is expressed as a function of the applied frequency f, which represents a typical relaxation process. At a temperature T, the dielectric loss factor ϵ'' has a normal Gaussian distribution as a function of the frequency f and has a maximum at frequency $f_{\epsilon''}$, where $f_{\epsilon''}$ is the resonance frequency that satisfies the resonance condition $2\pi f_{\epsilon''}\tau = 1$. The maximum dielectric loss factor at $f = f_{\epsilon''}$ is represented as follows:

$$\epsilon''_{max} = \frac{(\epsilon_s - \epsilon_0)}{2} \propto \frac{N}{T}. \tag{6}$$

Therefore, the maximum loss factor is proportional to the density of the dipole moments responsible for that relaxation process and it is also proportional to the reciprocal of the temperature.

If the relaxation time in Eq.(5) is substituted for τ in the resonance condition $2\pi f_{\epsilon''}\tau = 1$, the resonance frequency has the following form

$$f_{\epsilon''} = f_0 \exp\left(-\frac{E_M}{k_B T}\right). \tag{7}$$

The formula given in Eq.(7) indicates that the migration energy E_M is to be estimated from the Arrhenius relation of $\log(f_{\epsilon''})$ and $1/T$ if the temperature dependence of the resonance frequency is obtained experimentally in the measurements of the dielectric loss factor.

In the twin structure within the bulks corresponding to the crystal grains, there are two diffusion paths for the O^{2-} migrations - the path in the domains and the path along the domain walls. Moreover, in polycrystalline ceramic oxides, O^{2-} migrations occur in the grain boundaries as well. The migration energy and the optical phonon frequency vary with the zones because the ionic arrangements in these zones differ from each other. This implies that each zone has a relaxation time owing to the O^{2-} migrations peculiar to it. Theoretically, it is expected that a dielectric curve consisting of three relaxation peaks of different intensities at different resonance frequencies will be observed for the dielectric loss factor at each temperature in the polycrystalline ceramics of doped lanthanum gallates. As a result, the dielectric loss factor curves actually observed in experiments are not symmetric and must be distorted considerably.

The dielectric behaviors described until now are realized only in a system with a single relaxation process based on the Debye's theory [34,36-40]. This is a single relaxation system. However, dielectric relaxation processes usually include distributions of relaxation times [2,3,36,39,40]. In oxides, there are many different diffusion paths of O^{2-} migrations and the diffusion along each path requires the migration energy that is intrinsic to it. Even in the bulks, there must be many diffusion paths of O^{2-} migrations because lattice imperfections such as dislocations, impurities, vacancies, etc. disturb the regular arrangement of ions. Furthermore, in the boundaries that do not have any regularity in the arrangements of ions, there should be many diffusion paths for O^{2-} migrations and therefore the relaxation times must disperse very broadly as compared to the bulks. Therefore, the relaxation times disperse intrinsically in any zone in any ceramics oxides.

In a dielectric system that contains a distribution of relaxation times, the real and imaginary parts of the complex dielectric constant are given as follows:

$$\epsilon'(f) = \epsilon_0 + \frac{(\epsilon_s - \epsilon_0)}{2}\left(1 - \frac{\sinh\beta x}{\cosh\beta x + \cos(\beta\pi/2)}\right), \tag{8}$$

$$\epsilon''(f) = \frac{(\epsilon_s - \epsilon_0)}{2}\frac{\sin(\beta\pi/2)}{\cosh\beta x + \cos(\beta\pi/2)}, \tag{9}$$

where β is the parameter that represents the degree of the distribution of the relaxation times and $x = \log(2\pi f\tau)$ [36]. A single relaxation system corresponds to $\beta = 1$. Since the resonance condition is $2\pi f_{\epsilon''}\tau = 1$, which is also the case with the single relaxation system, the dielectric loss factor has a maximum value at $f = f_{\epsilon''}$,

$$\epsilon''_{max} = \frac{(\epsilon_s - \epsilon_0)}{2}\tan\left(\frac{\beta\pi}{4}\right) \propto \frac{N}{T}\tan\left(\frac{\beta\pi}{4}\right). \tag{10}$$

As the value of β decreases from 1, the distribution of the relaxation times becomes broad and the maximum dielectric loss factor is attenuated; however, the resonance frequency $f_{\epsilon''}$ remains unchanged.

As described in the previous section, an O^{2-} ion migrates with the assist of doped oxygen vacancies, which are created by impurities. These doped oxygen vacancies are first captured by the trapping centers at low temperatures. The oxygen vacancies dissociated thermally from the trapping centers with the energy E_O can move freely in the lattice. Here, E_O indicates the dissociation energy of an oxygen vacancy. Only these mobile oxygen vacancies can assist O^{2-} migrations. Since the number of doped oxygen vacancies is negligible as compared to the number of O^{2-} ions at the normal lattice sites at any temperature in an oxide, the number of O^{2-} ions that can migrate is in substance equivalent to the number of free mobile oxygen vacancies. If the number of doped oxygen vacancies is denoted by N_0, the density of the free mobile oxygen vacancies at T is N, which can be expressed as follows:

$$N = N_0\exp\left(-\frac{E_O}{k_B T}\right). \tag{11}$$

By substituting Eq.(11) for N in Eq.(10), the temperature dependency of the maximum dielectric loss factor can be given as follows:

$$\epsilon''_{max} \propto \frac{\exp(-E_O/k_B T)}{T}. \tag{12}$$

Therefore, in principle, the dissociation energy is obtained experimentally by measuring the ϵ''_{max} values as a function of T.

In the dielectric loss factor of most oxides, it is difficult to observe the relaxation peaks in practice. This is mainly because of the high values resulting due to the background of the loss factor. When there is no emergence of a dielectric relaxation peak in the loss factor, the approximation that the dielectric loss tangent tan δ is proportional to the loss factor ϵ'' is usually adopted, i.e., $\epsilon'' \propto \tan\delta = \epsilon''/\epsilon'$. This approximation is reasonably accepted because the frequency dependency of ϵ' is monotonous and has no extrema at any temperature [1,3,37,38,42]. In fact, resonance peaks due to dielectric relaxation processes are observed in the loss tangent of most oxides. However, in the relaxation processes in the bulks of the oxides, the resonance frequencies of the loss tangent are considerably high as compared to those of the loss factor [40,43,44]. Despite this, it appears that this approximation does not include any serious shortcomings in the estimations of the energy values required for the dielectric relaxation processes. Therefore, the experimental values of E_M and E_O obtained by using this approximation are generally accepted to be reliable.

In this approximation, we have the following relations; $f_{\tan\delta} \propto \exp(-E_M/k_\mathrm{B}T)$ and $(\tan\delta)_{max} \propto \exp(-E_O/k_\mathrm{B}T)/T$, where $f_{\tan\delta}$ is the resonance frequency in the loss tangent. Thus, the magnitudes of E_M and E_O for the O^{2-} migrations in the oxides are obtained experimentally using the Arrhenius relations of $\log(f_{\tan\delta})$ vs $1/T$ and $\log[T(\tan\delta)_{max}]$ vs $1/T$.

2. Complex-plane impedance analysis

One of the advantages of the ac measurements that can be applied to ionic conductors such as oxides is that complex-plane impedance analysis is possible. This is because impedance analysis provides very significant knowledge of ionic conduction. Usually, three different processes occur during the charge transport through a ceramic oxide. They are as follows: (i) bulk conduction (i.e., intra-grain conduction), (ii) conduction across the grain boundaries (i.e., inter-grain conduction), and (iii) transport across the electrode-specimen interface [45-48]. Impedance analysis distinguishes each of these processes separately. Since each circuit element corresponding to these processes is represented by an independent R-C combination, it is generally accepted that the conventional equivalent circuit of a ceramic oxide modeled in ac treatments comprises a series of three R-C parallel circuits: the first corresponds to intra-grain conduction, the second to inter-grain conduction, and the third to the transport in the interface between the oxide and the electrode. Figure 3(a) shows the representation of this conventional equivalent circuit. Since the modeling of an equivalent circuit is very important in the ac analyses, several attempts have been made to establish a more realistic equivalent circuit. As an example, a new theoretical treatment for the analysis and modeling of impedance spectroscopy was developed recently [49]. In this treatment, these processes can be observed with more accuracy by selecting the most realistic equivalent circuit. However, this method requires procedures that are currently not very simple. Therefore, we have employed the conventional equivalent circuit, which has been generally accepted, i.e., a series of three R-C parallel circuits, as shown in Fig. 3.

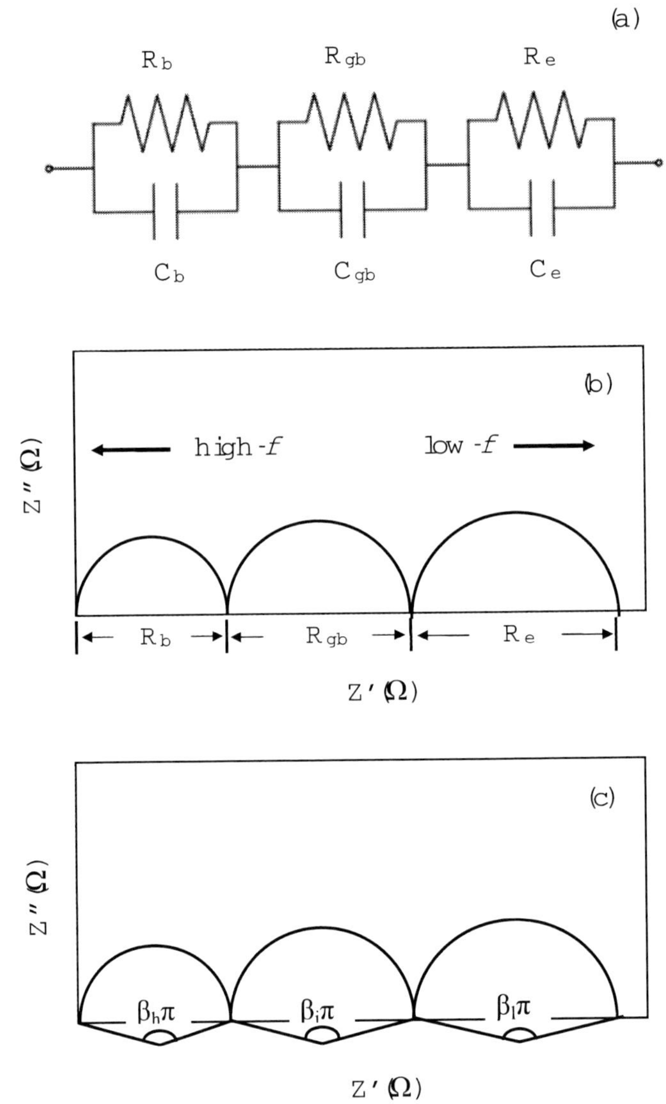

Figure 3. (a) The equivalent circuit of an oxides comprising of a series of three R-C parallel circuits; the suffix e denotes the parameters of the oxide-electrode interface, while the other suffixes have the same meanings as given in the text. (b) Three semicircles corresponding to the three R-C parallel circuits in a single relaxation system, where "$\longleftarrow$ high-f" indicates the increase direction of the applied frequency and "low-f $\longrightarrow$" expresses the decrease direction. (c) Three semicircular structure in a system with distributions of the relaxation times. Each arc intersects the real (Z') axis and the angle subtended by two intersections and the center is $\beta\pi$, where the suffixes have the same meanings as given in the text. The β values obtained in this manner are denoted such as $(\beta_i)_h$ and $(\beta_i)_i$ in the text in order to distinguish these values from the β values that are estimated in the results of dielectric relaxation processes.

The impedance of an R-C parallel circuit is expressed as follows:

$$Z = \frac{1}{1/R + j(2\pi f)C} = \frac{R}{1+(2\pi f)^2R^2C^2} - j\frac{(2\pi f)R^2C}{1+(2\pi f)^2R^2C^2} = Z' - jZ''. \quad (13)$$

Then, the real and imaginary parts of the impedance form a semicircle with a radius of $R/2$ in the complex plane, the highest point of the semicircle being at the frequency f_i,

$$(Z' - R/2)^2 + (Z'')^2 = (R/2)^2, \quad (14)$$

$$f_i = \frac{1}{2\pi RC}. \quad (15)$$

Since each R-C circuit has relations similar to Eqs.(14) and (15), three semicircles emerge in the complex plane in which the real parts Z' of the total impedance Z are plotted against the imaginary parts Z'' as a parametric function of the frequency of the applied ac electric field, f. Figure 3(b) shows three semicircular arcs corresponding to the three R-C parallel circuits. Usually, the highest-frequency arc passing through the origin of the complex plane corresponds to intra-grain conduction (the bulk conduction), the intermediate-frequency arc corresponds to inter-grain conduction (the boundary conduction), and the lowest-frequency arc corresponds to the interface process. In the measurement temperature region that we have usually employed, the lowest-frequency arcs of most oxides can hardly be recognized because frequencies below the lowest limit of our apparatus are required in the measurements of impedances corresponding to the lowest-frequency arcs.

Since the impedance analysis described above is based on a single relaxation process that takes place in each R-C circuit, as shown in Eq.(14), the center of each semicircle lies on the real axis (Z') in the complex-plane. However, most dielectric relaxation processes include the distributions of relaxation times, the degree of the distribution being characterized by the parameter β. In the single relaxation system, β is 1. However, in a system in which the relaxation times disperse, the capacitance has a dispersive function, as represented by the following equation,

$$C = C_0 + C_1 \int \frac{F(\tau)}{1+(2\pi f)^2\tau^2} d\tau, \quad (16)$$

where C_0 and C_1 are proportional to the static dielectric constant and the difference between the static and optical dielectric constants, respectively, and $F(\tau)$ is the distribution function of the relaxation times, which is subject to the following condition [8,36],

$$\int_0^\infty F(\tau)d\tau = 1. \quad (17)$$

These dispersions move the real semicircular arc from the semicircle of $\beta = 1$ downward in the complex-plane and therefore the center of the real arc deviates from the real axis [45-48].

Figure 3(c) shows three semicircular arcs with β_h, β_i and β_l, where the suffixes h, i and l indicate the parameters of the highest-, intermediate-, and lowest-frequency arcs, respectively. Each arc intersects the Z' axis and the angle subtended by two intersections and the center of the arc is $\beta\pi$. Therefore, the experimental β values are directly determined in the impedance analyses, although these values are assessed only indirectly in the measurements of the dielectric relaxation processes. Furthermore, the resistance values of the circuit elements are obtained from the intersections of the real axis and the semicircular arcs. On the real axis in the complex plane, the highest resistance of the highest-frequency arc is the bulk resistance R_b, the highest resistance of the intermediate-frequency arc is the sum of the bulk resistance R_b and the boundary resistance R_{gb} and hence the difference between the highest resistance values of the highest- and intermediate-frequency arcs is the boundary resistance R_{gb}. Theoretically, the sum of R_b and R_{gb} corresponds to the resistance of the polycrystalline ceramic oxide obtained by the direct current (dc) method. It is impossible to obtain the resistance values of R_b and R_{gb} separately by experimental means other than impedance analysis.

A semicircle in the results of impedance spectroscopy resulting from O^{2-} migrations in a lattice zone is always concomitant with a dielectric relaxation peak in the loss factor that is also caused by the O^{2-} migrations in that zone. However, the frequency ranges of the semicircle and the loss factor relaxation peak corresponding to that semicircle are considerably different. This is expected because the frequencies of the semicircle are subject to R and C whereas the frequencies of the dielectric relaxation peak are dependent on f_0 and E_M; the condition which the frequencies of the semicircle satisfy in the complex plane is quite different from the condition satisfied by the frequencies of the dielectric relaxation process. Further, there is another reason for the frequency difference between the semicircle in the complex plane and the relaxation process in the loss factor. In the case of the relaxation process in the bulks, it is theoretically shown that the frequency region of the impedance in which the relaxation peak appears is considerably higher than the frequency region of the loss factor for that relaxation peak [40].

3. Four-probe dc conductivity

Since the measurement of four-probe dc conductivity values is the fundamental technique employed in solid state physics to study the electric transport properties of solids, experiments on most oxides commence by measuring the dc conductivity values. The ionic conductivity due to the diffusion of ions, σ, is represented by the Nernst-Einstein formula in the following equation [8-10],

$$\sigma \cong \left(\frac{(q_L e)^2 \pi \nu a_L^2 N}{k_B T}\right) \exp\left(-\frac{Q}{k_B T}\right), \tag{18}$$

where ν is the ionic vibrational frequency, and $q_L e$, a_L and N are the electronic charge, jumping distance, and density of the diffusing ions that are responsible for ionic conduction, respectively. In Eq.(18), Q is the activation energy required for the migration of ions. In the case of ionic conduction due to O^{2-} migrations in doped lanthanum gallates, q_L is -2, a is the ionic spacing between the adjacent oxygen lattice sites, Q is E_M, and N is expressed as $N_0\exp(-E_O/k_B T)$. As explained previously, E_O and E_M are the energy values of the ionic conduction and these energy values consist of the components of the bulks and the boundaries. Therefore, the oxygen ionic conductivity of doped lanthanum gallates can be expressed as follows:

$$\sigma \cong \left(\frac{(-2e)^2 \pi \nu a_L^2 N_0}{k_B T} \right) \exp \left(-\frac{E_M + E_O}{k_B T} \right). \quad (19)$$

Therefore, E_{dc} is the sum of E_M and E_O. In the dc measurements, however, it is difficult to distinguish the migration energy and the dissociation energy within the bulks from the energy values in the boundaries on an experimental basis.

Experimental Details

1. Oxide specimens

In order to determine the reasons for the high ionic conductivity of doped lanthanum gallates, we first carried out ac measurements on a single crystal of $La_{0.95}Sr_{0.05}Ga_{0.9}Mg_{0.1}O_{3-\delta}$ solid solution, which is abbreviated here as LSGM, along with the dc measurements. After examining the experimental results obtained from the LSGM single crystal experiment very carefully, we carried out similar measurements for polycrystalline $La_{1-x}Sr_xGa_{1.1-x}Zr_{x-0.1}O_{3-\delta}$ ceramics (x = 0.2 - 0.5), i.e., LSGZ(x). This was done in order to investigate whether the factors that are responsible for high ionic conductivity as obtained from the results of the LSGM single crystal experiment are also observed in the polycrystalline ceramic oxides experiment.

The LSGM single crystal has been grown from the melt in argon and 1% oxygen using the Czochralski technique. The pulling rate was 1.2-2.5 mm/h [50]. A single crystal of the best quality and with dimensions of 15 mm $\phi \times$ 30 mm L was selected. The single crystal was produced by M. Berkowski (Institute of Physics, Polish Academy of Sciences, Al. Lotników, Warsaw 02-668, Poland). In order to determine the chemical compositions, chemical analysis of the single crystal was carried out [50]. The La content has been determined from sulfuric acid solution by direct titration with Trilon B (Ga has been masked by acetylacetone). The amount of Ga was determined by substitutional titration using Cu comlexonat and Trilon B (La was masked by NH_4F). For the Sr and Mg analysis, the crystal was melted using $Na_2CO_3 + Na_4B_2O_7 \times 10H_2O$ flux. The melt was dissolved in HCl (1:1). The Mg concentration was determined by atomic absorption spectroscopy using AAS-1N spectrometer (Carl Zeiss, Jena) with a propene/butane/air flame at wavelength 285.2 nm. Sr was determined by atomic-emission spectroscopy (Carl Zeiss, Jena) at wavelength 470.7 nm. This chemical analysis reveals that the experimental chemical composition of the single crystal is $La_{0.953}Sr_{0.054}Ga_{0.888}Mg_{0.105}O_{3-\delta}$, the cation concentration of which is practically identical to that of the starting compositions. The advantage of the single crystal produced in this manner is the good crystallinity, homogeneity, and phase purity. Since the electric transport properties of this high-quality single crystal are very sensitive to the types of impurities and their amounts, the chemical analysis was carried out. In the present study, the nominal compositions

Table 1. The sintering temperature T_s; the lattice parameters of the body centered orthorhombic lattice ($Imma$), a, b and c; and the perovskite cell parameters, $a_p = a/2$, $b_p = b/\sqrt{2}$, and $c_p = c/\sqrt{2}$, for every specimen of the LSGZ(x) polycrystalline ceramics, where x = 0.2, 0.3, 0.4 and 0.5. The temperatures of T_s in this table denote the sintering temperatures that maximize the ionic conductivity for each x.

x	specimen	T_s [°C]	a [Å]	b [Å]	c [Å]	$a_p = a/2$ [Å]	$b_p = b/\sqrt{2}$ [Å]	$c_p = c/\sqrt{2}$
0.2	LSGZ(0.2)	1520	7.828	5.524	5.559	3.914	3.906	3.931
0.3	LSGZ(0.3)	1520	7.883	5.561	5.586	3.942	3.932	3.950
0.4	LSGZ(0.4)	1520	7.928	5.610	5.603	3.964	3.966	3.962
0.5	LSGZ(0.5)	1470	7.984	5.642	5.643	3.992	3.989	3.990

of polycrystalline ceramic oxides are employed as the experimental chemical compositions.

The extremely high resolution of the powder diffraction patterns at room temperature allows all the reflections to be indexed according to the body-centered orthorhombic structure with the following lattice parameters: a = 7.794 Å, b = 5.499 Å, and c = 5.538 Å [50]. Several plates with $[001]_p$ directions were cut off from the crystal and polished to a thickness of 0.5 mm (see Fig. 2). The details of the twin structure of this crystal are investigated by white beam synchrotron x-ray diffraction (Laue technique) studies in the temperature range of 300 to 800 K [24,25].

The LSGZ(x) polycrystalline ceramic specimens (x= 0.5, 0.4, 0.3, and 0.2) were prepared by using a conventional solid-state reaction technique using La_2O_3, $SrCO_3$, Ga_2O_3, and ZrO_2 powders (3N). The mixtures were calcined in air at 1250°C for 12 h. After mixing the powders very carefully, this heat treatment was repeated. After grinding, the calcined powders were pressed into pellets and finally sintered in air for 24 h. At x = 0.3 and 0.2, the sintering temperature T_s was changed in order to maximize the ionic conductivity. The optimum sintering temperatures T_s are tabulated in Table 1 for all x components. The lattice structures and the lattice parameters were investigated using x-ray powder diffraction with Cu $K\alpha$ x-ray radiation (XRD) at room temperature (JEOL JDX-3530 X-ray diffractometer system). For the purpose of comparison, the main XRD reflections were indexed according to the body-centered orthorhombic lattice ($Imma$). The lattice parameters were refined by using the least squares method using the Rietveld analysis program. Table 1 summarizes the lattice parameters, a, b and c, for each specimen. The perovskite cell parameters - $a_p = a/2$, $b_p = b/\sqrt{2}$ and $c_p = c/\sqrt{2}$ - are also included in Table 1. In Fig. 4, the perovskite cell parameters are plotted against x. At x = 0.2 and 0.3, the differences in the perovskite cell parameters are never small; further, the line splitting is remarkable. Line splitting indicates the degree of the orthorhombic distortion from the ideal perovskite cubic cell. At x = 0.4 and 0.5, the line splitting is very small. This fact generally indicates that the deformation of the perovskite structure is small when x is 0.4 and 0.5. Therefore, it is reasonably accepted that the crystal structures of LSGZ(x) are nearly cubic at x = 0.4 and 0.5.

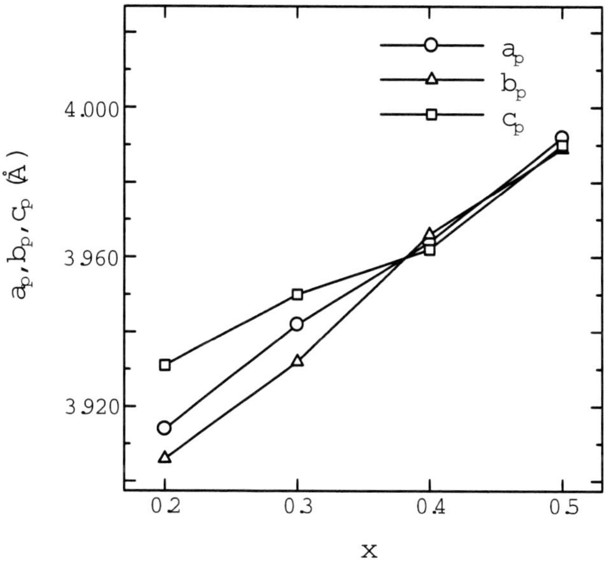

Figure 4. The plots of the perovskite cell parameters, a_p, b_p, and c_p, against x for LSGZ(x); here, the circles, triangles and squares are the plots of the lattice parameters, a_p, b_p and c_p, respectively.

2. Measurements of ac and dc methods

In the ac measurements, the four-probe pair method was employed in order to eliminate mutual inductances, unexpected capacitances, interferences between measured signals and unnecessary parameters because these factors have adverse effects on the measurements of the dielectric properties, especially in the high frequency range. Figure 5 illustrates the measurement principle of the four-probe pair method. The measured signal, i.e., the current, flows through the sample between the current probes HCUR and LCUR. The voltage drop of the measured signal that flows through the sample is detected by the potential probes HPOT and LPOT. H in HCUR and HPOT represents the drive voltage supplied from the internal vector voltmeter. Highly accurate ac measurements can be obtained by using this four-probe pair method. By using an Agilent 4294 A precision impedance analyzer, the capacitances and impedances of LSGM and LSGZ(x) were measured in the frequency range of 40 Hz -10 MHz. The electrodes for the ac measurements were made by coating Pt paste on two of the largest parallel surfaces of the specimens and by baking them at 900 °C for 0.5 h.

A Keithley 619 resistance bridge, an Advantest TR 6871 digital multimeter, and an Advantest R6161 voltage current source were employed for the four-probe dc conductivity measurements. Pt paste was also used for the electrodes in the dc experiments. In both the ac and dc measurements, the Pt-Pt/Rh thermocouple was used, and the temperature was controlled by using a Keithley 2000 multimeter. All ac and dc measurements were carried out in cooling runs as functions of the temperature over a range of 523K -1103 K in air.

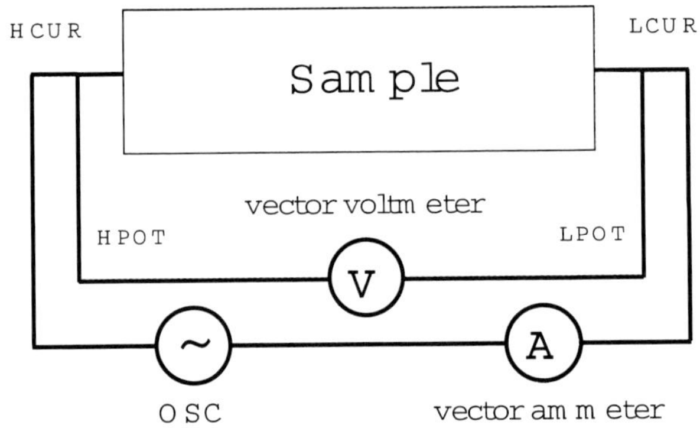

Figure 5. Illustration of four-probe pair method for ac measurements.

Experimental Results and Discussions

1. $La_{0.95}Sr_{0.05}Ga_{0.9}Mg_{0.1}O_{3-\delta}$ single crystal (LSGM)

1-a. Ionic Conduction in LSGM

Because ionic conduction in oxides is ascribed to the self diffusion of O^{2-} ions [1-4,15,16], Fig. 6 depicts the Arrhenius relation of $\log(\sigma_{dc}T)$ and $1/T$ for LSGM based on the Nernst-Einstein relation [8-10], where σ_{dc} is the four-probe dc conductivity normal to the $[001]_p$ crystallographic axis. The dc conductivity values of the LSGM single crystal are in the same orders as those of the other doped lanthanum gallates [1,3,15-21]. In the Arrhenius relation of Fig.6, there are marked variations due to the phase transitions at approximately 690 K and 870 K. Table 2 provides the magnitudes of the activation energy required for the dc conduction E_{dc} estimated from the Arrhenius relation in Fig. 6; E_{dc} = 1.00 $\pm$ 0.01 eV in the monoclinic phase, 0.82 $\pm$ 0.01 eV in the $R3c$ trigonal phase, and 1.01 $\pm$ 0.04 eV in the $R3\bar{c}$ trigonal phase. At $T < 550$ K in Fig. 6, the experimental plots deviate slightly from the linear Arrhenius relation within the monoclinic phase and this deviation must be due to the phase transition from the orthorhombic phase to the monoclinic phase. Most of the activation energy values of LSGM and LSGZ(x) obtained in the present study are summarized in Table 2.

According to the theoretical treatment of the complex plane impedance analysis given in the previous section, we have carried out this analysis at each temperature for LSGM. Figure 7 shows the results of impedance spectroscopy at 493, 593, and 693 K as examples. The ac field is applied in the $[001]_p$ direction normal to the largest surface of the plate in Fig. 2. In contrast with a ceramic oxide that theoretically contains three semicircular arc structure as mentioned before, the LSGM single crystal contains only the highest-frequency arc. This is a peculiar feature of this single crystal. In order to investigate the distribution of the relaxation times, the β value is estimated in the results of impedance spectroscopy at each temperature. The high-frequency arc intersects the real axis. Figure 7 (b) includes the plot of the center of the highest-frequency arc so as to estimate β; $\beta\pi$ is the angle subtended

Table 2. Activation energy values of ionic conduction in several phases of (a) the LSGM single crystal, (b) the LSGZ(x) polycrystalline ceramics, and (c) the LSGT(0.5) polycrystalline ceramic, where the results of LSGT(0.5) are the ones at $T <$ 960 K. In this table, $E_{\rm dc}$ is the activation energy for four-probe dc conduction, E_R is the energy determined from the temperature dependence of the electric resistance R obtained in the impedance analysis and E_M is the migration energy of an O^{2-} ion. The suffixes d and w represent the parameters of the domains and domain walls in the twin structure, and b and gb denote the grains and grain boundaries in polycrystalline ceramics. T_t for LSGZ(x) is the temperature at which two straight lines with different activation energy values cross over in the Arrhenius relation of log$(\sigma_{\rm dc}T)$ and $1/T$ as shown in Fig.16.

(a) LSGM single crystal				
phase	$E_{\rm dc}$ [eV]	E_R [eV]	$(E_M)_d$ [eV]	$(E_M)_w$ [eV]
Monoclinic ($T <$ 600 K)	1.00 ± 0.01	0.92 ± 0.01	0.97 ± 0.01	0.82 ± 0.02
$R3c$ trigonal (600 K $< T <$ 840 K)	0.82 ± 0.01	0.75 ± 0.01	0.79 ± 0.02	0.65 ± 0.03
$R\overline{3}c$ trigonal (920 K $< T$)	1.01 ± 0.04			

(b) LSGZ(x) polycrystalline ceramics				
specimen	$E_{\rm dc}$ [eV] $T < T_t$	$E_{\rm dc}$ [eV] $T > T_t$	$(E_R)_b$ [eV]	$(E_R)_{gb}$ [eV]
LSGZ(0.2)	0.89 ± 0.02	0.73 ± 0.02	0.83 ± 0.01	1.16 ± 0.01
LSGZ(0.3)	0.92 ± 0.02	0.78 ± 0.01	0.90 ± 0.02	1.14 ± 0.02
LSGZ(0.4)	0.94 ± 0.02	0.83 ± 0.03	0.93 ± 0.02	1.15 ± 0.03
LSGZ(0.5)	1.01 ± 0.03	0.91 ± 0.02	0.95 ± 0.02	1.14 ± 0.02
specimen	T	$(E_M)_d$ [eV]	$(E_M)_b$ [eV]	$(E_M)_w$ [eV]
LSGZ(0.2)	$T <$ 740 K	0.75 ± 0.02	1.01 ± 0.06	0.53 ± 0.02
	$T >$ 740 K	0.67 ± 0.02	1.01 ± 0.06	0.51 ± 0.05
LSGZ(0.5)		0.86 ± 0.06	1.15 ± 0.03	0.31 ± 0.04

(c) LSGT(0.5) polycrystalline ceramics ($T <$ 960 K)				
$E_{\rm dc}$ [eV]	$(E_R)_b$ [eV]	$(E_M)_b$ [eV]	$(E_R)_{gb}$ [eV]	$(E_M)_{gb}$ [eV]
0.93 ± 0.05	0.82 ± 0.04	0.78 ± 0.06	1.14 ± 0.05	0.99 ± 0.06

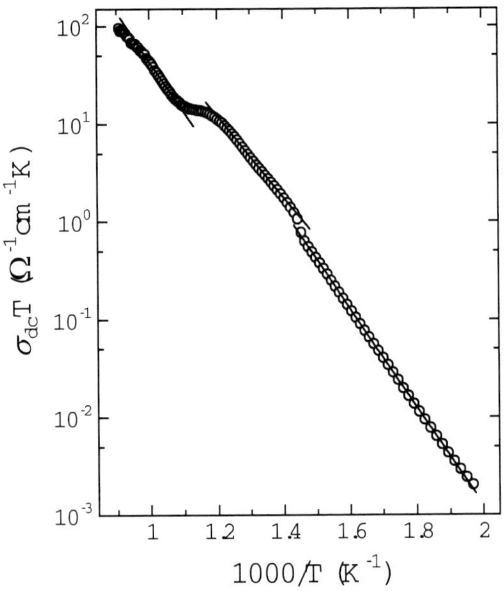

Figure 6. Arrhenius relation of log($\sigma_{dc}T$) and $1/T$ for the LSGM single crystal, where σ_{dc} is the four-probe dc conductivity. The open circles denote the experimental plots and the straight line in each phase yields the activation energy for dc conduction E_{dc}.

by two intersections and the center of the arc, where one intersection is the origin of the complex plane. Though the values of β are obtained directly in the impedance analysis, they are also indirectly estimated as parameters used in the curve-fitting treatment in the analysis of the dielectric relaxation processes, which will be mentioned later. Therefore, in the present review, the values of β obtained by the impedance analysis are denoted such as $(\beta_i)_h$ in order to distinguish from the β values estimated by curve-fitting of the dielectric relaxation processes, where the suffixes i and h in $(\beta_i)_h$ mean the impedance analysis and the highest-frequency arc. For the LSGM single crystal, the impedance spectroscopy yields $(\beta_i)_h = 0.91 \pm 0.02$ in the temperature region where the impedance analysis is possible.

Next, the resistance value R is obtained from the real axis (Z') intercept in the result of impedance spectroscopy at each temperature. Since the electric conductivity is proportional to the reciprocal of resistance, Fig. 8 shows the Arrhenius relation of log(T/R) and $1/T$. There are two linear portions corresponding to the monoclinic and $R3c$ trigonal phases. These straight lines yield $E_R = 0.92 \pm 0.01$ eV for the monoclinic phase and 0.75 ± 0.01 eV for the $R3c$ trigonal phase (see Table 2), where E_R is the activation energy required for the electric conduction in each phase. The magnitudes of E_R are marginally smaller than that of E_{dc} in both the monoclinic and $R3c$ trigonal phases. Since it is expected that the ionic transport across the domain walls is very different from that along the walls, the difference between E_{dc} and E_R must be mainly due to the fact that the ac electric field is applied parallel to the domain walls of the single crystal plate whereas the dc field is normal to the ac field.

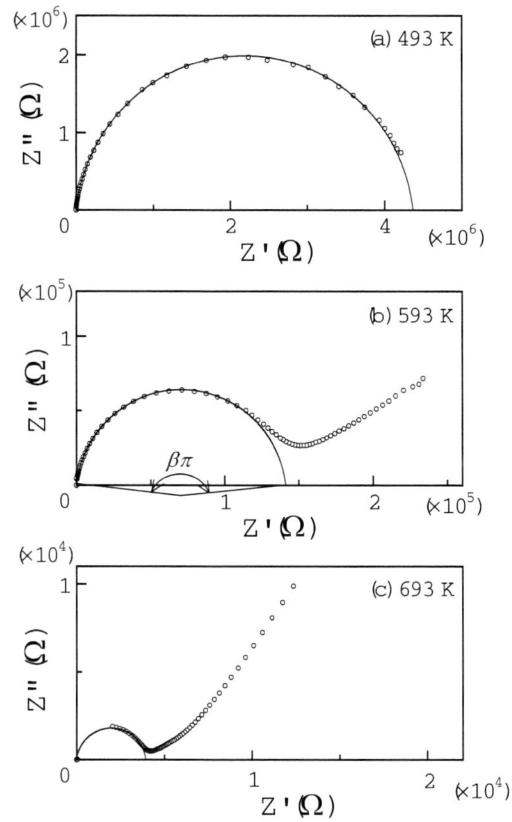

Figure 7. Impedance spectroscopy in the complex plane for LSGM at (a) 493 K, (b) 593 K, and (c) 693 K. At 593 K, the center of the semicircular arc is plotted with the angle $\beta\pi$.

1-b. Dielectric relaxation process in LSGM

According to the dielectric theory described in the previous section, it would be most appropriate if the dielectric relaxation process could be analyzed by using the experimental results of the dielectric loss factor ϵ''. In LSGM, however, no dielectric relaxation peak appears in the loss factor; instead, it appears in the loss tangent like most other oxides. Based on the explanation provided in the previous section, we have used the approximation that the dielectric loss tangent is proportional to the loss factor. In Fig.9, we have plotted the dielectric loss tangent tan δ against the applied frequency f as a parametric function of T for LSGM over the temperature range of 633 K to 993 K. A dielectric relaxation peak appears at each temperature, but it is somewhat distorted. This is a peculiar feature of doped lanthanum gallates.

At the monoclinic-to-trigonal transition point, a drastic change takes place in the loss tangent. It is surmised that this change must be mainly a consequence of the large dipole moments due to the ionic displacement around the O^{2-} ions at the saddle points during the migration processes of the O^{2-} ions in the $R3c$ trigonal phase as compared to the dipole

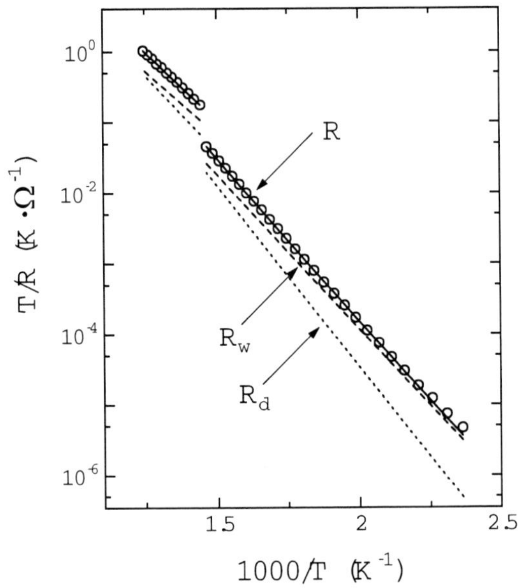

Figure 8. Arrhenius relation of $\log(T/R)$ vs $1/T$ for LSGM, where the open circles denote the experimental plots of the electric resistance R obtained in the impedance spectroscopy. The Arrhenius relations of $\log(T/R_d)$ vs $1/T$ and $\log(T/R_w)$ vs $1/T$ are also demonstrated, where R_d (dotted line) and R_w (dashed line) are the values of the resistance in the domains and along the domain walls in the twin structure, respectively, which are estimated by curve-fitting. The solid line represents the theoretical resistance of $R = R_d R_w/(R_d + R_w)$, the values of which agree well with the experimental plots.

moments in the monoclinic phase. The critical radii of the cation triangles, r_{crit}, in the $R3c$ trigonal phase are generally greater than those in the monoclinic phase [50]. These critical radii correspond to the effective ionic radii of the saddle points. This fact explains the drastic change in the loss tangent at the monoclinic-to-trigonal transition point. The large values of r_{crit} in the $R3c$ trigonal phase easily increase the displacement of ions, thereby resulting in large dipole moments which yield the large loss factors, as shown in Fig. 9. As calculated previously [51-53], exact assessments of the energy values of the ions in oxides require extremely complicated theoretical calculations which involve important physical parameters of each ion in oxides such as the wave functions of shell electrons, electronic and ionic polarizabilities, and so on. Nevertheless, the size of the space through which an ion passes is the most important parameter that mainly dominates the migration of an ion. Therefore, the large critical radii result in low migration energy of the O^{2-} ions in the $R3c$ trigonal phase as compared to those in the monoclinic phase. This is evident from the difference between the energy values of the monoclinic and $R3c$ trigonal phases, which are tabulated in Table 2.

The dielectric properties obtained in the present study have been analyzed within the framework of the Debye's model [34,36-40]. However, instead of a single relaxation process, the distribution of the relaxation times has been included in the numerical analysis of

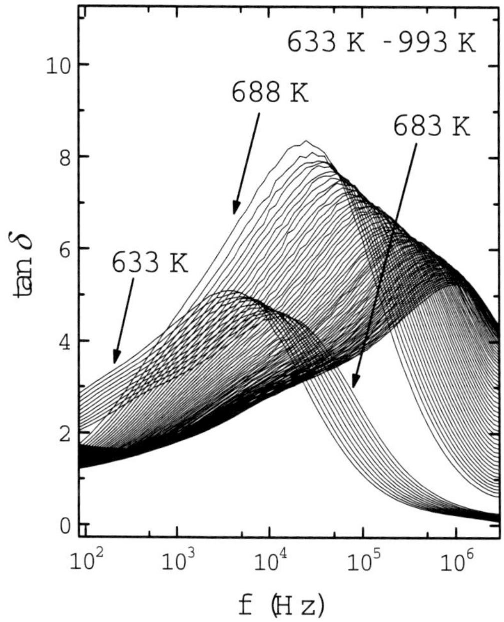

Figure 9. Frequency dependencies of the dielectric loss tangent tan δ at 5 K increments as a parametric function of T in the range of 633 - 993 K for LSGM. A drastic change occurs when the temperature changes from 683 to 688 K; this is due to the monoclinic-to-trigonal ($R3c$) phase transition.

each dielectric relaxation process by employing the parameter β. The dielectric loss tangent curve at each temperature in Fig. 9 is distorted and therefore does not exhibit a normal Gaussian distribution. This must be due to the overlapping of two relaxation processes. In such a case, as described in our previous studies [4,5,7], iterations of the least squares method can divide the distorted loss tangent curve into individual relaxation process by using the formula of the experimental loss tangent (tan $\delta)_{\exp}$ as follows:

$$(\tan\delta)_{\exp} = \frac{A_h}{T}\frac{\sin(\beta_h\pi/2)}{\cosh(\beta_h x_h) + \cos(\beta_h\pi/2)} + \frac{A_i}{T}\frac{\sin(\beta_i\pi/2)}{\cosh(\beta_i x_i) + \cos(\beta_i\pi/2)}, \quad (20)$$

where the suffixes h and i represent the parameters of the high- and intermediate-frequency relaxation processes, respectively. The constant term A in each process is proportional to the density of the O^{2-} ions that are able to migrate.

At the resonance frequency of the loss tangent $f_{\tan\delta}$ that satisfies the condition $2\pi f_{\tan\delta}\tau = 1$, the maximum loss tangent at T is given as (tan $\delta)_{\max} = (A/T)\exp(\beta\pi/4)$. The least squares method is iterated until the values of $(f_{\tan\delta})_h, A_h, \beta_h, (f_{\tan\delta})_i, A_i$, and β_i yield the curve that fits best to the experimental plots. Figure 10(a) shows the dielectric loss tangent tan δ as a function of the applied frequency f at 693 K along with the two theoretical curves obtained by the iteration treatments and their total curve. At high temperatures, the tail of the low-frequency peak emerges at low frequencies, which is plainly recognized at 843 K in Fig. 10(b). In such a case, another term representing

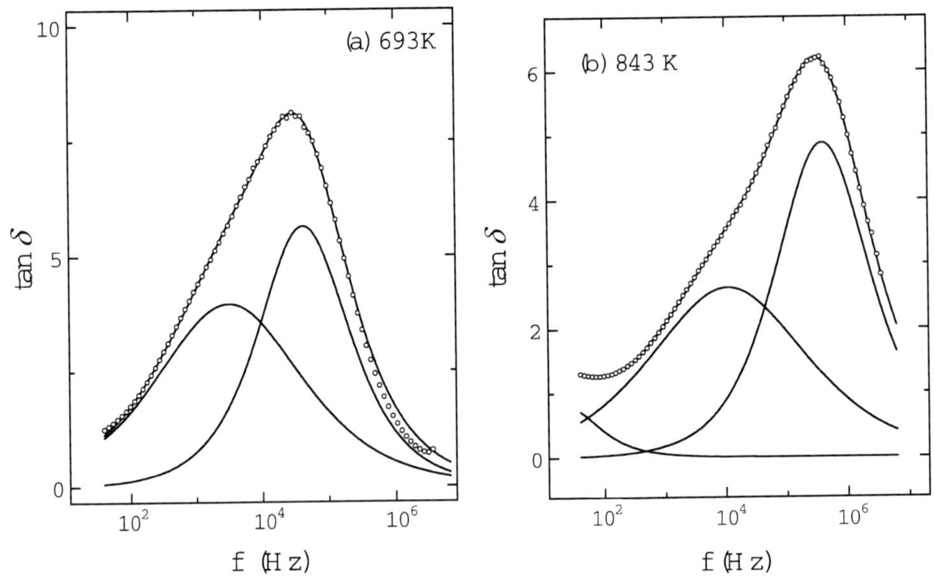

Figure 10. Experimental plots of tan δ (open circles) against f with three theoretical curves (solid lines) obtained by iterations of the least squares method; (a) the high- and intermediate-frequency relaxation peaks and the resultant curve of these relaxation peaks for LSGM at 693 K and (b) similar result at 843 K, which includes another relaxation peak at low frequencies.

the low-frequency curve should be included in the experimental loss tangent (tan $\delta)_{\rm exp}$ in Eq.(20), i.e., $(A_l/T)\sin(\beta_l\pi)/[\cosh(\beta_l x_l) + \cos(\beta_l\pi/2)]$, where the suffix l indicates the parameters of the low-frequency peak. Figure 10(b) includes three curves, i.e., the high-, intermediate-, and low-frequency peaks, and the resultant curve of these three relaxation peaks besides the experimental plots.

The β values estimated from the dielectric relaxation process by iterations of the least square method are denoted such as $(\beta_r)_i$, as well as the case of the impedance analysis. The suffix r in $(\beta_r)_i$ indicates the dielectric relaxation process and the suffix i is described above. Then, $(\beta_r)_h = 0.94 \pm 0.04$ and $(\beta_r)_i = 0.72 \pm 0.02$. It is noteworthy that the value of $(\beta_i)_h$ obtained in the impedance analysis lies between that of $(\beta_r)_h$ and $(\beta_r)_i$, i.e., $(\beta_r)_i < (\beta_i)_h < (\beta_r)_h$. There must be some reason why such a relative relation of these β values is realized. This will be described later. Since the value of $(\beta_r)_i$ is small as compared to $(\beta_r)_h$, the ionic conduction responsible for the intermediate-frequency relaxation process must involve many different paths of O^{2-} migrations as compared to the ionic conduction that corresponds to the high-frequency peak. From the dielectric results of the relaxation processes in the loss tangent, it is evident that the electric conduction in the LSGM single crystal is the combination of the conduction responsible for the high-frequency relaxation process and that responsible for the intermediate-frequency process. The total theoretical curve in Fig. 10(a) deviates slightly from the experimental plots at the high frequency at approximately $f = 10^6$ Hz. This deviation might be caused by the electrode effects that modify the bulk dispersion especially at high frequencies.

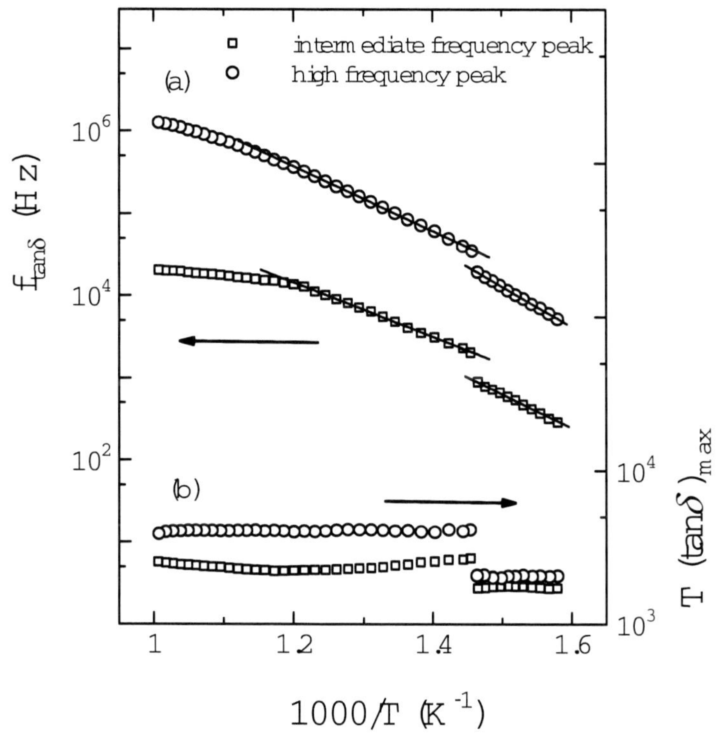

Figure 11. Arrhenius relations of (a) log($f_{\tan\delta}$) vs $1/T$ and (b) log[$T(\tan\delta)_{max}$] vs $1/T$ for LSGM, where the open circles and open squares represent the experimental plots for the high-frequency dielectric relaxation peak and the intermediate-frequency peak, respectively. Arrhenius relations of log($f_{\tan\delta}$) vs $1/T$ yield $(E_M)_d$ and $(E_M)_w$ in the monoclinic and $R3c$ trigonal phases.

Figure 11(a) shows the relation between $f_{\tan\delta}$ and T for the high- and intermediate-frequency relaxation processes in the Arrhenius representation. Since the resonance condition yields $f_{\tan\delta} = f_0\exp(-E_M/k_BT)$, the migration energy of the O^{2-} ions E_M and the optical phonon frequency f_0 in each phase can be estimated in Fig. 11(a). In the monoclinic phase, $(E_M)_h = 0.97 \pm 0.01$ eV, $(f_0)_h = 1.60\times10^{12}$/s, $(E_M)_i = 0.82 \pm 0.02$ eV, and $(f_0)_i = 5.71\times10^9$/s, while in the $R3c$ trigonal phase, $(E_M)_h = 0.79 \pm 0.02$ eV, $(f_0)_h = 6.90\times10^{10}$/s, $(E_M)_i = 0.65 \pm 0.03$ eV, and $(f_0)_i = 8.131\times10^8$/s. These migration energy values are summarized in Table 2. These numerical results indicate that the LSGM single crystal consists of two lattice parts: one has a high ionic vibrational frequency and the other has low ionic vibrational frequency. The high-frequency dielectric relaxation process results from the oxygen ionic conduction with rather high migration energy in the lattice part of the high ionic vibrational frequency, whereas oxygen ionic conduction with low migration energy in the lattice zone of the low ionic vibrational frequency results in the low-frequency relaxation process.

As described in the previous section, the oxygen vacancies dissociated thermally from the trapped states assist O^{2-} migrations and the dissociation energy E_O is to be obtained experimentally from the following relation: $(\tan\delta)_{\max} \propto \exp(-E_O/k_BT)/T$ [1,2,4,7]. Figure 11(b) also includes the plots of $\log[T(\tan\delta)_{\max}]$ against $1/T$ for both the high- and low-frequency relaxation processes. As compared to the other doped lanthanum gallates [1-3], the results in Fig. 11(b) exhibit highly complicated behaviors. In particular, extremely large variations occur around the phase transition points. Such variations may be due to the occupation and release of oxygen vacancies between the domains and domain walls at the phase transitions, which require reconstructions of the domain wall structure [25]. Therefore, strong interactions are expected between the domain walls and the point defects such as oxygen vacancies. Similar interactions are observed experimentally between the domain walls and the oxygen vacancies with the atomic scale mechanism of the twin memory effect [54,55]. However, with regard to the LSGM single crystal, it is impossible to estimate the real values of E_O in Fig. 11(b).

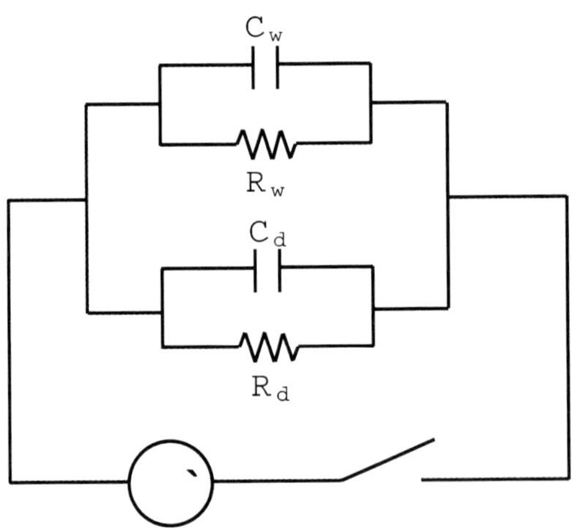

Figure 12. The ac equivalent circuit in the twin structure of LSGM when the ac field is applied along the $[001]_p$ direction.

The LSGM single crystal contains only the highest-frequency semicircle in the impedance spectroscopy, but it contains two dielectric relaxation processes in the loss tangent. This is a very significant feature that is peculiar to the LSGM single crystal. This feature can be easily understood from the illustration of the twin structure in Fig. 2. In this structure, the domain walls are parallel to the $[001]_p$ direction from the rear to the front of the crystal plate of LSGM [24,25]. Furthermore, there are two diffusion paths of the O^{2-} ions when the ac electric field is applied parallel to the $[001]_p$ direction that is normal to the largest surface of the plate, i.e., the paths within the domains and along the domain walls. These diffusion paths constitute a parallel R-C circuit, which is illustrated in Fig. 12. This is the equivalent circuit of the twin structure, which is modeled in the ac treatments.

Therefore, the impedance Z of the equivalent circuit in Fig. 12 is given as follows:

$$\begin{aligned}\frac{1}{Z} &= \frac{1}{R_d} + j(2\pi f)C_d + \frac{1}{R_w} + j(2\pi f)C_w \\ &= \frac{R_d + R_w}{R_d R_w} + j(2\pi f)(C_d + C_w) \\ &= \frac{1}{R} + j(2\pi f)C, \end{aligned} \quad (21)$$

where the suffixes d and w denote the parameters of the domains and the domain wall zones, respectively. As a result, the LSGM single crystal comprises one semicircular structure with a radius given by $R/2 = R_d R_w / 2(R_d + R_w)$ because the impedance in Eq.(21) has the same form as that in Eq.(13). This argument accounts for the impedance spectroscopy in Fig. 7 perfectly. However, there are two relaxation processes in the dielectric loss tangent. One relaxation process is due to the O^{2-} migrations along the domain walls, while the other is due to the O^{2-} migrations within the domains. The resultant electric current due to the O^{2-} migrations in these different lattice zones leads to the ionic conduction of the LSGM single crystal.

1-c. Oxygen ionic conduction within domains and domain wall zones in twin structure

At present, the structures in the domain walls are still unknown and therefore it is not clear whether the domain walls have a disordered structure or an ordered structure such as the crystallographic shear planes in certain oxygen-deficient transition metal oxides [56-58]. However, the ionic arrangements in the wall zones between the adjacent twin domains must be incoherent as compared to the ionic arrangements within the domains. Since the oxygen vacancies relax the lattice distortion due to incoherence, it is expected that the oxygen vacancies segregate preferentially near the domain walls. The incoherence at the domain walls induces elastic softening, thereby resulting in low migration energy values required for ionic conduction along the domain walls as compared to that in the domains [29]. Furthermore, the incoherency at the domain walls has another important function; the normal phonon modes at the domain walls are broken down by this incoherency. Consequently, the frequencies of the ionic vibrations in the domain wall zones must be low. Based on this speculation, it is reasonably accepted that the low-frequency dielectric relaxation process results from the ionic conduction along the domain walls, whereas the domain conduction leads to the high-frequency relaxation process. Because of this reason, the suffixes h and i used earlier are replaced with d and w, respectively. As shown in Eq.(21), the semicircular arc in the impedance spectroscopy results of Fig. 7 is the result of the combination of the ionic conduction in the domains with $(\beta_r)_d$ and the ionic conduction along the domain walls with $(\beta_r)_w$. Therefore, it is appropriate that the value of $(\beta_i)_h$ obtained from the semicircular arc in Fig. 7 lies between that of $(\beta_r)_d$ and $(\beta_r)_w$, as described previously, i.e., $(\beta_r)_w < (\beta_i)_h < (\beta_r)_d$.

Since the values of r_{crit} in the $R3c$ trigonal phase are large as compared to those in the monoclinic phase as mentioned previously, the low energy value of $(E_M)_d$ and the large dielectric loss tangent in the $R3c$ trigonal phase as compared to that in the monoclinic phase are suitably explained.

From Eq.(19), the resistance for the oxygen ionic conduction is proportional to the resistivity relaxation time τ_r and inversely proportional to the product of the ionic vibrational frequency ν and the density of the free mobile oxygen vacancies that assist the O^{2-} migrations N. The ionic vibrational frequency ν is equal to f_0 described before. Since the resistivity relaxation time includes the exponential function exp(E_M/k_BT), the resistance in the domains is represented by the following expression,

$$
\begin{aligned}
R_d &= (R_0)_d \exp\left(\frac{(E_M)_d + (E_O)_d}{k_BT}\right) \\
&\propto \frac{(\tau_r)_d}{(\nu)_d (N_0)_d} \exp\left(\frac{(E_O)_d}{k_BT}\right). \qquad (22)
\end{aligned}
$$

Then, the pre-exponential factor $(R_0)_d$ is inversely proportional to the product of $(\nu)_d$ and $(N_0)_d$. The resistance along the domain walls has a similar formula. In the present study, it is impossible to obtain R_d and R_w separately, but the migration energy values $(E_M)_d$ and $(E_M)_w$ are estimated independently. Although the dissociation energy values $(E_O)_d$ and $(E_O)_w$ are impossible to obtain as described earlier, the dissociation energy is generally very low as compared to the migration energy [1-3]. Therefore, by neglecting the contributions of the dissociation processes, we have tried to estimate the magnitudes of the pre-exponential factors $(R_0)_d$ and $(R_0)_w$. This has been done by using the experimental values of the resistance obtained by the impedance analyses, which is the resultant resistance $R = R_dR_w/(R_d + R_w)$.

Figure 8 compares the calculated resultant resistance R (solid line) with the experimental plots. In the assessments of the resultant resistance values, the following values for the pre-exponential factor of the resistance are used. i.e., $(R_0)_d = 2.43 \times 10^{-3}\Omega$ and $(R_0)_w = 2.30 \times 10^{-2}\Omega$ in the monoclinic phase, and $(R_0)_d = 1.81 \times 10^{-2}\Omega$ and $(R_0)_w = 1.20 \times 10^{-1}\Omega$ in the $R3c$ trigonal phase. Figure 8 includes the relations of log(T/R_d) vs $1/T$ (dotted line) and log(T/R_w) vs $1/T$ (dashed line). From Fig. 8, it is understood that the resultant resistance R is considerably low as compared to R_d and R_w. The frequency values of the ionic vibrations obtained in Fig.11(a) enable the numerical assessment of $[(\nu)_w(N_0)_w]/[(\nu)_d(N_0)_d] = (R_0)_d/(R_0)_w$, which yields $(N_0)_w/(N_0)_d \cong 28$ in the monoclinic phase and $(N_0)_w/(N_0)_d \cong 12$ in the $R3c$ trigonal phase.

Since the magnitudes of $(N_0)_w/(N_0)_d$ have been calculated by neglecting the dissociation processes of the oxygen vacancies from the trapped states, the assessed ratios of $(N_0)_w/(N_0)_d$ include some uncertainties. In spite of this, the estimated values of the $(N_0)_w/(N_0)_d$ ratio strongly suggest that the oxygen vacancies segregate preferentially at the domain walls to achieve stabilization. Consequently, the electric resistance for the oxygen ionic conduction along the domain walls is low as compared to that in the domains. The ratio of the number of oxygen vacancies in the domains to that in the domain walls is in good agreement with the estimation obtained by Lee et al. [29]. Moreover, the low resistance along the domain walls which correspond to one element of the parallel circuit in Fig. 12 contributes effectively to the significant reduction of the resultant resistance.

Ionic conduction in the twin structures cannot be explained only on the basis of impedance analysis because this analysis cannot distinguish electric conduction within the domains from the conduction along the domain walls. It is possible to directly obtain the energy values required for ionic conduction in these different zones separately from the measurements of the dielectric relaxation processes, but it is impossible to obtain information about the resistance values in the different zones. However, the present study indicates that a speculation of ionic conduction in twin structures is possible with high precision if the experimental results by both these ac methods are available together.

As described previously, it is still unknown whether the domain wall structure is ordered or disordered. In the case of the ordered structure, the oxygen vacancies that constitute the domain walls together with the other ions are very stable and cannot migrate very easily because their movement partially breaks the periodicity of the ordered structure. In particular, the dissociation of an oxygen vacancy from the very stable ordered structure requires high energy. Furthermore, the small value of β in the wall zones, i.e., $(\beta_r)_w$ = 0.72 $\pm$ 0.02, as compared with the value of β in the domains, i.e., $(\beta_r)_d$ = 0.94 $\pm$ 0.04, indicates that there are many various migration paths of O^{2-} ions along the domain walls. This is an important feature of disordered structures.

2. $La_{1-x}Sr_xGa_{1.1-x}Zr_{x-0.1}O_{3-\delta}$ polycrystalline ceramics, LSGZ(x)

2-a. XRD and impurity phase in LSGZ(x)

The main XRD patterns of the specimens of LSGZ(x) as well as those of most other doped lanthanum gallates at room temperature indicate a phase with an orthorhombic structure at x = 0.2 and 0.3 and a cubic perovskite structure at x = 0.4 and 0.5, as described previously. Table 1 summarizes the lattice parameters of the body-centered orthorhombic and the perovskite cell parameters at each value of x along with the sintering temperature T_s, x being 0.5, 0.4, 0.3, and 0.2. As x decreases, the atomic ratios [La^{3+}]/[Sr^{2+}] at the A site in the perovskite structure and [Ga^{3+}]/[Zr^{4+}] at the B site increase. Since the ionic radii of La^{3+} and Sr^{2+} at the A-site are 1.36 Å and 1.44 Å, respectively, while those of Ga^{3+} and Zr^{4+} at the B-site are 0.62 Å and 0.72 Å [59], respectively, every lattice parameter reduces with the decrease in x.

There exists an extremely weak satellite line due to an impurity phase around 2θ = 30° in the XRD. Figure 13(a) shows the XRD pattern of LSGZ(0.5) sintered at 1470 °C. The satellite line of the impurity phase is clearly recognized at $2\theta \cong 30°$. In the XRD pattern of strontium- and magnesium-doped $LaGaO_3$ ceramics, the reflection peaks due to the impurity phases of $LaSrGa_3O_7$ at the triplet junctions of the crystal grains and $LaSrGaO_4$ within the intra-grains are observed at $2\theta \cong 30°$ [16,22,49]. Therefore, it is presumed that the impurity phase observed at $2\theta \cong 30°$ in LSGZ(x) may be either $LaSrGaO_4$ or $LaSrGa_3O_7$. Figure 13(b) demonstrates the XRD of LSGZ(0.2) in which this impurity line is significantly attenuated as compared to LSGZ(0.5). The intensity

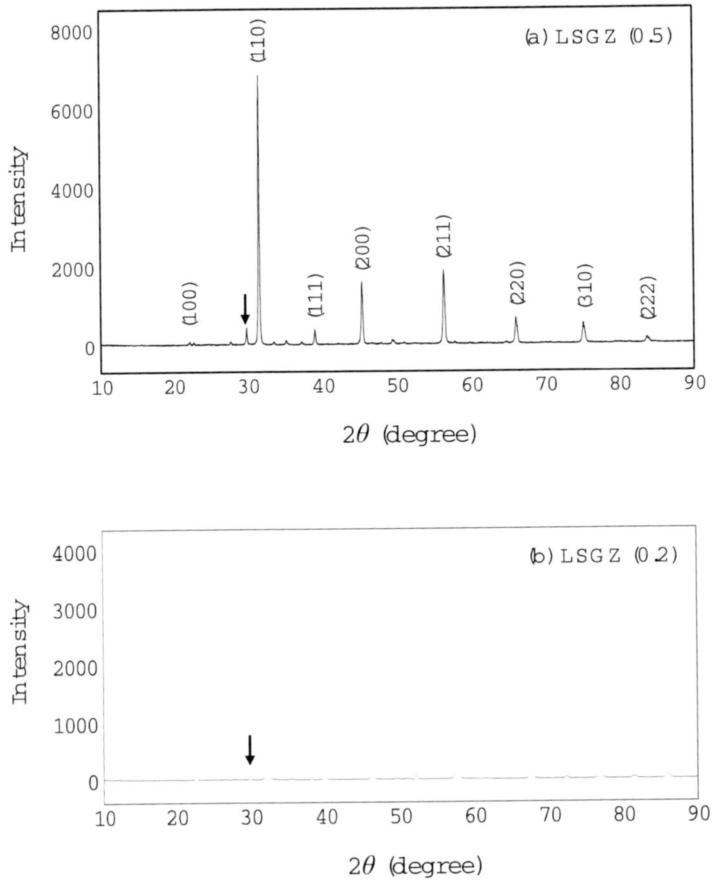

Figure 13. (a) XRD patterns of LSGZ(0.5) at room temperature and (b) LSGZ(0.2). The arrow indicates x-ray reflection due to the impurity phase. Indexes of main reflection peaks are indicated in Fig. 13(a).

of the impurity phase around 2θ = 30° for each specimen is normalized by the ratio of $I_{ip}/I_{(110)}$, where I_{ip} is the intensity of the impurity phase and $I_{(110)}$ is the intensity of the main peak at $2\theta \cong 33°$ in the XRD due to the (110) plane in the orthorhombic lattice. The $I_{ip}/I_{(110)}$ ratio changes by the factor x. Furthermore, this ratio also varies by T_s even if x is fixed. Table 3 summarizes the $I_{ip}/I_{(110)}$ ratios for all the prepared specimens.

At x = 0.3 and 0.2, the sintering temperature T_s is changed from 1420 to 1570°C with increments of 50°C. The four-probe dc conductivity values of these specimens $\sigma_{\rm dc}$ have been measured as a function of temperature. On the basis of the Nernst-Einstein relation [8-10], Fig. 14 depicts the Arrhenius relations of $\log(\sigma_{\rm dc}T)$ and $1/T$ as a parametric function of T_s for all the specimens of x = 0.3. Each curve is nearly parallel in the entire temperature region of the dc measurements with the exception of the specimen for which T_s = 1420°C. The results in Table 3 and Fig. 14 indicate that the dc conductivity certainly corresponds to the $I_{ip}/I_{(110)}$ ratio; the dc conductivity increases as the $I_{ip}/I_{(110)}$ ratio decreases in the measured temperature region. This is evident from the result in Fig. 15

Table 3. Values of T_s, $I_{ip}/I_{(110)}$ and σ_{dc} at 1073, 973, 873 and 773 K for every specimen of LSGZ(x); here, I_{ip} is the intensity of the x-ray reflection due to the impurity phase around $2\theta = 30°$ in XRD and $I_{(110)}$ is the intensity of the main peak at $2\theta \cong 33°$ due to the (110) plane in the body-centered orthorhombic lattice.

				dc conductivity σ_{dc} [$\Omega^{-1}cm^{-1}$]			
x	specimen	T_s [°C]	$I_{ip}/I_{(110)}$	1073 K	973 K	873 K	773 K
0.2	LSGZ(0.2)	1420	0.042	2.01×10^{-2}	9.82×10^{-3}	3.78×10^{-3}	9.44×10^{-4}
@		1470	0.026	2.83×10^{-2}	1.43×10^{-2}	5.75×10^{-3}	1.60×10^{-3}
@		1520	0.022	2.96×10^{-2}	1.50×10^{-2}	6.25×10^{-3}	1.75×10^{-3}
@		1570	0.024	2.68×10^{-2}	1.38×10^{-2}	5.63×10^{-3}	1.60×10^{-3}
0.3	LSGZ(0.3)	1420	0.031	1.50×10^{-2}	7.13×10^{-3}	2.55×10^{-3}	6.10×10^{-4}
@		1470	0.026	1.99×10^{-2}	9.44×10^{-3}	3.49×10^{-3}	8.87×10^{-4}
@		1520	0.012	2.18×10^{-2}	1.02×10^{-2}	3.86×10^{-3}	1.03×10^{-3}
@		1570	0.030	1.59×10^{-2}	7.63×10^{-3}	2.89×10^{-3}	7.84×10^{-4}
0.4	LSGZ(0.4)	1520	0.017	6.64×10^{-3}	3.02×10^{-3}	1.07×10^{-3}	2.63×10^{-4}
0.5	LSGZ(0.5)	1470	0.066	2.53×10^{-3}	1.03×10^{-3}	3.31×10^{-4}	7.26×10^{-5}

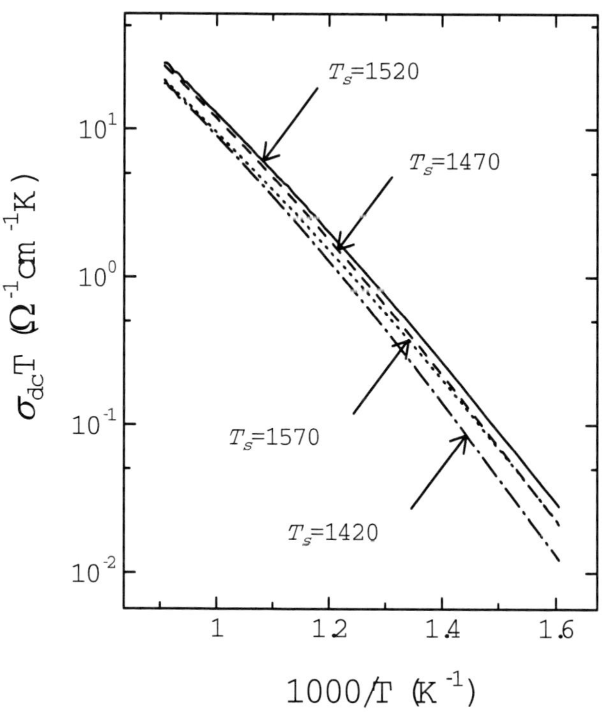

Figure 14. Arrhenius relations of $\log(\sigma_{dc}T)$ and $1/T$ for all specimens of LSGZ(0.3) as a parametric function of the sintering temperature T_s.

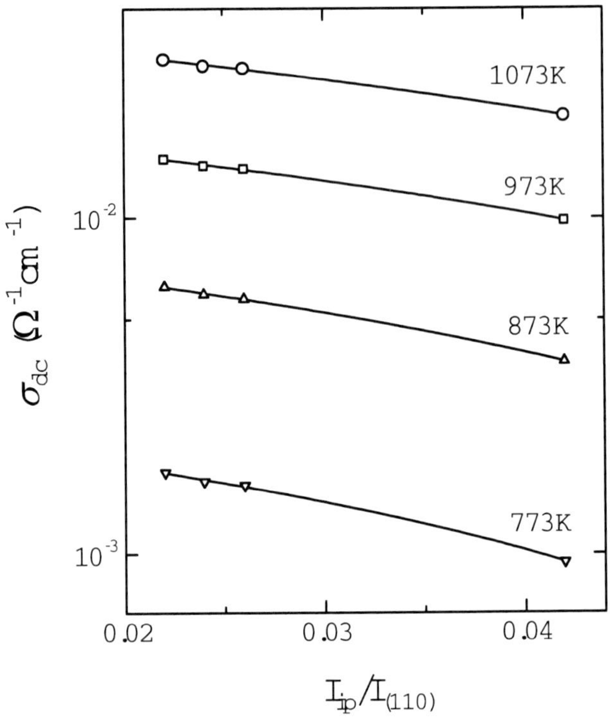

Figure 15. Relations of log(σ_{dc}) and $I_{ip}/I_{(110)}$ at 773, 873, 973 and 1073 K for LSGZ(0.2).

that demonstrates the relations of log(σ_{dc}) and $I_{ip}/I_{(110)}$ at 773, 873, 973, and 1073 K for LSGZ(0.2). LSGZ(0.3) also exhibits a relation that is very similar to that shown in Fig. 15. It appears that the impurity phase is responsible for the residual ionic resistance; however, it does not have any direct profound effects on the conduction mechanism owing to the O^{2-} migrations. It should be emphasized that the suppression of the impurity phase clearly increases ionic conductivity; however, this increment is marginal. Hereafter, ac and dc measurements have been carried out in the present study for the specimens of $x = 0.2$, 0.3, and 0.4 that were sintered at T_s = 1520°C. As for $x = 0.5$, the conductive behaviors of the sample sintered at 1470°C have been measured. Table 1 lists these sintering temperatures.

2-b. Ionic conductivity in LSGZ(x)

Figure 16 depicts the Arrhenius relations of log($\sigma_{dc}T$) and $1/T$ for all the specimens. There are three noteworthy features of the result in Fig.16, which can be stated as follows: (i) the dc ionic conductivity increases with the decrease in x, (ii) in each specimen, two straight lines cross over at a temperature denoted by T_t, (iii) there is a decrease in T_t from 840 K to 810 K as x decreases from 0.5 to 0.2, and (iv) the value of E_{dc} in every sample is somewhat lower at $T > T_t$ as compared to the value of E_{dc} at $T < T_t$. The experimental values of E_{dc} for all the specimens are summarized in Table 2.

As a routine work, the impedance analyses of LSGZ(x) have been carried out on the

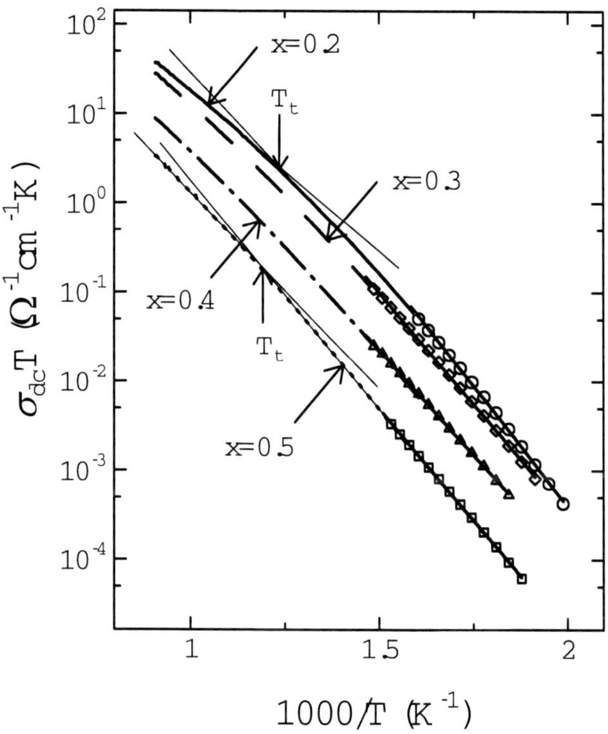

Figure 16. Arrhenius relations of log($\sigma_{\rm dc}T$) and $1/T$ for all the specimens of LSGZ(x) that were sintered at their optimum temperatures. Each relation consists of two straight lines that cross over at T_t. On the plots of the relations of $x = 0.2$ and 0.5, the arrows indicating T_t are included as examples. The plots indicate the electric conductivity converted from $(R_b + R_{gb})$. The squares, triangles, diamonds, and circles represent the experimental plots of $x = 0.5$, 0.4, 0.3 and 0.2, respectively.

basis of the conventional R-C equivalent circuit. Figure 17 shows the results of impedance spectroscopy at 523, 593, and 663 K for LSGZ(0.3). Each result of the spectroscopy contains two semicircular arcs, i.e., the highest-frequency arc due to the intra-grain conduction (the bulk conduction) and the intermediate-frequency arc resulting from the inter-grain conduction (the boundary conduction). The bulk resistance R_b and the boundary resistance R_{gb} are estimated by the usual means described previously. Since impedance analysis is possible at $T < T_t$ for every specimen, these resistance values are the ones at temperatures below T_t. Figure 16 shows the Arrhenius plots of the conductivity values converted from $(R_b + R_{gb})$. These plots are in good agreement with the four-probe dc conductivity values in the temperature regions where impedance analyses are possible.

Even though the values of R_b and R_{gb} are obtained, it is impossible to assess the bulk and the boundary conductivity values because the volume fractions of the bulks and boundaries and so on are unknown. However, the resistance values normalized per unit volume of the specimens are obtained. Figure 18 shows the plots of log(lT/SR_b) and log(lT/SR_{gb}) against $1/T$, where l and S are the length and surface area of the specimen

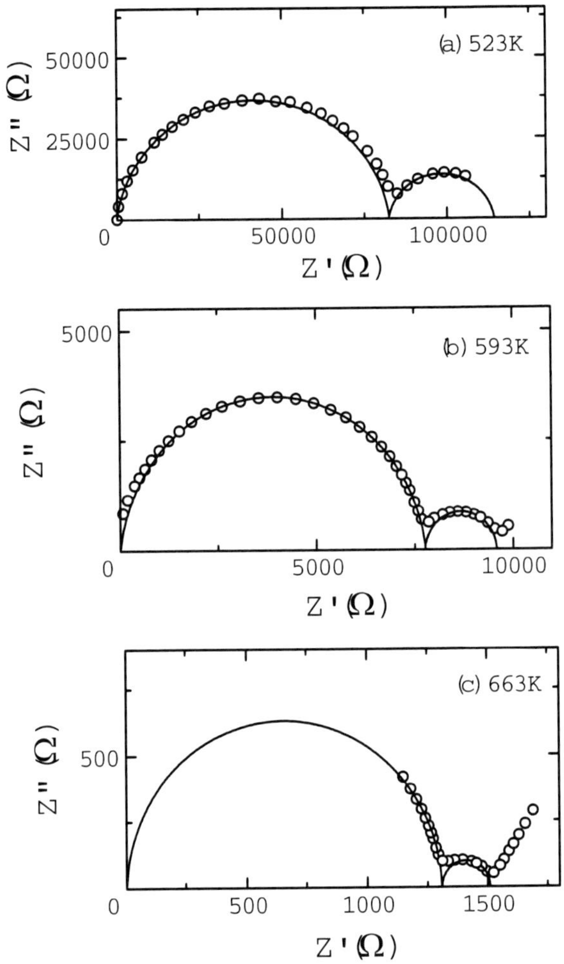

Figure 17. (a) Impedance spectroscopy in the complex-plane for LSGZ(0.3) at 523 K, (b) 593 K, and (c) 663 K.

for the ac measurements, respectively. The normalized boundary resistance values for all the specimens are nearly equal irrespective of x, while the normalized bulk resistance decreases with the decrease in x. This fact indicates that one of the main reasons for the increase in the ionic conductivity due to the decrease in x must be the reduction in the bulk resistance.

The Arrhenius relation of log(lT/SR) and $1/T$ yields the activation energy of the ionic conduction. Table 2 includes $(E_R)_b$ and $(E_R)_{gb}$ for all the specimens, where $(E_R)_b$ and $(E_R)_{gb}$ are the activation energy values of the ionic conduction in the bulks and boundaries, respectively. $(E_R)_b$ decreases with the decrease in x. This is identical to the behavior of the dc activation energy E_{dc}. A decrease in x implies that the number of Sr^{2+} and Zr^{4+} ions that are substituted for La^{3+} and Ga^{3+} decrease. Therefore, a decrease in x might enlarge the critical radii of the cation triangle of the perovskite lattice r_{crit} because of the ionic radii of these ions indicated in the previous subsection [18,22,23,59]. As

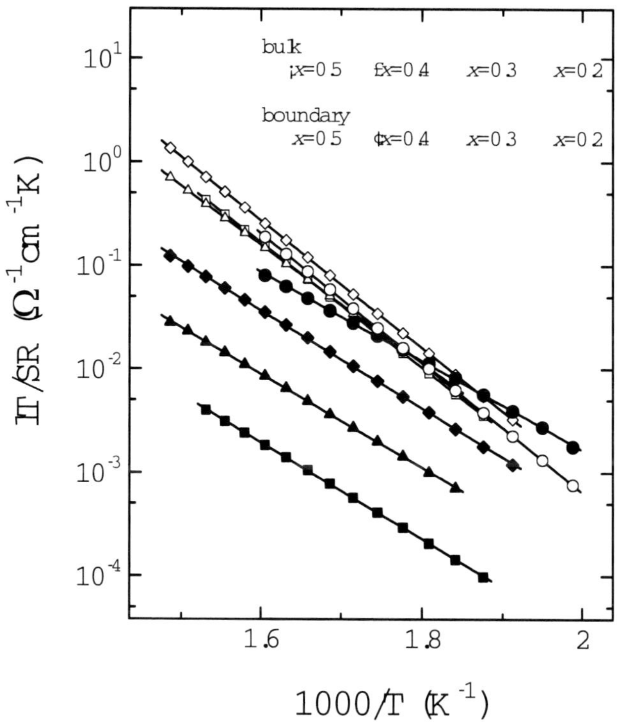

Figure 18. Arrhenius plots of $\log(lT/SR_b)$ vs $1/T$ and $\log(lT/SR_{gb})$ vs $1/T$ for LSGZ(x), where l and S are the length and surface area of the specimen for the ac measurements. The squares, triangles, diamonds, and circles indicate the experimental plots of x = 0.5, 0.4, 0.3, and 0.2. The solid symbols denote the experimental plots of lT/SR_b, while the open symbols denote the experimental plots of lT/SR_{gb}.

indicated in Table 1, however, the decrease in x also reduces the volume of the perovskite unit cell. Since there is a possibility that these two effects induced by the decrease in x offset each other, the decrease in x is unlikely to effectively contribute to the reduction in the activation energy required for ionic conduction. There must be an important reason for this reduction - the twin structure in the bulks for a small value of x. As described in the previous section, the twin structure in the LSGM single crystal decreases the activation energy of ionic conduction and increases the electric conductivity. When x decreases, LSGZ(x) is distorted from the cubic to the orthorhombic structure, which contains the twin structure that is peculiar to the lanthanum gallates doped with a small amount of impurity ions. Therefore, it appears that the formation of the twin structure at a small value of x can account for the reduction in the activation energy due to the decrease in x. The possibility that the bulks contain the twin structure when the value of x is small will be examined in the next subsection. With regard to $(E_R)_{gb}$, no clear x dependence is observed, as shown in Table 2.

The impedance analyses yield $(\beta_i)_h$ = 0.92 $\pm$ 0.03 and $(\beta_i)_i$ = 0.88 $\pm$ 0.03 for LSGZ(0.2). The magnitudes of $(\beta_i)_h$ and $(\beta_i)_i$ for the other LSGZ(x) specimens are similar to those of LSGZ(0.2).

2-c. Dielectric relaxation processes in LSGZ(0.5) and LSGZ(0.2)

The dielectric loss tangent tan δ is plotted against the applied frequency f as a parametric function of temperature for LSGM(0.5) at increments of 10 K in Fig. 19(a) and for LSGM(0.2) in Fig. 19(b). Although both the specimens exhibit complicated dielectric loss tangent curves, their general features are essentially similar. The curves of the tan δ - f relation at 633 K in Fig. 19(a) and at 703 K in Fig. 19(b) are indicated by a bold line in order to show the frequency dependencies of the loss tangent curves clearly. As described in the previous section, a complicated curve consisting of several dielectric relaxation processes is divided into individual processes by iterations of the least squares method based on the theoretical formula of the experimental loss tangent, i.e., Eq.(20). For both the specimens in Fig. 19, it seems best to divide the experimental curve into three relaxation processes. In such a case, another term should be included in Eq.(20) in a manner similar to the analysis of Fig. 10(b).

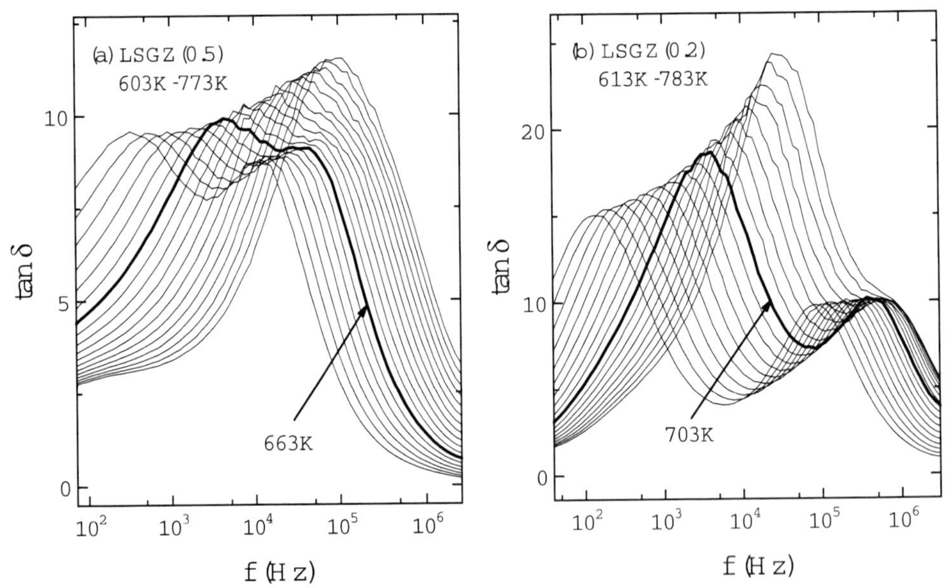

Figure 19. (a) Frequency dependencies of the dielectric loss tangent tan δ at 10 K increments as a parametric function of T for LSGZ(0.5) in the temperature range of 603-773 K and (b) LSGZ(0.2) in the range of 613 - 783 K. The curves at 633 K (a) and 703 K (b) are denoted by bold lines in order to show the frequency dependencies of the loss tangent curves clearly.

In Figs. 20(a) and 20(b), the experimental values of the dielectric loss tangent are plotted against the applied frequency along with three relaxation processes that are split by the least squares method and the total curve of these three processes for LSGZ(0.5) at 663 K and for LSGZ(0.2) at 703 K. It is evident that each specimen contains three dielectric relaxation processes, i.e., the high-, intermediate-, and low-frequency relaxation processes. Although these three frequency peaks in each specimen overlap partially, each resonance frequency can be distinguished directly. In Fig. 20(a), however, it is difficult to estimate the relative intensity ratios of these peaks accurately because the two maxima in the experimental curve of the dielectric loss tangent are too close with respect to the frequencies. It is even more difficult to trace the variations of their intensity ratios as a function of temperature. Therefore, it is difficult to estimate the value of E_O in each relaxation process of LSGZ(0.5); however, the resonance frequencies of the three relaxation processes can be determined precisely. This enables the estimations of the magnitudes of E_M required for the ionic conduction responsible for these relaxation processes.

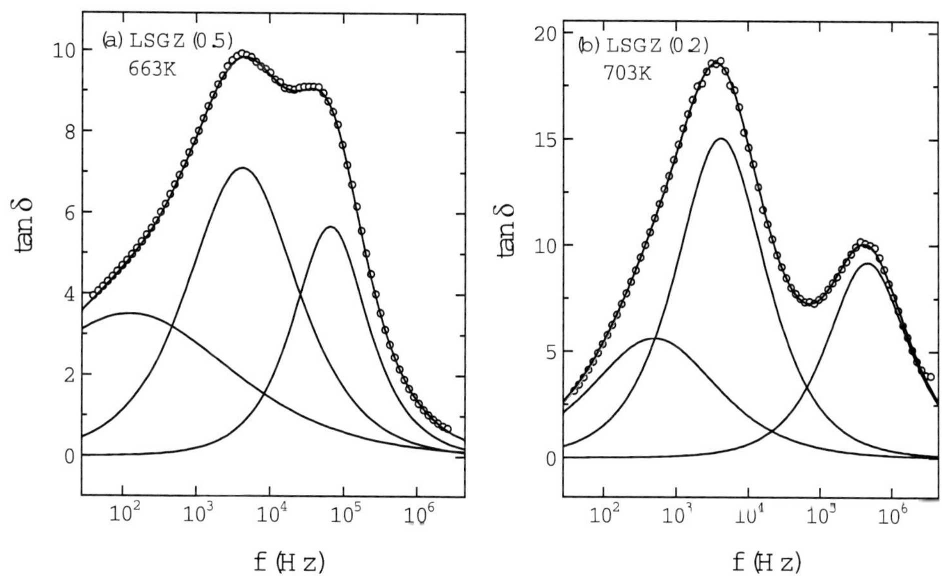

Figure 20. (a) Experimental plots (open circles) of the dielectric loss tangent against the applied frequency along with three relaxation peaks that are split by the least squares method and their resultant curve for LSGZ(0.5) at 663 K and (b) for LSGZ(0.2) at 703 K. Three curves that are split theoretically by the least squares method and their resultant curve are denoted by solid lines.

Figurc 21(a) depicts the Arrhenius relations of $\log(f_{\tan\delta})$ and $1/T$ for the individual frequency peaks of LSGZ(0.5) over the temperature range of 613 - 783 K. Their activation energy values are $(E_M)_h$ = 0.86 ± 0.06 eV, $(E_M)_i$ = 1.15 ± 0.03 eV, and $(E_M)_l$ = 0.31 ± 0.04 eV (see Table 2). In LSGZ(0.5), $(E_M)_l$ is very small as compared to $(E_M)_h$ and $(E_M)_i$, the values of which are rather comparable with the migration energy values in $La_{0.9}Sr_{0.1}Ga_{0.9}Mg_{0.1}O_{3-\delta}$ polycrystalline ceramics [3]. Furthermore, the relaxation times involved in the low-frequency relaxation process disperse significantly in comparison with the other two peaks because $(\beta_r)_l$ = 0.39 ± 0.06, $(\beta_r)_i$ = 0.76 ± 0.05 and $(\beta_r)_h$ = 0.96 ± 0.01. The magnitudes of $(\beta_r)_i$ and $(\beta_r)_h$ in LSGZ(0.5) correspond rather well to the β values for the oxygen ionic conductions in the boundaries and bulks of

$La_{0.9}Sr_{0.1}Ga_{0.9}Mg_{0.1}O_{3-\delta}$ [3]. With regard to LSGZ(0.5), these facts might imply that the origin of the low-frequency relaxation process is unclear at the present stage, but the high- and intermediate-frequency peaks result from the O^{2-} migrations in the bulks and boundaries.

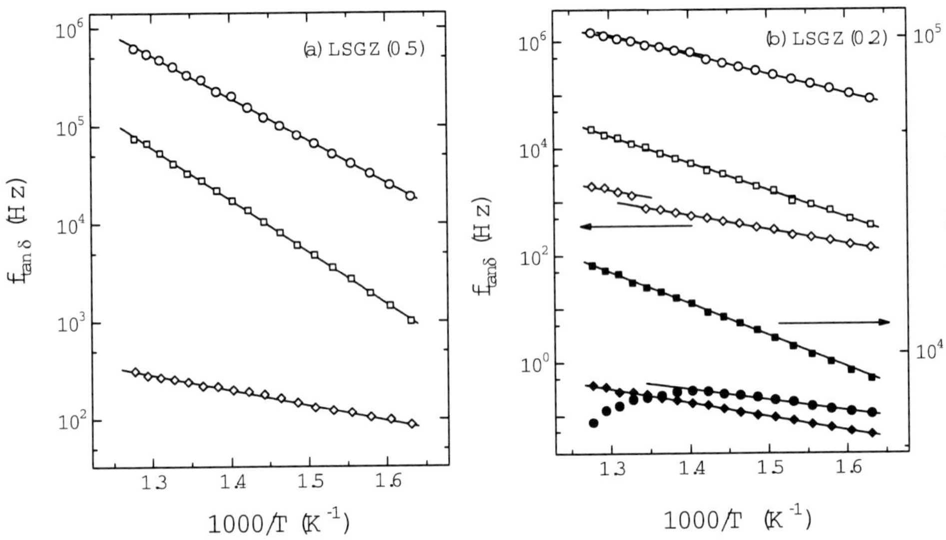

Figure 21. (a) Arrhenius relations of log($f_{\tan\delta}$) vs $1/T$ for the high-, intermediate- and low-frequency relaxation peaks of LSGZ(0.5) and (b) Arrhenius relations of log($f_{\tan\delta}$) vs $1/T$ and log[$T(\tan\delta)_{max}$] vs $1/T$ for each frequency relaxation peak of LSGZ(0.2). The circles, squares and diamonds are the plots of the high-, intermediate- and low-frequency relaxation peaks. The open and solid symbols are the plots related to the resonance frequencies and the maximum loss tangent, respectively.

With regard to LSGZ(0.2), Fig. 21(b) shows the Arrhenius plots of log($f_{\tan\delta}$) against $1/T$ in the same temperature region as Fig. 21(a). There is a crossover of two straight lines around 740 K in the high-frequency relaxation process: $(E_M)_h$ = 0.75 ± 0.02 eV below 740 K and 0.67 ± 0.02 eV above 740 K. Furthermore, the Arrhenius line of the low-frequency relaxation process contains the step function around 740 K: $(E_M)_l$ = 0.53 ± 0.02 eV below 740 K and 0.51 ± 0.05 eV above 740 K. These energy values are listed in Table 2. The crossover and the step function around 740 K clearly imply that the rclaxation processes responsible for the high- and low-frequency relaxation processes originate in the same phenomenon. These variations around 740 K must be the consequence of a phase transition. However, the Arrhenius relation of the intermediate-frequency relaxation process does not include any variation around 740 K. This fact indicates that the intermediate-frequency relaxation process might result from the boundary conduction because a phase transition may hardly have any significant effects in the boundaries. The Arrhenius relation of the intermediate-frequency peak yields $(E_M)_i$ = 1.01 ± 0.06 eV. The crossover temperature T_t of the dc conductivity observed in Fig. 16 and the phase transition temperature (740 K) recognized in Fig. 21(a) are not in agreement. This disagreement must be due to the fact that the dc conductivity includes the component of the boundary conduction that has the high activation energy.

In LSGZ(0.2), $(\beta_r)_l$ = 0.65 $\pm$ 0.10, $(\beta_r)_i$ = 0.84 $\pm$ 0.04, and $(\beta_r)_h$ = 0.94 $\pm$ 0.03. The β value of the intermediate-frequency relaxation process is comparable with the value of the relaxation processes in the boundaries in polycrystalline ceramics of other doped lanthanum gallates. In comparison with LSGZ(0.5), $(\beta_r)_l$ is considerably large, while the value of $(E_M)_l$ is very high in LSGZ(0.2).

With regard to LSGZ(0.2), when the β values obtained by the iteration treatments are compared with those values obtained by the impedance analyses, two features can be observed. i) The values of $(\beta_i)_i$ and $(\beta_r)_i$ are in agreement within the experimental errors. Therefore, in LSGZ(0.2), the intermediate-frequency dielectric relaxation process and the intermediate-frequency semicircle might originate from the same phenomenon. ii) The value of $(\beta_i)_h$ lies between $(\beta_r)_h$ and $(\beta_r)_l$ estimated by the iteration treatments, i.e., $(\beta_r)_l < (\beta_i)_h < (\beta_r)_h$. The relative relation of these β values is very similar to that observed in the LSGM single crystal.

2-d. Low-frequency relaxation processes in LSGZ(0.5) and LSGZ(0.2)

If the β values of the LSGM single crystal obtained by the iteration treatments are compared with the β values of LSGZ(0.2), it turns out that $(\beta_r)_h$ of LSGZ(0.2) and $(\beta_r)_d$ of LSGM are in good agreement. Furthermore, $(\beta_r)_l$ of LSGZ(0.2) coincides well with $(\beta_r)_w$ of LSGM within the experimental errors. These numerical comparisons of the β values clearly indicate that both the LSGM and LSGZ(0.2) specimens consist of several zones of different structures and that some zones in these oxides have common structures.

Figure 21(b) also includes the Arrhenius relations of log[$T(\tan\delta)_{max}$] and $1/T$ for the individual relaxation processes of LSGZ(0.2). The Arrhenius relations of the intermediate- and low-frequency processes do not include any anomaly around 740 K, unlike the Arrhenius relations of the resonance frequencies. With regard to the high-frequency relaxation process, a decrease in [$T(\tan\delta)_{max}$] is observed when the temperature exceeds 720 K. Such a phenomenon is also observed in the LSGM single crystal that contains the twin structures [6]. These Arrhenius relations yield $(E_O)_h$ = 0.067 $\pm$ 0.002 eV below 720 K, $(E_O)_i$ = 0.19 $\pm$ 0.03 eV, and $(E_O)_l$ = 0.087 $\pm$ 0.001 eV. The activation energy of the boundary conduction obtained by the impedance analysis for LSGZ(0.2) is $(E_R)_{gb}$ = 1.16 $\pm$ 0.01 eV, which is close to the value of the sum of $(E_M)_i$ and $(E_O)_i$ within the experimental errors., i.e., 1.20 $\pm$ 0.06 eV, as predicted in Eq.(19).

As indicated previously, the value of $(\beta_r)_l$ of LSGZ(0.2) is quite different from that of LSGZ(0.5). Furthermore, $(E_M)_l$ in LSGZ(0.2) is much higher than $(E_M)_l$ in LSGZ(0.5). Because of this considerable difference in the behaviors of the low-frequency relaxation processes of LSGZ(0.5) and LSGZ(0.2), it is difficult to believe that these relaxation processes of LSGZ(0.5) and LSGZ(0.2) are caused due to the same phenomenon. As indicated by the Arrhenius relations of log($f_{\tan\delta}$) against $1/T$ in Fig. 21(b), the high-frequency relaxation process and the low-frequency process change simultaneously around 740 K in LSGM(0.2). It should be noted that the twin structures are observed experimentally in the bulks of polycrystalline doped lanthanum gallate ceramics when the amount of the impurity ions is small [32,33]. As described several times in the past, the twin structure contains two diffusion paths for the O^{2-} migrations; the path in the domains and the path along the domain walls. These paths correspond to the parallel circuit of two independent R-C combinations in Fig. 12, which yields high ionic conductivity. The increase in the dc ionic conductivity as x decreases from 0.5 to 0.2 must be mainly due to formation of the twin structure in the non-cubic ferroelastic phases of the compositions with low doping levels.

The dielectric relaxation behaviors of the high- and low-frequency relaxation processes in LSGZ(0.2) are consistent with the behavior of the ionic conduction due to the O^{2-} migrations in the twin structure. This deduction is also ensured by not only the arguments about the β values in the LSGM single crystal and LSGZ(0.2) but also the Arrhenius relations in Fig. 21(b), which have been explained above. In conclusion, in LSGZ(0.2), the high-frequency relaxation process results from the O^{2-} migrations in the domains of the twin structure, the low-frequency relaxation process is due to the ionic conduction along the domain walls, and the intermediate-frequency peak is ascribed to the O^{2-} migrations in the grain boundaries. In heavy doped lanthanum gallates like LSGZ(0.5) with a cubic crystal structure, the twin structure cannot exist according to the laws of symmetry.

It is probable that the existence of the twin structure within the bulks in LSGZ(0.2) might be the main reason for the difference in the behaviors of the low-frequency relaxation processes of LSGZ(0.2) and LSGZ(0.5). Therefore, it is possible that the ionic conduction across the electrode-specimen interface in LSGZ(0.5) may result in the low frequency peak. If this is realized, even LSGZ(0.2) is expected to contain another frequency peak due to this ionic conduction. However, it is practically impossible to divide one experimental curve of the dielectric loss tangent of LSGZ(0.2) into four relaxation processes, even if the least squares method is iterated. The good agreement between the experimental plots and the total of three relaxation peaks divided by the least squares method in Fig. 20(b) might indicate that the relaxation process due to the ionic conduction across the interfaces would contribute marginally to the total dielectric relaxation curve in LSGZ(0.2). As shown in Fig. 10(b), the low-frequency peak is partially observed at 843 K in the LSGM single crystal. This peak must emerge for the same reason as the low-frequency peak in LSGZ(0.5); the ionic conduction across the electrode-specimen interface must be responsible for both the low-frequency relaxation processes of LSGM at 843 K and LSGZ(0.5).

Taking into consideration the abovementioned speculation, the equivalent circuit of LSGZ (x) modeled in the ac treatments is represented as shown in Fig. 22, when x is 0.2 and 0.3. The resultant circuit of the twin structure of the bulk element, i.e., the parallel circuit of two independent R-C combinations, and the R-C parallel circuit in the boundary element are connected in series. Since there are no twin structures within the intra-grains when $x = 0.4$ and 0.5, the equivalent circuit is represented by Fig.3(a).

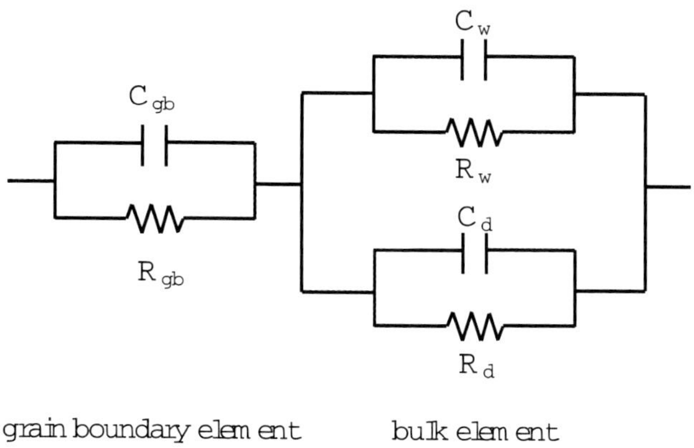

Figure 22. The conventional equivalent circuit of LSGZ(x) when x = 0.2 and 0.3.

2-e. LSGT(0.5) and LSGZ(0.5)

The research and development of LSGZ(x) is on the line of the study about LSGT(x), i.e., $La_{1-x}Sr_xGa_{1.1-x}Ti_{x-0.1}O_{3-\delta}$, and the details of LSGT(0.5) have been reported in the previous literature [7]. Therefore, it is quite appropriate to compare the conductive behaviors of LSGZ(0.5) in the present review with those of LSGT(0.5) in order to investigate the effect on ionic conduction due to the replacement of Ti^{4+} with Zr^{4+}. In Figure 23, $\log(\sigma_{dc}T)$ is plotted against $1/T$ for both the specimens; here, the plots of LSGT(0.5) are the same as those in the previous report [7]. The ionic conductivity of LSGT(0.5) is approximately three times as high as that of LSGZ(0.5). The energy value of E_{dc} required for dc conduction of LSGT(0.5) at $T < 960$ K and that of LSGZ(0.5) at $T < T_t$ are listed in Table 2. The difference between these E_{dc} values of LSGT(0.5) and LSGZ(0.5) is never so large if the experimental errors are considered.

In the previous study [6], the impedance spectroscopy results of LSGT(0.5) comprised two semicircular arcs at each temperature, which is also the case with LSGZ(0.5) in the present study; further, the bulk and boundary resistance values, R_b and R_{gb}, were obtained as usual. The values of $(E_R)_b$ and $(E_R)_{gb}$ of LSGT(0.5) were estimated by using the temperature dependencies of R_b and R_{gb}. In Table 2, these energy values of LSGT(0.5) are compared with the values of $(E_R)_b$ and $(E_R)_{gb}$ of LSGZ(0.5). The magnitudes of $(E_R)_{gb}$ of both the specimens are almost the same, although the value of $(E_R)_b$ of LSGZ(0.5) is higher than that of LSGT(0.5). Therefore, this difference in $(E_R)_b$ is mainly responsible for the difference in E_{dc} of these specimens. Despite the fact that the differences in the activation energy values required for ionic conduction in these specimens are not so big, the ionic conductivity in LSGZ(0.5) is low as compared to that in LSGT(0.5).

As described in the introduction, there might be some problems involved with the Ti^{4+}/Ti^{3+} couples in LSGT(0.5) in a strong reducing atmosphere, thereby resulting in high electronic conductivity due to the polaronic conduction of the $3d$ carriers [34,35]. Although

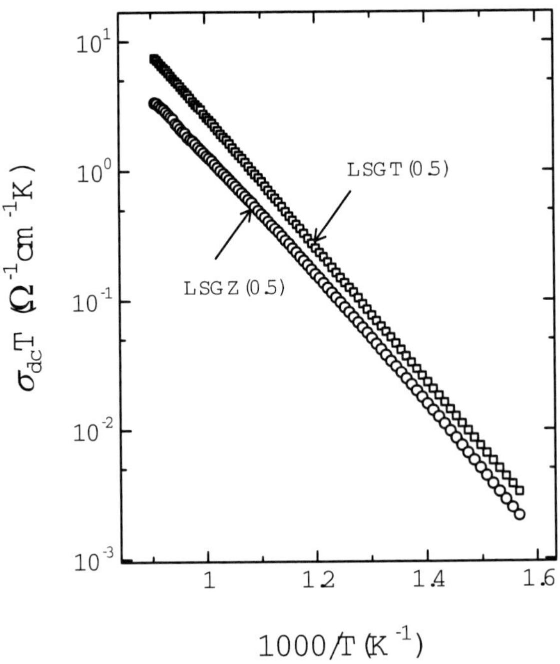

Figure 23. Arrhenius relations of log($\sigma_{dc}T$) and $1/T$ for LSGT(0.5) (open squares) and LSGZ(0.5) (open circles).

the electric conductivity in LSGZ(0.5) is somewhat low as compared to that of LSGT(0.5), LSGZ(0.5) might probably be free from such a serious problem if it is employed as an electrolyte in SOFC. Owing to this reason, the LSGZ(x) system must be stable even in a strong reducing atmosphere irrespective of the value of x. Since the ionic conductivity of LSGZ(0.2) is more than ten times greater than that of LSGZ(0.5) as shown in Fig. 16, LSGZ(0.2) ceramics are considered to be possible candidates for use as electrolytes in an SOFC because LSGZ(0.2) excludes electronic conduction completely; moreover, its ionic conductivity is considerably higher than that of LSGT(0.5).

Summary

The main points of the present review can be summarized as follows.

1. Oxygen ionic conduction in oxides results from the self diffusion of O^{2-} ions. The elementary process of the oxygen diffusion involves the migration of an O^{2-} ion from a lattice site to the next vacant lattice site across a saddle point in a diffusion path. Therefore, the presence of oxygen vacancies is the minimum requirement for O^{2-} migrations. When an O^{2-} ion passes through a saddle point, the displacement of ions around the O^{2-} ion at the saddle point results in a relaxation process that is characterized by a relaxation time. Even in the bulks that correspond to the crystal grains, there exist many different diffusion paths because of the lattice imperfections

involved in the bulks. This implies that a relaxation process due to O^{2-} migrations involves the distribution of relaxation times. The degree of distribution of the relaxation times is represented by the parameter β.

2. The oxygen ionic conductivity values of doped lanthanum gallates are considerably high as compared to those of most other oxides. Therefore, doped lanthanum gallates are considered to be strong potential candidates for use as electrolytes in an SOFC. From the scientific point of view, the investigation of the reason why doped lanthanum gallates have high ionic conductivity values constitutes an important research subject. In fact, no parameter involved in the Nernst-Einstein relation can account for the high ionic conductivity of doped lanthanum gallates. This is because the differences between the values of these parameters for doped lanthanum gallates and other oxides are not significant although these differences are expected to lead to the high ionic conductivity of doped lanthanum gallates. In order to study this research subject in greater detail, it is important to clarify the dynamics of O^{2-} migrations in doped lanthanum gallates.

3. The ac measurements provide very significant knowledge about the migration dynamics of the O^{2-} ions in oxides. Using impedance analysis, the resistance values of the bulks and the boundaries in polycrystalline oxide ceramics can be distinguished. Moreover, the β values in the bulks and the boundaries are obtained directly using impedance analysis. Relaxation processes are observed in the dielectric properties due to the displacement of ions when O^{2-} ions pass through saddle points because the product of an electronic charge and the displacement of an ion results in a dipole moment. In the measurements of dielectric relaxation processes, the activation energy values required for O^{2-} migrations are obtained directly, but the β values are assessed indirectly because these values are obtained as parameters used in the curve fitting process. However, if these two techniques in the ac method are used together, a speculation as to the electric conduction in oxides is possible with high precision.

4. The present study commenced by elucidating the electric transport properties of an LSGM single crystal produced by using the Czochralski method, i.e., $La_{0.95}Sr_{0.05}Ga_{0.9}Mg_{0.1}O_{3-\delta}$, using mainly the ac method along with the dc measurements. This was followed by the investigation of the ionic conduction of polycrystalline $La_{1-x}Sr_xGa_{1.1-x}Zr_{x-0.1}O_{3-\delta}$ ceramics, i.e., LSGZ(x).

5. The LSGM single crystal formed with the twin structure, which has been shown using high-resolution white-beam synchrotron x-ray diffractions, contains only one semicircular arc in the impedance spectroscopy. Further, two dielectric relaxation processes emerge in the relation of the loss tangent tan δ and the applied ac field frequency f. From the arguments based on the experimental results of the activation energy of the resistance obtained by using impedance analysis, the activation energy values required for these two dielectric relaxation processes, and the β values, it has been revealed that the combination of O^{2-} migrations within the domains and along the domain walls in the twin structure results in the ionic conduction in the LSGM single crystal. The conventional equivalent circuit of the twin structure modeled in the ac treatments is a parallel circuit of two independent R-C combinations; one corresponds to the ionic conduction due to O^{2-} migrations in the domains, while the other corresponds to the migrations along the domain walls. This parallel circuit significantly reduces the resultant resistance in the twin structure.

6. In polycrystalline LSGZ(x) (x = 0.2, 0.3, 0.4, and 0.5), impedance spectroscopy for every specimen contains two semicircular structure at each temperature and three dielectric relaxation processes emerge in the loss tangent. These features are identical to those of the other doped lanthanum gallates ceramics. However, the behavior of the low-frequency relaxation process of LSGZ(0.2) is very different from that of LSGZ(0.5). Furthermore, the ionic conductivity of LSGZ(0.2) is more than ten times greater than that of LSGZ(0.5). Considering the fact that the bulks comprise the twin structures in the doped lanthanum gallates in the case of small values of x, the experimental results of the activation energy values and the β values in LSGZ(0.2) have been discussed in terms of the ionic conduction due to O^{2-} migrations in the twin structure within the bulks and across the grain boundary. It is surmised that the low-frequency relaxation process observed in LSGZ(0.2) results from the O^{2-} migrations along the domain walls. The equivalent circuit in the bulks, which consists of a parallel circuit of two independent R-C combinations, plays a very important role in the high ionic conduction in doped lanthanum gallate ceramics. With respect to the cubic compositions that are realized when x is large as in the cases of LSGZ(0.4) and LSGZ(0.5), the bulks in the ceramics do not contain the twin structure. Therefore, the low-frequency relaxation process in these cases is ascribed to the charge transport in the specimen-electrode interface.

Acknowledgements

The authors are very grateful to M. Berkowski (Institute of Physics, Polish Academy of Science) for the growth of the LSGM single crystal, and S. Mochizuki and Y. Morishita (Yokohama National University) for their assistance in these research projects. D. I. Savystkii is deeply thankful to U. Bismayer (University of Hamburg) for the collaboration in the investigation of the structures of the LSGM single crystals. The ac and dc measurements for the LSGM single crystal have been carried out by the collaboration of M. Kurumada and Y. Morishita, while the measurements of LSGZ(x) were carried out by the collaboration of M. Kurumada, S. Mochizuki, and Y. Morishita. Both these experiments were carried out in Yokohama National University with financial support from AGC Seimi Chemical Co., Ltd. The authors are deeply indebted to T. Inagaki and Y. Fujie (AGC Seimi Chemical Co., Ltd.) for their encouragement.

References

[1] Iguchi, E., S. Nakamura, S., Munakata, F., Kurumada, M., and Fujie, Y. *J.mAppl. Phys.* 2003, vol. 93, p. 3662.

[2] Iguchi, E. and Mochizuki, S. *J. Appl. Phys.* 2004, vol. 96, p. 3889.

[3] Kurumada, M., Hara, H., Munakata, F., and Iguchi, E. *Solid State Ionics* 2005, vol. 176, p. 245.

[4] Kurumada, M., Hara, H., and Iguchi, E. *Acta Materialia* 2005, vol. 53, p. 4839.

[5] Komine, S., Iimure, T., and Iguchi, E. *Solid State Ionics* 2005, vol. 176, p. 2523.

[6] Kurumada, M., Savytskii, D. I., and Iguchi, E. *J. Appl. Phys.* 2006, vol. 100, p. 014107.

[7] Iguchi, E., Kurumada, M., and Mochizuku, S. Defects and Diffusion in Ceramics - An Annual Retrospective, Trans Tech Publications Ltd., Zurich, 2005, vol. VII, pp.115-127.

[8] Dekker, A. J. *Solid State Physics*, Prentice-Hall, New York, 1957, Chaps. 6 and 7.

[9] Kittel, C. *Introduction to Solid State Physics*, John Wiley & Sons, New York, 1971, 4th edit. Chap. 19.

[10] Swalin, R. A. *Thermodynamics of Solids,* Wiley, New York, 1970, Chap. 15.

[11] Ando, K. and Oishi, Y. *J. Nucl. Sci. Technol.* 1983, vol. 20, p. 973.

[12] Ishihara, T., Maysuda, H., and Takita, Y. *J. Am. Ceram. Soc.* 1994, vol. 116, p. 3801.

[13] Feng, M. and Goodenough, J. B. *J. Solid State Inorg. Chem.* 1994, vol. T31, p. 663.

[14] Huang, P. and Petric, A. J. *Electrochem. Soc.* 1997, vol. 143, 1644.

[15] Stevensen, J. W., Armstrong, T. R., McCready, D. E., Pederson, L. R., and Weber, W. J. J. *Electrochem. Soc.* 1997, vol. 144, p. 3613.

[16] Huang, K., and Goodenough, J. B. *J. All. Compounds* 2000, vol. 303-304, p.454.

[17] Marti, W., Fischer, P., Altorfer, F., Scheel, H. J., and Tadin, M. *J. Phys.: Condens. Matter* 1994, vol. 6, p. 127.

[18] Ishihara, T., Matsuda, H., and Takita, Y. *Solid State Ionics* 1995, vol. 79, p. 147.

[19] Kim, J. H. and Yoo, H. I. *Solid State Ionics* 2000, vol. 140, p. 105.

[20] Anderson, P. S., Marques, F. M. B., Sinclair, D. C., and West, A. R. *Solid State Ionics* 1999, vol. 118, p. 229.

[21] Westphal, D., Mather, G. C., Marques, F. M. B., Jakobs, S., and Guth, U. *Solid State Ionics* 2000, vol. 136/137, p. 19.

[22] Anderson, P. S., Mather, G. C., Marques, F. M. B., Sinclair, D. C., and West, A. R. *J. Eur. Cerami. Soc.* 1999, vol. 19, p. 1665.

[23] Lybye, D., Poulsen, F. W., and Mongensen, M. *Solid State Ionics* 2000, vol. 128, p. 91.

[24] Savytskii, D. I., Trots, D. M., Vasylechko, L. O., Tamura, T., and Berkowski, M. *J. Appl. Crystallogr.* 2003, vol.36, p.1197.

[25] Savytskii, D. I., Vasylechko, L., Bismayer, U., Paulmann, C., and Berkowski, M. in *NATO Science Series*, edited by N.Sammes et al. Springer, 2005, vol.202, p.135.

[26] Bueble, S., Knorr, K., Brecht, E., and Schmahl, W. W. *Surf. Sci.* 1998, vol. 400, p. 345.

[27] Bueble, S. and Schmahl, W. W. *Mater. Struct.* 1999, vol. 6, p. 140.

[28] Harrison, R. I., T. Redfern, S. A., and Salje, E. K. H. *Phys. Rev. B* 2004, vol. 69, p. 144101.

[29] Lee, W. T., Salje, E. K. H., and Bismayer, U. *J. Phys.: Condens. Matter* 2003, vol. 15, p. 1353.

[30] Calleja, M., Dove, M. T., and Salje, E. K. H. *J. Phys.: Condens. Matter* 2003, vol. 15, p. 2301.

[31] Bartels, M., Hangen, V., Burianek, M., Getzlaff, M., Bismayer, U., and Wiesendanger, R., *J. Phys.: Condens. Matter* 2003, vol. 15, 957 (2003).

[32] Drennan, J., Zelizko, V., Hay, D., Ciacchi, F. T., Rajendran, R., and Badwal, S. P. S. *J. Mate. Chem.* 1997, vol. 7, p. 79.

[33] Mathews, T. and Sellar, J. R. *Solid State Ionics*, 2000, vol. 135, p. 411.

[34] E. Iguchi, *Recent Research Development in Physics & Chemistry of Solids,* Transword Research Network, Triavandrum, 2002, pp.159-179.

[35] Iguchi, E., Hashimoto, Y., Kurumada, M., and Munakata, F. *J. Appl. Phys.* 2003, vol. 94, p. 758.

[36] Fröhich, H. *Theory of Dielectric*, Claredon, Oxford, 1958, pp. 70-106.

[37] Iguchi, E., Kubota, N., Nakamori, T., Yamamoto, N., and Lee, K. *J. Phys. Rev. B* 1991, vol. 43, p. 8646.

[38] Iguchi, E. and Akashi, K. J *Phys. Soc. Jpn.* 1992, vol. 61, p. 3385.

[39] Maiti H. S. and Basu, R. N. *Mater. Res. Bull.* 1986, vol. 21, p. 1107.

[40] Gerhardt, R. *J. Phys. Chem. Solids* 1994, vol. 55, p. 1491.

[41] Austin I. G. and Mott, N. F. *Adv. Phys.* 1969, vol. 18, p. 41.

[42] Gehlig, R. and Salje, E. K. H. *Philos. Mag. B* 1983, vol. 47, p. 229.

[43] Iguchi, E. and W. H. Jung, W. H., *J. phys. Soc. Jpn.* 1994, vol. 63, p. 3078.

[44] Iguchi, E., Nakamura, N., and Aoki, A. *J. Phys. Chem. Solids* 1997, vol. 58, p. 755.

[45] Bauerle, J. E. *J. Phys. Chem. Solids* 1969, vol. 30, p. 2657.

[46] MacDonald, J. R. *J. Chem. Phys.* 1974, vol. 61, p. 3977.

[47] Franklin, A. D. *J. Am. Cerm. Soc.* 1975, vol. 58, p. 465.

[48] MacDonald, J. R. *Impedance Spectroscopy,* Wiley, New York, 1987.

[49] Abram, E. J., Sinclair, D. C., and A. R. West, A. R. *J. Electroceram.* 2003, vol. 10, p. 165.

[50] Vasylechko, L., Vashook, V., Savytskii, D. I., Senyshyn, A., Niewa, R., Knapp, M., Ullmann, H., Berkowski, M., Matkovski, M., and Bismayer, U. *J. Solid State Chem.* 2003, vol. 172, p. 396.

[51] Dienes, G. J., Welch, D. O., Fischer, C. R., Hatcher, R. D., Lazareth, O., and Samberg, M. *Phys. Rev. B* 1975, vol. 11, p. 3060.

[52] Iguchi, E., Tamenori, A., and Kubota, N. *Phys. Rev. B* 1992, vol. 45, p. 697.

[53] Iguchi, E. and Nakatsugawa, H. *Phys. Rev. B* 1995, vol. 51, p. 10956.

[54] Hayward, S. A. and Salje, E. K. H. *Miner. Mag.* 2000, vol. 64, p. 195.

[55] Salje, E. K. H., Hayward, S. A., and Lee, W. T. *Acta Crystallogr., Sect A: Found. Crystallogr.* 2005, vol. 61, p. 3.

[56] Iguchi, E. and Tilley, R. J. D. Philos. Trans. R, *Soc. Lond. Ser. A,* 1977, vol.286, p. 55.

[57] Shimizu, Y. and Iguchi, E. *Phys. Rev. B*, 1978, vol. 17, p. 2505.

[58] Aizawa, K., Iguchi, E., and Tilley, R. J. D. *Proc. R. Soc. Lond. A*, 1984, vol. 394, p. 299.

[59] Shannon, R. D. *Acta Crystallogr. A*, 1976, vol. 32, p. 751.

In: Diffusion and Reactivity of Solids
Editor: James Y. Murdoch, pp. 163-207

ISBN: 978-1-60021-890-3

Chapter 4

NANOSIZED MATERIALS AS ELECTRODES FOR LITHIUM ION BATTERIES

Jesús Santos-Peña[1], Julián Morales[1], Enrique Rodríguez-Castellón[2] and Sylvain Franger[3]

[1] Departamento de Química Inorgánica e Ingeniería Química, Edificio Marie Curie, Campus de Rabanales, Universidad de Córdoba, 14071 Córdoba, Spain

[2] Departamento de Química Inorgánica y Cristalografía, Universidad de Málaga, Spain

[3] Laboratoire de Physico-Chimie de l'Etat Solide, UMR CNRS 8182, ICMMO, Université Paris XI, 91405 Orsay, France

Abstract

In this work we show some results on the research of nanosized materials with potential applications in lithium ion batteries. The study is focussed on positive electrodes such as olivine $LiFePO_4$ and α-$LiFeO_2$ as well as negative electrodes based on iron containing spinels. For the positive electrodes, the nanosized nature was found to enhance the efficiency of the lithium extraction/insertion reaction, due to a reduced path length for the transport of electrons and lithium ions. Moreover cycling properties were improved in the nanomaterials due to the combination of faster reaction kinetics and increased electrolyte-electrode interface. Capacities as high as 140 mAh/g were observed for $LiFePO_4$ when is modified by adding of conductive systems such as copper or carbon. α-$LiFeO_2$ nanobelts showed better electrochemical properties than other lithium ferrite polymorphs. For the spinels, capacities as high as 1400 mAh/g were found. However, the nanometric character induces the formation of a solid electrolyte interface that decreases the reversibility of the reaction with lithium. The three systems are examples of the applicability of nanodesign in the search for new electrodes for rechargeable batteries.

Introduction

Portable electric power sources play a very important role in the life of individuals in modern society. In today's industrialised life, we regularly see portable telephones, portable computers, portable CD and DVD players, portable tools and a wide variety of objects that are not directly connected to standard voltage outlets. For obvious reasons, it would also be desirable to have electric vehicles (automobiles and scooters), particularly in urban areas. Today, the Li-ion battery provides the highest energy capacity per unit weight among rechargeable systems, which include lead acid, Ni-Cd and Ni-MH batteries. Also, the Li-ion battery is subject to none of the ecological problems associated with heavy metals. Its excellent energy to weight ratio, which ranges from 100 to 150 Wh/kg, has helped it to consolidate as the best choice for portable telephones and computers [1]. The secondary lithium battery has evolved through a modification in technology over the past thirty years. Initially, the device was based on a layered chalcogenide (TiS_2, MoS_2) as cathode, an organic solvent containing a dissolved lithium salt as electrolyte, and lithium metal as anode [2]. This system met with severe safety problems after multiple charge/discharge cycles, and never gathered a substantial industrial market, so, it was replaced in the 1990s by the lithium ion battery, which is today's fastest growing battery system. In the Li-ion battery, the cathode is a lithiated transition metal oxide ($LiCoO_2$, $LiNiO_2$, $LiCo_{1-x}NiO_2$, $LiMn_2O_4$), the electrolyte is usually an organic solvent containing a dissolved lithium salt and the anode is carbon (usually in graphitic form). Although it provides important advantage over aqueous secondary batteries (NiCd, Ni/MH and lead acid) in terms of stored energy capacity and shelf life, their specific energy is only marginally higher (about twice than in previous systems). In fact, one can see that their increased energy is mainly the result of the voltage of the lithium ion systems (3.5—4.0 V) with respect to the aqueous systems (1.2—2.0 V).

Although lithium ion batteries feature very on low self discharge and excellent efficiency, they exhibit a moderate charging rate (2—4 hours to recharge a completely discharged battery) by effect of intrinsic problems due to the solid state diffusion of lithium from the positive electrode and into the negative electrode. One way of alleviating this problem is by reducing the particle size of the active material [3—6]. This, however, may also reduce the efficiency of the charging process or increase self discharge in the system.

The diffusion coefficient of Li^+ in a solid electrode is typically of $10^{-12} cm^2 \cdot s^{-1}$. Based on Fick's second law and this realistic diffusion coefficient, a particle *ca.* 100 nm in size would be completely intercalated or deintercalated within 15 min (4C). Therefore, the short diffusion path lengths of nanomaterials are beneficial for electrode kinetics. Other interesting properties of nanosized materials in lithium batteries include enhanced reactivity towards lithium [4,7], suppression of phase transformations —which increases electrochemical reversibility— [8,9] and the ability to use defects and high surface areas to obtain high capacities for lithium [10—14]. In some cases, the nanometric size of particles boosts reactivity by inducing changes in electrode conductivity [13,15]. One can also expand the cycling life for negative electrodes undergoing volume changes during cycling by preparing the material as nanoparticles [see for example 3,13]. This has the disadvantage that the increased reactivity associated to a high surface/volume ratio can affect self-discharge and result in poorly cycling cells. Nanoparticles of some compounds in charged state can react with the organic electrolyte and induce excessive heat in the battery system or even lead to catastrophic failure [16]. Furthermore, the low degree of packing of these systems reduces volumetric energy densities.

There is abundant literature on the use of nanosized electrodes in lithium ion batteries. One sound example of the advantages of nanomaterials is the capacity provided by the positive electrode $LiCoO_2$ material currently used in Sony batteries. Polycrystalline $LiCoO_2$ can furnish a capacity of only 20 $mAh{\cdot}g^{-1}$ upon charging at 1000 $mA{\cdot}g^{-1}$; as nanocrystalline powder [17], however, it has the ability to deliver an average capacity of 120 $mAh{\cdot}g^{-1}$. Nanosized RuO_2, which provides 1100 $mAh{\cdot}g^{-1}$ over several cycles, has been reported to exhibit excellent reversibility in the electrochemical reaction; this has been ascribed to excellent mass and charge transport properties of the nanometric phases involved in the reaction [18]. "Conductive" $LiFePO_4$ is a very promising active compound for positive electrodes in secondary lithium batteries. The polyanionic array endows it with excellent structural stability in the charged state, making it a safe material for batteries. Also, its low cost is an attractive quality for this application. The rate capabilities of $LiFePO_4$ based batteries can also be enhanced by reducing the particle size of electrode materials; in fact, such rate capabilities are known to be limited by slow diffusion of lithium ion [14,19,20]. The use of nanostructured cathodes can therefore improve the intercalation behavior through shortened diffusion distances with nanoparticles.

These are several examples of the utility of nanosized systems in lithium ion batteries, an area of increasing interest [1,5,21,22], including electrodes based on nanosized $LiNi_{0.5}Mn_{1.5}O_4$ [23], vanadium oxide [14,24-27], $Li_4Ti_5O_{12}$ [14,28], $LiFePO_4$ [20,21,29], transition metal chalcogenides [30-32], TiO_2 [33,34], silicon [35-38], C/Si composites [36,39-41], tin based systems [3,37,42-44], intermetallics [45-49], carbon nanotubes [50-52], Fe_xO_y [4,7,8,53-55], NiO [4,7,55], Cu_xO [4,7,55-59] and Co_xO_y [4,7,45,55,60-66]. In this chapter, we discuss the results obtained so far for nanosized materials with promising properties as positive or negative electrodes for lithium ion batteries and examine the effect of their nanometric particle size on the electrochemical performance of the corresponding cells.

Lithium Iron Phosphate

Research into new cathodes for lithium ion batteries has expanded considerably since the pioneering studies of Padhi et al. on the olivine structure $LiFePO_4$ [19]. This compound, when optimised, provides a capacity close to 170 $mAh{\cdot}g^{-1}$ by effect of the complete extraction of lithium (i.e. 1 mole of lithium per formula). However, the reaction is hindered by the insulating character of the phosphate and the diffusion coefficient of lithium ($\approx 10^{-14}$ $cm^2{\cdot}s^{-1}$) [19,20,29,67–69]. One way of circumventing this shortcoming is by thoroughly mixing the compound with a conductive additive such as carbon [20,29,68–77] or a metal [78,79]. This field of research has recently been extended to the use of conductive organic polymers (e.g. polypyrrole) [80]. Also, the electrochemical performance of the phosphate can be improved by reducing its particle size to the nanometric level [14, 20, 29, 71, 79, 81–83]. This approach somewhat contradicts the shrinking core shell mechanism proposed to explain the transformation of $LiFePO_4$ into $FePO_4$ during the charging process [20,29,67–69]. However, it seems rather plausible that the kinetics of Li ions being removed from the particle core should improve as particle size is decreased.

Figure 1 illustrates the electrochemical performance of various $LiFePO_4$ compounds as a function of particles size. The faradic output, which corresponds to the insertion process ($\eta = \frac{x}{1}$, where x is the lithium content in the Li_xFePO_4 material), is indeed clearly dependent on this dimensional parameter. The insertion efficiency is seemingly optimal when crystallite size is within the range of maximum length of the lithium diffusion pathway (L), the latter being imposed by the current density used for cycling.

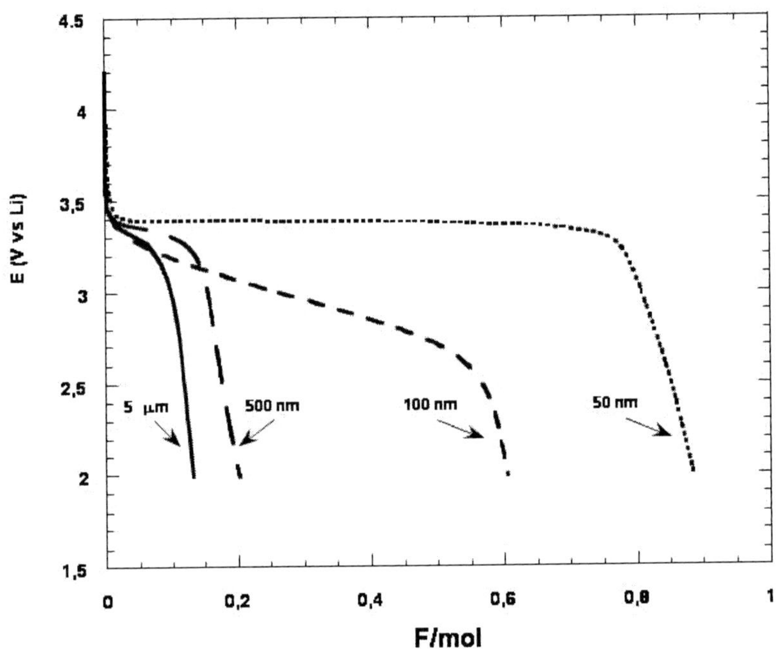

Figure 1. Dependence of a $LiFePO_4$ based cell first discharge on the pristine particle size.

Parameter L values can be estimated from the integrated form of the Fick's first law (under one-way and semi-infinite diffusion conditions) :

$$J_{Li} = -\widetilde{D}_{Li}\frac{dC_{Li}}{dL} \Rightarrow \frac{dx}{dt} = -\frac{\widetilde{D}_{Li}}{L}\frac{dx}{dL} \Rightarrow L = \sqrt{2\widetilde{D}_{Li}t}$$

Table I. Calculated values discussed in the text.

Cycling rate	L_{max} (nm)	Ratio L_{max}/ particle radius			
		1 µm	0,5 µm	100 nm	10 nm
C/20	850	0.85	> 1	> 1	> 1
C/10	604	0.60	> 1	> 1	> 1
C/5	425	0.43	0.86	> 1	> 1
C	190	0.19	0.38	> 1	> 1
2C	134	0.13	0.26	> 1	> 1
4C	94	0.09	0.18	0.94	> 1

Table I shows selected some calculated values of the maximum length of the lithium diffusion pathway L, as a function of the cycling rate (the mean diffusion coefficient was taken to be $\tilde{D}_{Li} = 5.10^{-14}$ cm^2.s^{-1}) and compares them with common particle radii.

Ensuring good electrochemical performance even at high rates (> C-rate) therefore entails using actually nanosized particles (< 100 nm); otherwise, only a small fraction of each crystallite will obviously be accessible to lithium ions and a reduced overall efficiency ($\eta \ll 1$). Therefore, a need exists for new $LiFePO_4$ synthetic routes providing as small particles as possible.

The most common method for obtaining $LiFePO_4$ is a solid–state reaction involving the decomposition of various oxysalts. The reaction takes place at moderate temperatures, where the size of the as–formed phosphate particles is in the micron range [81]. Prosini et al. [82] found chemical lithiation by lithium iodide of nanosized anhydrous $FePO_4$ and subsequent heating under an Ar/H_2 atmosphere at 550 °C to give nanoparticulated $LiFePO_4$ with very good electrochemical activity; however, the procedure requires a pristine nanomaterial. Huang et al. [71] succeeded in limiting the size of $LiFePO_4$ by introducing oxidized carbon particles as nucleating agent. The particles obtained at 700 °C were 100–200 nm in size. Recently, Hsu et al.[83] used a method involving the calcination of a mixture of citric acid, iron oxalate, di-ammonium hydrogen phosphate and lithium carbonate to obtain nanometric phosphates over the temperature range 400–900 °C. Interestingly, it was necessary to decompose the oxalates and citrate up to 700 °C in order to obtain an electroactive powder. Below such a temperature, organic compounds are seemingly incompletely decomposed, so the carbon coating required to ensure electronic conductivity in the phosphate is not formed. Singhal et al. [14] precipitated a precursor by using lithium oxalate, iron hydroxide and $NH_4H_2PO_4$. Heating at 350 °C and annealing between 550 and 800 °C provided $LiFePO_4$ in particles sizes over the range 5–50 nm.

In this work, we used a method that produces nanoparticulated spinels with excellent performance in lithium ion batteries. The method involves mechanical grinding of lithium, nickel and manganese salts in the presence of a large excess of hydrated oxalic acid [23] to obtain $LiNi_xMn_{2-x}O_4$ phases. The slurry thus obtained provides a nanomaterial upon calcination at low and moderate temperatures. We adapted this procedure to the phosphate phases and obtained electroactive nanosized $LiFePO_4$ at 550 °C. We tested various conductive additives in order to alleviate the insulation problems of this nanosized phosphate. Initially, we coated the phosphate particles with copper obtained by *in situ* reduction of CuCl.

Lithium iron phosphate is typically prepared by heating a mixture of precursors above 500 °C. Usually, iron(II) is protected from oxidation in an inert atmosphere during the thermal treatment. The main difference of our method from other solid–state methods using related precursors is that it includes previous grinding and the use of a large excess of hydrated oxalic acid, two factors that seemingly have a significant effect on the results. Thus, they facilitate the formation of a highly homogeneous nanometric mixed oxalate precursor [84]. Also, oxalic acid released while heating isolates nanoparticles from one another, thereby hindering particle growth or agglomeration. In addition, hydration water facilitates nucleation and growth of the reaction products. Briefly, lithium iron phosphate was obtained as follows: $FeC_2O_4{\cdot}2H_2O$ (1.80 g, Fluka), $(NH_4)_2HPO_4$ (1.32 g, Fluka)), $LiC_5H_7O_2$ (1.60 g, Strem Chemicals) and $H_2C_2O_4{\cdot}2H_2O$ (8 g, Fluka) were ball–milled in a Retsch mill at 50 rpm for 30 min. The resulting slurry was supplied with 2 ml of ethanol and air–dried. The powder collected was subsequently calcined at 550 °C under an argon atmosphere for 2 h. We chose this moderate temperature in order to ensure adequate crystallinity in the compound and avoid sintering, which increases particle size. After cooling to room temperature, a brown solid, nanometric $LiFePO_4$, was obtained.

For coating with copper, the nanomaterial was immersed in a solution containing freshly prepared CuCl in methanol [85]. The Cu:$LiFePO_4$ weight ratio used was 5:95. Because this compound is readily oxidized, the process was carried out inside a glove box (Mbraun 150). The brown solid darkened upon addition of several milligrams of $NaBH_4$ under continuous stirring. The solid was collected by filtration after 2 h of stirring and thoroughly washed with methanol, dried in an inert atmosphere and stored in tightly sealed glass containers under an argon atmosphere for transportation. Like Croce et al.[78], we chose copper for this purpose on the grounds of its low toxicity and cost; however, we used a different approach involving obtaining Cu^0 *in situ* by chemical reduction of Cu^ICl. The reaction of this salt with $NaBH_4$ has been confirmed to yield nanoparticulated copper [85] and proceeds according to

$$2\,Cu^ICl + 2NaBH_4 \rightarrow 2NaCl + B_2H_6 + 2Cu + H_2 \quad (1)$$

Formally, the copper thus obtained should be more effectively dispersed than submicrosized commercial copper [78] and hence increase the conductivity of the Cu_{nano}–$LiFePO_4$ composite.

We checked the purity of the $LiFePO_4$ obtained by XRD (Figure 2). The scanning conditions for structural refinement included a 15–125 ° (2θ) recording range, a 0.02° step size and 8 s per step. The peaks associated to the compound matched those of the orthorrombic phase (*S.G. Pnmb*). Rietveld refinements, performed using the GSAS software suite [86], yielded the following cell parameter values: a = 10.298 Å, b = 5.995 Å and c = 4.695 Å with $\alpha = \beta = \gamma = 90°$. The fitted parameters were Rwp = 17%, Rp = 11.9 and Rb = 5.4 %. Also, there were few peaks at 30.35, 43.39 and 57.56 °(2θ) that could not be indexed in the olivine-type structure; rather they were indexed in a cubic structure with a lattice constant of 8.34 Å, which is consistent with either maghemite, γ-Fe_2O_3, or magnetite, Fe_3O_4. The presence of one of these phases made the sample magnetic (as checked with a magnet stick). Based on the heating temperature used, only the magnetite phase could be stable; its content as calculated from the Rietveld refinement was *ca.* 6%. The magnetite may have been formed from traces of oxygen contained in the argon flow or occluded in particle pores [81]. Other authors [74] have found $LiFePO_4$ to yield Fe_2O_3 and $Li_3Fe_2(PO_4)_3$ directly during oxidation in the air at 400 °C. Interestingly, Fe_2O_3 was found to be an impurity in a related solid-state reaction [83]. This phase seemingly appears when no citric acid –the carbon precursor used in their preparation– is present. This phenomenon was also observed by Dokko et al. [87], who added ascorbic acid to the reaction mixture in order to remove an amorphous layer of iron oxide that appeared at 400°C. Based on these findings, our precursors can be deemed ineffective to obtain carbon-rich materials, as confirmed by elemental analyses. Otherwise, Rietveld refinements on the XRD patterns for the Cu_{nano}–$LiFePO_4$ material confirmed that the olivine-type structure was maintained during the reaction and that cell parameters were only slightly different (viz. a = 10.302 Å, b = 5.998 Å and $c = 4.695$ Å with $\alpha = \beta = \gamma = 90°$. No copper was detected in the XRD pattern. Again, peak for iron oxide were observed.

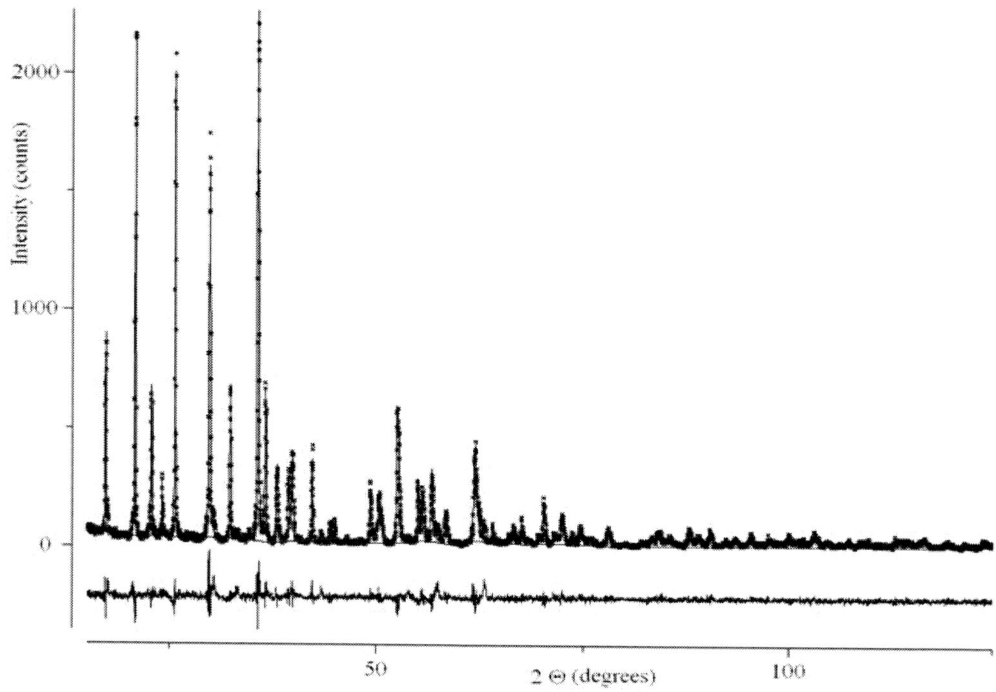

Figure 2. XRD patterns for nanosized $LiFePO_4$. Experimental (dotted), calculated (line) and the difference (bottom).

Transmission electron microscopy (TEM) images were obtained on a Jeol 2010 microscope operating at 200 keV. Samples were prepared in an ultrasonic bath, using hexane as dispersing agent and a grid covered with a carbon perforated film. TEM images provided direct evidence of crystallinity in the material. The sample consisted of nanoparticles (Figure 3a) with reasonably well–defined edges and sizes below 80 nm. Figure 3b shows an HRTEM image of a crystallite exhibiting structural uniformity and a lattice spacing of 0.25 nm, which corresponds to the (311) lattice planes. Our particles were better-defined than those formed by chimie douce methods [82] or a related solid-state reaction at 550 °C [83]. TEM images (Figure 3c) revealed that stirring of the phosphate in the copper(I) solution resulted in rounder particle edges and hindered the observation of well-defined particles. The crystallographic orientation of the particles was seemingly unaffected; also, the d-spacing as calculated from the fringes was again consistent with the value for the (311) planes. No copper could be identified from the TEM images. So far, all attempts at detecting isolated or agglomerated copper particles by combining microdiffraction and energy dispersive microanalysis have failed. The diffuse diffraction peak in Cu_{nano}–$LiFePO_4$ nanocomposites must be due to particle microstructure degradation caused by stirring rather than to diffusion of Cu atoms into the phosphate framework.

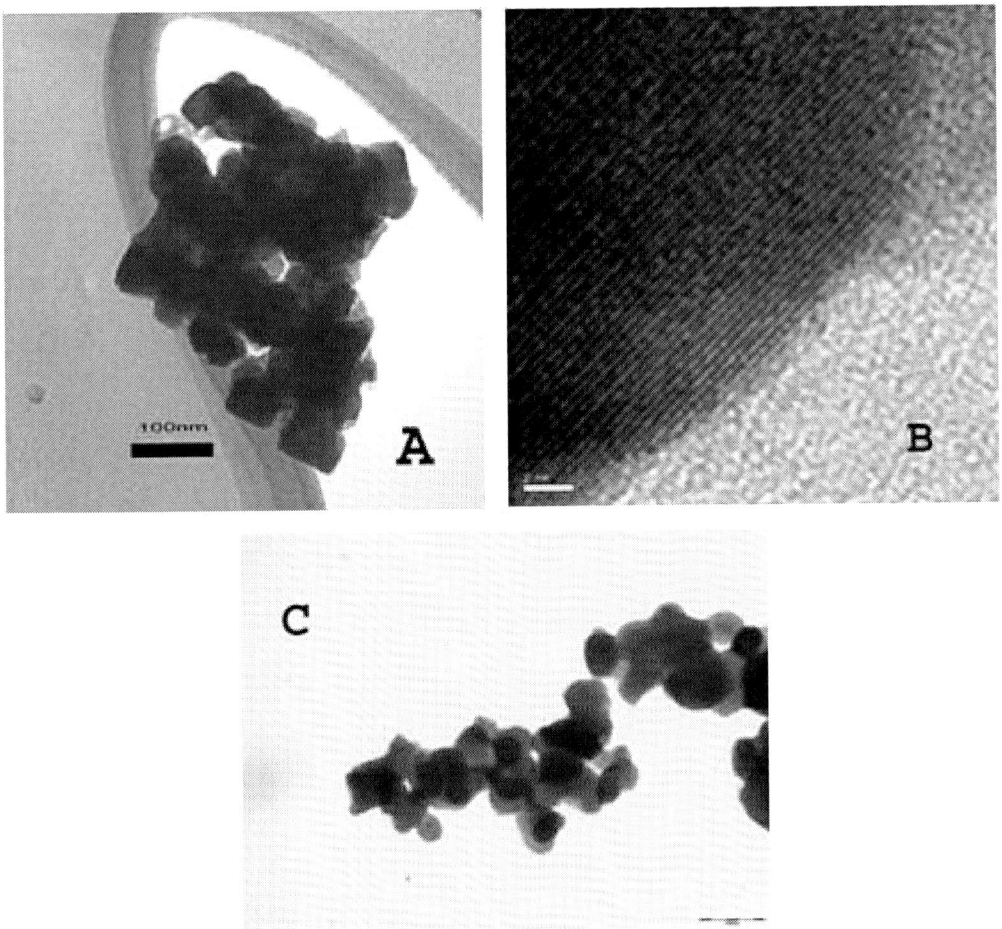

Figure 3. TEM images of (a) and (b) $LiFePO_4$; (c) Cu_{nano}–$LiFePO_4$ composite. The bar corresponds to 100 nm for (a) and (c), and to 2 nm for (b).

The oxidation states of iron and phosphorus, and the Fe/P atomic ratio, were checked by XPS. X-ray photoelectron spectra were obtained on a Physical Electronics PHI 5700 spectrometer, using non-monochromatic Mg Kα radiation (300 W, 15 kV, 1253.6 eV) and a multi-channel detector. Spectra were recorded in the constant pass energy mode at 29.35 eV, using a 720 μm diameter analysis area. Binding energy (BE) values were referred to the C 1s peak (284.8 eV) from the adventitious contamination layer during data processing of the XPS spectra. Recorded spectra were always fitted using Gauss–Lorentz curves in order to more accurately determine the binding energy of the different element core levels. The error in BE was estimated to be *ca.* ± 0.1 eV. The spectra for the C 1s and Cu 2p regions were initially recorded for 10 min. Such a short acquisition time was intended to avoid the previously reported reduction of Cu^{2+} species to Cu^{+} by the action of the X-ray excitation source as much as possible [88]. For recording of XPS spectra, the containers were opened in the air and the pellets rapidly transferred to the preparation chamber of the XPS spectrometer. In this way, the exposure time of the sample to air was minimized.

The corresponding data are shown in Table II. The Fe/P ratio obtained, 1.02, was consistent with the compound stoichiometry. The low intensity/background ratio of the Li 1*s* emission peak resulting from its low scattering coefficient towards X-rays precluded the accurate determination of its binding energy and the estimation of the element content.

Table II. XPS binding energies for various atoms in the $LiFePO_4$ and Cu_{nano}-$LiFePO_4$ samples.

Sample	Atom	Emission peak	Binding energy (eV)	Assignation
$LiFePO_4$	Fe	$2p_{3/2}$	711.4	Fe^{2+}
		$2p_{1/2}$	724.6	
	P	$2p$	133.7	$[PO_4]^{3-}$
	O	$1s$	531.4	$O^{=}$
Cu_{nano}–$LiFePO_4$	Fe	$2p_{3/2}$	711.2	Fe^{2+}
		$2p_{1/2}$	724.4	
	P	$2p$	133.7	$[PO_4]^{3-}$
	O	$1s$	531.3	$O^{=}$
	Cu	$2p_{3/2}$	930.0	Cu^0
		$2p_{3/2}$	932.9	Cu^I-O
		$2p_{3/2}$	934.6	Cu^{II}-O
		$2p_{1/2}$	949.8	Cu^0
		$2p_{1/2}$	952.8	Cu^I-O
		$2p_{1/2}$	954.5	Cu^{II}-O

Figure 4a shows the spectrum for Fe. The shape and binding energies of the Fe 2*p* emission peaks are consistent with those obtained by other authors for phospholivines [89] and hinder the identification of iron oxides. Furthermore, the P 2*p* binding energy corresponds to the PO_4^{3-} group [90] and rules out the presence of iron phosphide [91], a product formed through carbothermal reduction during the preparation of a phospholivine. Therefore, no carbothermal reaction took place during the calcining process. The presence of copper was confirmed from XPS measurements that also ruled out the formation of metallic iron (by reduction of Fe^{II} from the phosphate) or residual chloride. The binding energy values obtained are listed in Table II. The fact that the P and Fe peaks retained their positions (Figure 4b) provides direct evidence of the stability of $LiFePO_4$ upon contact with $NaBH_4$.

One unexpected finding was a complex Cu 2p profile (Figure 5) that was resolved into the contributions of three different oxidation states (Cu^{II}, Cu^I and Cu^0). The presence of divalent copper is also supported by the typical shake-up observed at 940 and 943 eV. These data suggest a pronounced reactivity of Cu towards oxygen as the likely result of the nanometric size of the particles. Despite the precautions exercised to minimize contact of the composite material with the air, the element was partially oxidized –at least at a surface level. The Cu^{II}:Cu^I:Cu^0 atomic proportion was found to be 2:1:1. Therefore, only 1.25% by weight of the Cu_{nano}–$LiFePO_4$ mixture consisted of metallic copper, so the improvement in conductivity should have been less than expected for total conversion (5 wt %). This accounts for the inability to detect copper nanoparticles by XRD or from the HRTEM images.

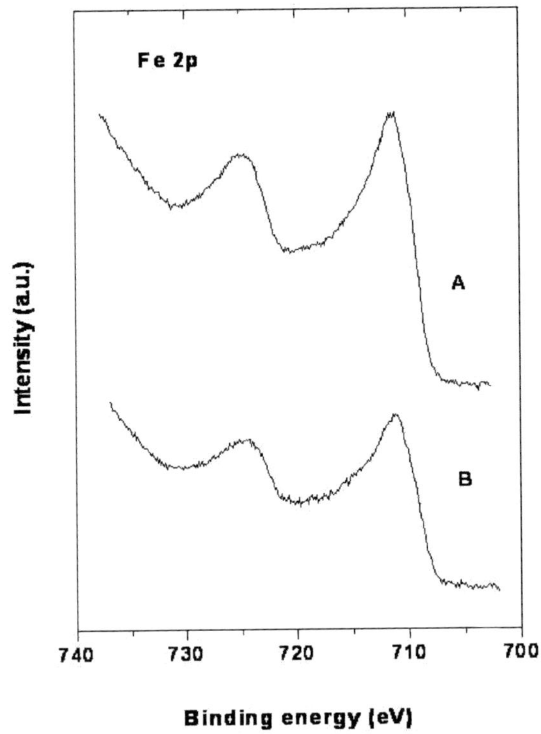

Figure 4. XPS spectrum for iron in (a) $LiFePO_4$ and (b) Cu_{nano}–$LiFePO_4$.

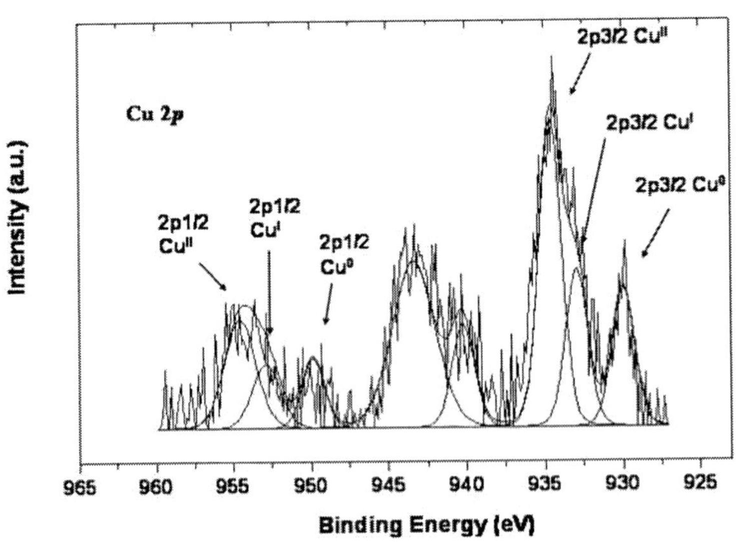

Figure 5. Cu 2p spectrum for the Cu_{nano}-$LiFePO_4$ composite.

Peak deconvolution is also shown.

Electrochemical measurements were made on CR2032 coin cells, using an electrode mixture consisting of active material, carbon black and Teflon in a 85:12:3 proportion by weight. In the general procedure for all electrodes tested in this work, an amount of a few milligrams of the electrode mixture was pressed onto stainless steel grid of 14 mm diameter under a pressure of 4 ton·cm^{-2}. The electrolyte used was 1 M $LiPF_6$ in 2:1 EC:DMC as solvent. Both the reference electrode and the counter-electrode consisted of a lithium disk. Potentiostatic cycles were carried out in a MacPile II (BioLogic) potentiostat at 10 mV·2 min^{-1}. Galvanostatic discharges were performed on a Arbin system under a C/10 or C/5 regime (i.e. one mole of lithium was transferred in 10 or 5h, respectively).

Figure 6a shows selected potentiostatic charge–discharge curves for a cell made from nanosized $LiFePO_4$. A broad peak centered at 3.55 V was obtained in the first cycle. This oxidation peak corresponds to the extraction of lithium from the structure according to

$$LiFe^{II}PO_4 \rightarrow Li^+ + e^- + Fe^{III}PO_4 \tag{2}$$

The reverse reaction (*viz.* the insertion of lithium into the heterosite structure) provided a broad peak at 3.33 V. The difference between both potentials, ΔE_p, was 220 mV and testifies to the kinetic limitation of these electrodes by effect of their high resistivity. [70,76,78]. Moreover, the reduction peak (denoted by I_c) was weaker than the oxidation peak (denoted by I_a), as expected for a slow transfer process. Peak intensity decreased in the third cycle, which suggests that oxidation and reduction were much more strongly hindered, and ΔE_p rose to 360 mV. In the fifth cycle, the peaks were clearly shifted from their original positions and much broader, consistent with the decreased electrochemical activity.

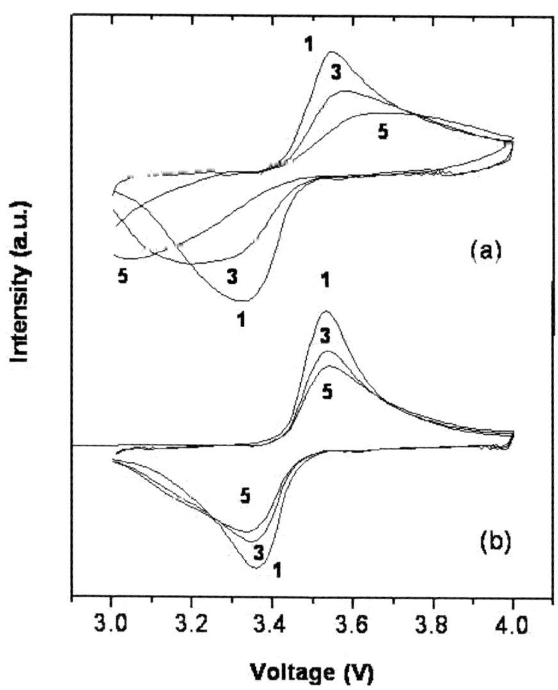

Figure 6. Voltammetric cycling of Li/$LiPF_6$(EC,DEC)/lithium iron phosphate electrodes over the 3.0–4.0 V voltage range. (a) Nanosized $LiFePO_4$, (b) Cu_{nano}–$LiFePO_4$. Scan rate 5 mV·min^{-1}.

Figure 7.a shows selected galvanostatic charge/discharge cycles for the nanosized $LiFePO_4$ cell. A typical flat voltage plateau close to 3.5 V was obtained during the first charge and discharge. The first capacity was 120 mAh·g^{-1}; this, however, is 50 mAh·g^{-1} less than the theoretical capacity associated to total delithiation of the phosphate. These values depart markedly from those obtained by Hsu et al. [83] for nanoparticles prepared at the same temperature, which they found to be electrochemically inactive. Furthermore, our capacity values are higher than those found for an hydrothermally synthetized $LiFePO_4$ containing amorphous iron oxides [87] and a nanostructured phosphate prepared by Singhal et al.[14] Note that, in the latter case, the authors used 20 wt% rather than 12% carbon black to prepare the electrode mixture.

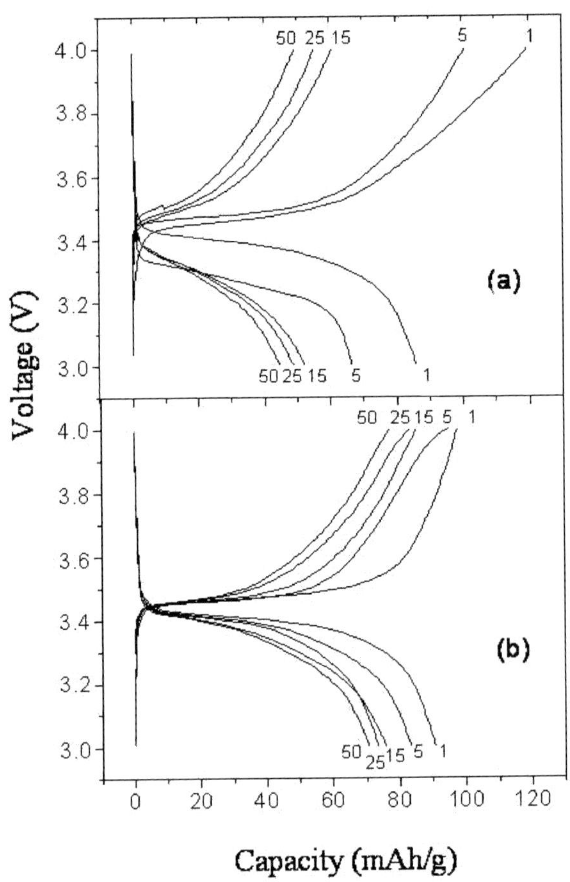

Figure 7. Galvanostatic cycles under a C/10 regime for cells with cathode based on (a) nanosized $LiFePO_4$ and (b) Cu_{nano}–$LiFePO_4$.

Consistent with the potentiostatic results, the capacity was barely retained upon cycling (Figure 8) and, after five cycles, faded by 42%. This behaviour is typical of a phosphate-electrode system and illustrates the inadequate electronic conductivity of nanoparticles. Moreover, as suggested by Yamada et al. [81], the presence of iron oxide impurities may detract from cell performance. As noted earlier, the nanoparticles obtained in this work, while somewhat more electroactive than other $LiFePO_4$ nanoparticles obtained using more complex methods, exhibit an electrochemical response that differs markedly from theoretical calculations in addition to quite poor cell performance. This is widely ascribed to the lack of a carbon or phosphide network ensuring electron conduction between particles. Based on the results of Hsu et al. [83], the electroactivity of nanosized phosphates is determined not only by the presence of the conductive network, but also by particle shape, which was well-defined in our case. In order to improve the electrical conductivity of the electrode, we added copper to our original material.

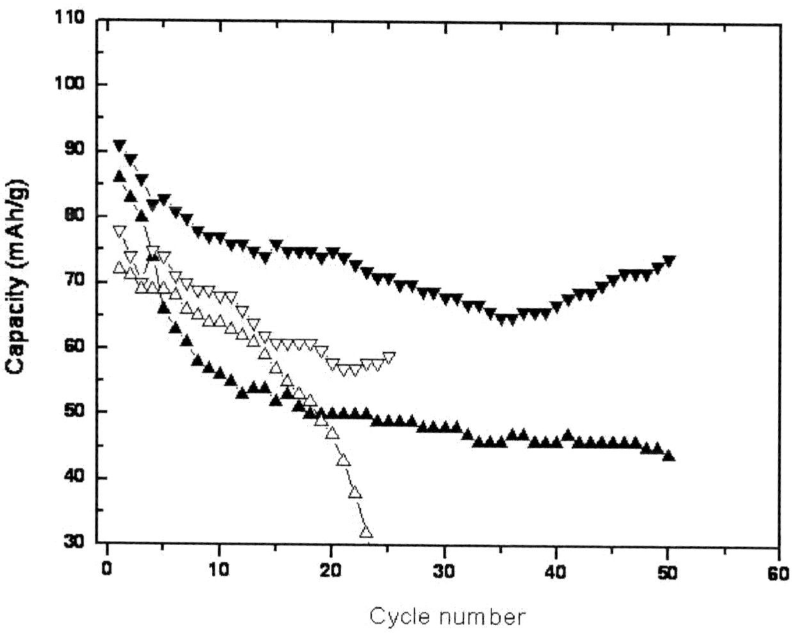

Figure 8. Variation of cell capacity as a function of the number of cycles. (▲) $LiFePO_4$ under C/10, (▼) Cu_{nano}-$LiFePO_4$ under C/10, (△) $LiFePO_4$ under C/5 and (▽) Cu_{nano}-$LiFePO_4$ under C/5.

Notwithstanding the low copper content, the favourable effect of the element on the electrochemical performance of the cell was clearly apparent. Figure 6b shows the potentiostatic cycles for the cell based on the Cu_{nano}–$LiFePO_4$ mixture. Now, the oxidation and reduction peaks were sharper than those for untreated $LiFePO_4$ (Fig.6a), the effect being stronger for the reduction peak. On further cycling, the positions of the peaks remain unchanged, even though their area decreased gradually. This suggests that electron charge transfer is facilitated by a lower resistivity in the nanocomposite. Copper oxidation was not apparently detected in these samples, probably because it may have been masked by the main reaction (the Cu potentials are 3.36V–3.51V for Cu^{2+}/Cu and Cu^{+}/Cu, versus Li^{+}/Li, respectively). Galvanostatic cycling (Figure 7b) also provided direct evidence of the favourable effect of copper on the electrochemical properties of the electrode. In the first charge, the compound provided a flat voltage profile and a capacity of 100 $mAh \cdot g^{-1}$; however, 10 $mAh \cdot g^{-1}$ was lost after discharge. The voltage plateau remained remarkably stable on cycling, but its width decreased with cycling. The composite retained part of the capacity and voltage profiles over several cycles (Figure 7b); thus, 70% of the original capacity was retained after 50 cycles. In order to further check the performance of our composites, we tested the electrodes under a faster regime (*viz.* C/5). As shown in Figure 8, the faster cycling was, the lower capacity was in the first cycle. Once again, however, the copper composite exhibited greater retention (75% in capacity in the 25th cycle versus 50% for the copper-free material).

These promising results led us to examine the use of inexpensive techniques which were bound to provide reliable results. Thus, we explored the use of metal deposition systems, which are commonly used to prepare samples prior to their examination by electron microscopy, and selected gold as the specie to be deposited. We also prepared two additional composite materials, namely: (i) copper–$LiFePO_4$ mixed by hand ; and (ii) C-coated $LiFePO_4$ obtained by mechanochemical activation and pyrolysis of an organic precursor (typically a sucrose). The electrochemical performance of the composites is discussed in terms of the additive distribution and its potential reaction with the electrolyte during charge/discharge cycles.

$LiFePO_4$ was finely dispersed on aluminium foil prior to coating. To this end, phosphate particles were ultrasonicated in acetone, dropped on aluminium and finally transferred to an oven at 40 °C. This procedure was repeated several times until a deposit of 20–30 mg of phosphate was obtained. Gold was deposited with a BALTEC SCD-005 coater at 34 mA for 160 s, which produced a coating of theoretical thickness 25 nm. A working distance of 50 mm was used. This sample was named Au_v–$LiFePO_4$. A second composite, named Cu_m–$LiFePO_4$, was prepared by manually mixing copper (*Merck*, $\phi < 63$ microns) with the nanophosphate in a 1:99 weight ratio. Finally, the C_{mc}–$LiFePO_4$ sample was prepared by mixing sucrose with the phosphate in a C:phosphate weight ratio 5:95 and subsequently heating at 600 °C *in vacuo* for 30 min. Coating had no effect on the cell dimensions of the phosphate. Gold was presumably present in small amounts and particle sizes so it was not detected by XRD; by contrast, the metal in the Cu composite was indeed identified, probably because of its large particle size and high crystallinity despite its low content.

Table III. XPS binding energies (eV) for various atoms in the phosphates prepared in this work.

Sample	Fe 2*p*	P *2p*	C *1s*	Cu 2*p*	Au *4f*
C_{mc}-$LiFePO_4$	711.2	133.8	284.8, 286.2, 287.2, 288.6	-	-
Au_v-$LiFePO_4$	711.1	133.3	-	-	54.0
Cu_m-$LiFePO_4$	711.2	133.6	-	932.6, 934.5 2.5, 954.4	-

In order to check the presence and chemical state of the conductive additives, we made XPS measurements of the three treated samples and compared their spectra with that for pristine $LiFePO_4$. Table III shows the binding energies (BE) of the constituent elements. No significant shifts were observed with the different treatments in any sample, the BE values for the P and Fe 2p peaks being consistent with that for the phosphate group [90] and Fe^{2+}, respectively [89]. Also, P/Fe ratios were close to 1 except for the C_{mc}–$LiFePO_4$ sample, which had a value of 1.42. As with the Cu_{nano}–$LiFePO_4$ composite, no iron phosphide entities were identified by XRD. This suggests the formation of a phosphorous-rich iron phosphide at the surface level by effect of the reducing conditions created by sucrose [91].

The C 1s spectrum for the C_{mc}–$LiFePO_4$ sample was fitted to four components. The main component was observed at 284.8 eV and arose mainly from carbon in the pyrolized sucrose. The other three peaks were weak and assigned to C bound to O, either via a single bond (peak at 286.3 e V) or as a carboxylate and/or carbonate resulting from surface carbonation (peaks at 287.2 and 288. 2 eV). The presence of traces of the organic precursors used in the synthetic procedure may have been the origin of the latter signals, at least at the surface level [20]. Analysing the Cu environment by XPS had the inherent difficulty that peak shape was affected by the exposure time to the X-ray sources. With a short scanning time (10 min), the Cu 2p spectrum exhibited two contributions that can be assigned to Cu^0 and Cu^{2+} species. Under these conditions, the Cu^0/Cu^{2+} ratio was close to 0.8; however, increasing the exposure time to 60 min raised the ratio to 2.3, which reflects a reducing effect of the X-ray beam, as noted earlier. In any case, Cu added to the phosphate was oxidized at the surface level. However, the oxide film must be very thin because it was easily removed without the need for argon sputtering. One interesting finding was the presence of a high content of metal at the surface level. Thus, the Cu:Fe and Au:Fe atomic ratios were 0.13:1 and 0.18:1, respectively, and several times higher than those based on stoichiometric calculations for Cu or a limited deposition time for Au. This reveals a preferential location of these phases on the surface of the phosphate nanoparticles and can help understand the electrochemical properties of the Au_v–$LiFePO_4$ composite.

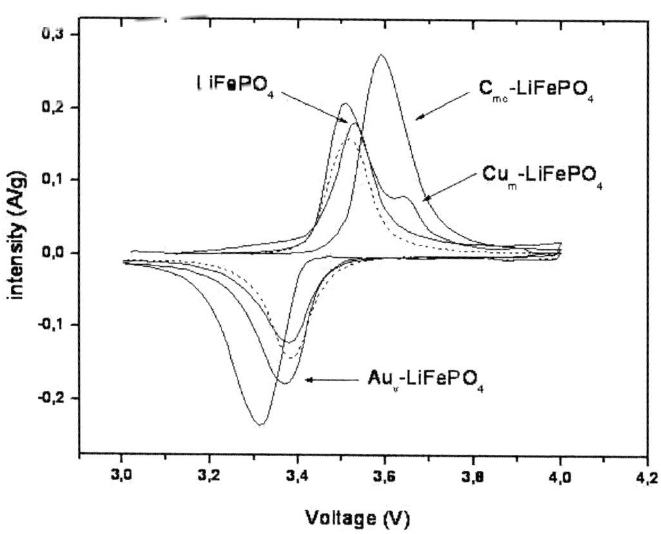

Figure 9. First voltammetric cycling of Li/$LiPF_6$ (EC,DMC)/phosphate based electrodes over the voltage range 3.0-4.0 V.

Figure 9 shows the first voltammetric charge–discharge curves for the cells based on these nanocomposites. The peaks for Cu_m–$LiFePO_4$ were stronger than that for the pristine material. Moreover, the potential difference (ΔE_p) decreased from 220 mV to 130 mV. Both results suggest an improved kinetics of electron transfer and the corresponding lithium insertion/de-insertion process.

The anodic scan for the Cu_m–$LiFePO_4$ sample exhibited another peak at 3.65 V that was seemingly irreversible as it was not detected in the cathodic scan. Although the peak voltage was consistent with that for Cu^+/Cu vs. Li^+/Li (3.56 V), the presence of Cu above the peak, as revealed from *ex-situ* XRD patterns, rules out this assignation. Based on two additional findings (*viz.* an increase in peak intensity and voltage as the Cu content was increased), the peak was ascribed to partial oxidation of the electrolyte catalyzed by the element. The decomposition products probably coat the particles and form a solid electrolyte interface (SEI) that hinders reaction development on further cycling –in fact, the peak was only observed in the first scan. Therefore, this unwanted reaction barely affects the overall electrolyte integrity and, as shown below, the beneficial effect of Cu is preserved [78,79].

The effect of sputtered gold is controversial because ΔE_p (130 mV) was similar to the value for the other two additives even though peak intensity was lower. Somehow, gold favours the electron transfer (through decreased polarization between the oxidation and reduction peaks), but seemingly hinders lithium extraction/insertion. Whereas no significant changes in peak position or intensity other than the above-mentioned disappearance of the peak at 3.65 V were observed on further cycling, the peaks for Au_v–$LiFePO_4$ underwent dramatic changes in both position ($\Delta E_p \approx 250$ mV at the fifth cycle) and intensity (which decreased on cycling). The C_{mc}–$LiFePO_4$ cells exhibited a peculiar behaviour. Thus, the oxidation and reduction peaks were very strong, and the I_a/I_c ratio was close to one. Therefore, the carbon coating on the phosphate particles was effective and increased the electrochemical activity of the material. However, ΔE_p was close to 280 mV, which is higher than the value for pristine $LiFePO_4$. In fact, a similar value was obtained by Franger et al. for C_{mc}–$LiFePO_4$ prepared from micrometric phosphate and glucose heated at 600 °C [20]. Therefore, our ΔE_p value is typical for composites obtained under these synthetic conditions and independent of the particle size of the pristine phosphate. ΔE_p decreased to 190 mV in the second cycle. A similar decrease was previously observed in other composites [78]. On further cycling, the peaks retained their positions but weakened; however, they continued to be stronger than for the other composite-based cells. Therefore, the role played by the additive is strongly dependent on its chemical nature.

Table IV lists selected capacity values provided by the cells under a galvanostatic regime (C/10). The capacities associated to the oxidation and reduction peaks in the first charge/discharge cycle decreased in the sequence C_{mc}–$LiFePO_4 >$ Cu_m–$LiFePO_4 >$ $LiFePO_4 >$ Au_v–$LiFePO_4$. The C_{mc}–$LiFePO_4$ composite yielded 140 $mAh\cdot g^{-1}$ in the first charge, consistent with previously reported values [20]. The good electrochemical performance of this composite is also reflected in Figure 10, which shows the variation of the discharge capacity as a function of the number of cycles. After the second cycle, the discharge capacity of the C_{mc}–$LiFePO_4$ cell levelled off at *ca.* 120 $mAh\cdot g^{-1}$. The copper-based composite exhibited a first charge capacity of 130 $mAh\cdot g^{-1}$ (i.e. 76% of the theoretical value) versus 52% for the as-prepared $LiFePO_4$ nanomaterial. Croce et al. [78] previously proposed copper as an excellent alternative to carbonaceous materials. In our case, the cell provided up to 100

$mAh{\cdot}g^{-1}$ over at least 20 cycles. Surprisingly, the worst performer was the coated sample (Au_v–$LiFePO_4$). The capacity delivered by this electrode was even lower than that of the untreated compound. The origin of this behavior is unclear as coating agents such as Au are also good electronic conductors. The main difference lies in the way the conductor was incorporated and might account for the disparate electrochemical response observed. In the evaporation mode for adding the electronic conductor, phosphate particles may be coated with a layer of Au that must be crossed by Li ions during the insertion/deinsertion process. This barrier may hinder the electrochemical reaction, as reflected in the decreased discharge capacity delivered by the cell. However, if Cu and $LiFePO_4$ are mixed by hand, the surface of the active particles remains essentially unaltered and the role played by the additive is limited to improving the electronic conductivity of the electrode. This decreases the cell impedance.

Table IV. Some electrochemical parameters of the phosphate based cells tested in this work

Sample	1st charge capacity ($mAh{\cdot}g^{-1}$)	1st discharge capacity ($mAh{\cdot}g^{-1}$)	10th charge capacity ($mAh{\cdot}g^{-1}$)	10th discharge capacity ($mAh{\cdot}g^{-1}$)
$LiFePO_4$	89	88	79	78
C_{mc}-$LiFePO_4$	139	125	121	120
Cu_m-$LiFePO_4$	128	110	102	102
Au_v-$LiFePO_4$	72	52	51	51

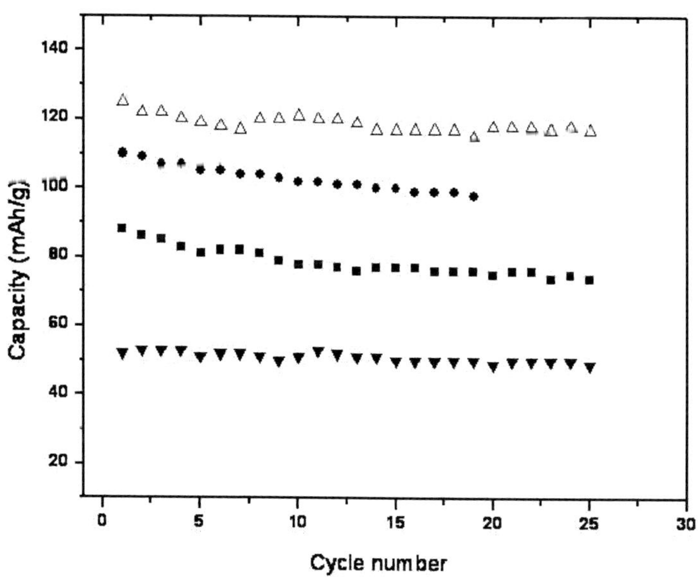

Figure 10. Variation of the cell discharge capacity under a regime C/10 for cells based on the phosphates prepared in this work. (■) $LiFePO_4$, (●) Cu_m-$LiFePO_4$, (▼) Au_v-$LiFePO_4$ and (△) C_{mc}-$LiFePO_4$.

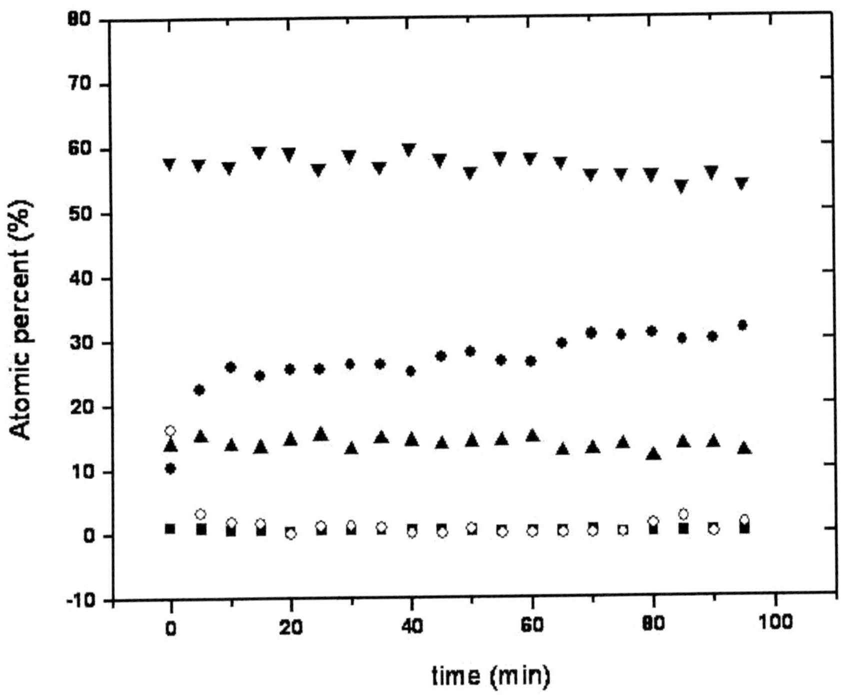

Figure 11. XPS depth profiles for (○) C 1s, (■) Au 4f, (●) Fe 2p, (▲) P 2p and (▼) O 1s.

In order to shed more light on the gold composite, we recorded XPS depth profiles for carbon, iron, phosphorus, oxygen and gold by etching the surface with Ar^+ and determined the ensuing surface atomic ratios. Figure 11 shows the variation of the surface composition as a function of the etching time. Adventitious carbon was removed from the surface within a short time (about 5 min). Besides, the variation of the iron, phosphorus and oxygen surface contents was barely dependent on the sputtering time, and the relative concentrations of these elements were similar even after 100 min.

Figure 12 shows the variation of the XPS spectra for Au 4f as a function of the sputtering time. The intensity of the gold peak decreased with increasing exposure time to Ar^+; however, more than 100 min was needed to decrease its relative surface content from 1.2% to 0.18% (Fig. 12). These results depart from those recently reported for $LiNi_{0.5}Mn_{1.5}O_4$ prepared in pellet form. The gold content measured at the surface level under the same sputtering conditions was markedly higher (around 40%); after a few minutes of etching, however, it dropped to negligible levels [92]. This means that, when a powdered sample is sputtered, the gold not only coats particle surfaces, but also deposits as uniformly dispersed nanoclusters. Access of Li ions to those sites of the phosphate surface in contact with Au nanoclusters must therefore be hindered, and the reactivity of the electrode towards lithium (and hence cell capacity) decreased as a result.

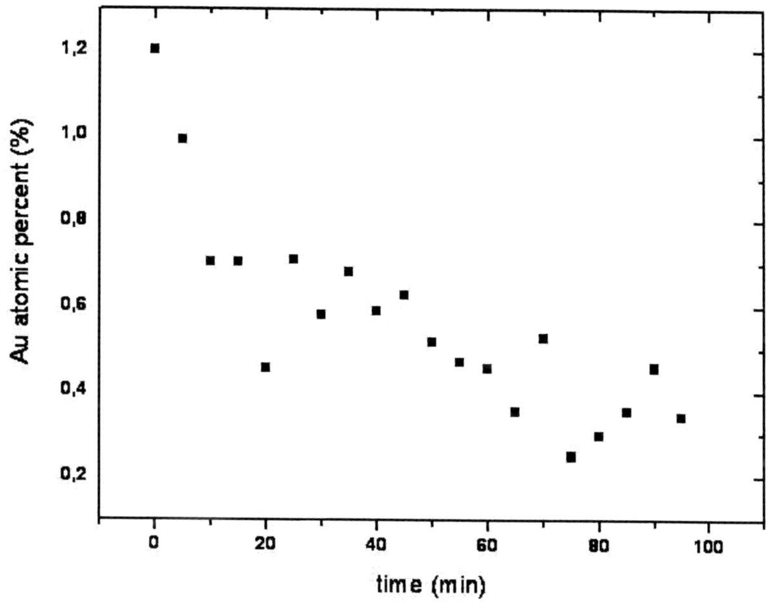

Figure 12. Variation of the Au 4f XPS spectrum as a function of the Ar^+ sputtering time.

Therefore, well-crystallized nanometric $LiFePO_4$ of uniform particle size (ca. 80 nm) was synthesized at 550 °C by using a simple method involving a homogeneously mixing of iron oxalate, diammonium hydrogen phosphate, lithium acetylacetonate and excess oxalic acid plus grinding. In contrast to the typically inactive or poorly active nanomaterials provided by alternative solid-state reactions at the same temperature, our nanomaterial was electroactive in lithium batteries. The initial charge capacity of the cell, 120 $mAh{\cdot}g^{-1}$, was not completely recovered during the first discharge owing to the poor electronic properties of the material. Initially, we tried to improve the cycling efficiency by introducing copper (hypothetically in nanosized form) thoroughly mixed with the phosphate via a chemical reaction. Although copper was partially oxidized, a small fraction (less than 1.5% by weight) sufficed to increase the reversibility of the lithium extraction/insertion process by a factor close to 2 relative to the copper–free material. The capacities delivered by the Cu_{nano}–$LiFePO_4$ composite in the 50th cycle were close to 80 $mAh{\cdot}g^{-1}$. We subsequently tested evaporated gold, co-ground metallic copper and carbon obtained by *in-situ* pyrolysis of sucrose as conductive additives. Voltammetric cycles revealed that the three additives boost the electron transfer kinetics. However, galvanostatic experiments showed that only the addition of C and Cu increased cell capacity and the ability to retain it. XPS measurements of the Au_v–$LiFePO_4$ composite showed that gold not only coats particle surfaces, but also deposits as homogeneously dispersed nanoclusters. This treatment seems to hinder the diffusion of lithium ions and hence the decrease in capacity. Apart from using the C_{mc}–$LiFePO_4$ composite material, which is well-optimized for commercial applications, metallic copper incorporation proved the most suitable procedure for enhancing the electrochemical performance of our new nanosized lithium iron phosphate. In any case, the electrochemical properties shown there are acceptable for a nanophosphate prepared with a fast one-step method using a moderate temperature.

An optimized C–$LiFePO_4$ composite cathodic material obtained by mechanochemical activation [20] has been tested with nanocrystalline $Li_4Ti_5O_{12}$ as anode. The operating cell can be described by the following electrochemical equation:

$$Li_{4+x}Ti_5O_{12} + Li_{1-x}FePO_4 \leftrightarrow Li_4Ti_5O_{12} + LiFePO_4 \quad \text{with } 0 < x < 1 \tag{3}$$

This process involves the exchange of lithium ions between the two electrodes via the non-aqueous electrolyte. The charge–discharge profile of such a cell is expected to include a flat plateau around 2 V since both the anodic and the cathodic material exhibit two-phase electrochemical lithium insertion. This behaviour was indeed observed in our experimental system (Figure 13). A mean available voltage of 1.85 V was obtained from it.

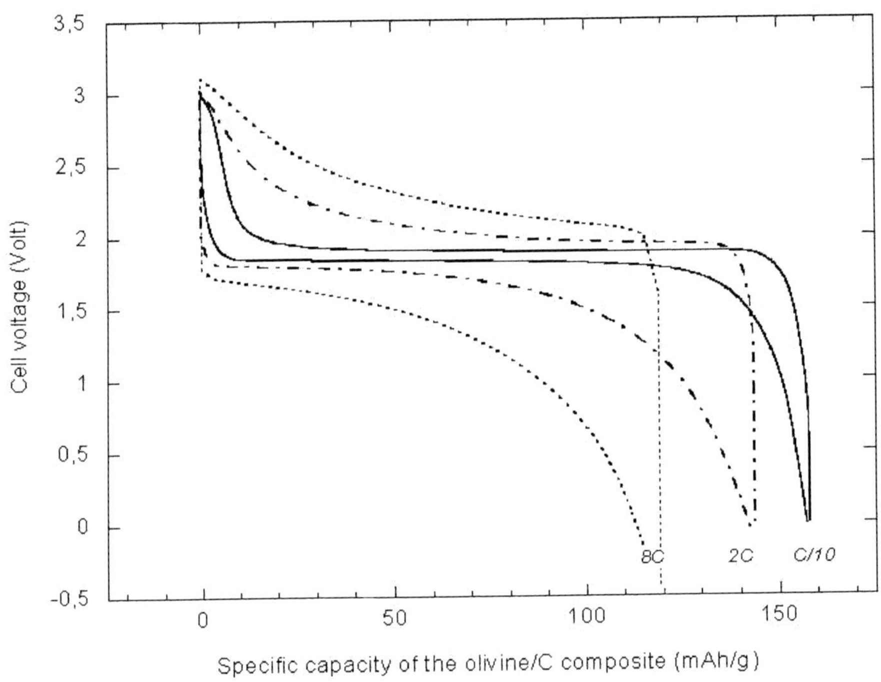

Figure 13. Galvanostatic cycle of the $Li_4Ti_5O_{12}/C_{mc}$-$LiFePO_4$ cell under different rates, at room temperature.

At a slow scan rate (C/10), this cell delivers a specific capacity of about 160 $mAh \cdot g^{-1}$ referred to the cathode, which corresponds to 95 % of the theoretical capacity for pure lithium iron phosphate. The fine, homogeneous carbon coating formed by thermal decomposition of sucrose on the $LiFePO_4$ particles, which improves the general conductivity of this compound, facilitates the achievement of good capacities even at room temperature and relatively high rates.

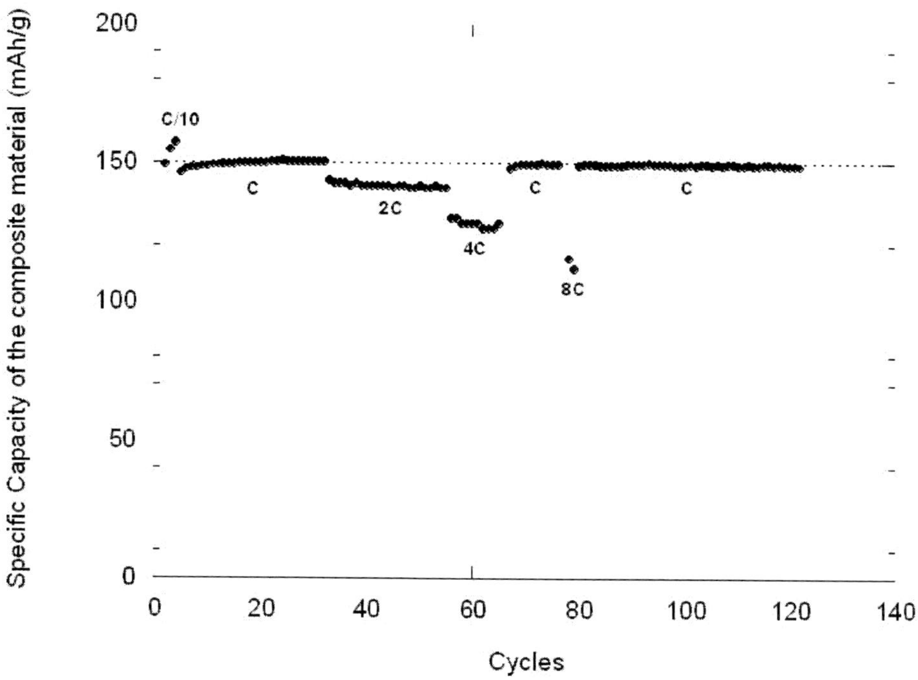

Figure 14. Capacity (referred to the cathodic material) of the $Li_4Ti_5O_{12}$/C-$LiFePO_4$ cell in function on the charge/discharge regime, at room temperature.

In order to confirm the good capabilities of our $Li_4Ti_5O_{12}$|C–$LiFePO_4$ battery, we tested it under variable cycling conditions including relatively low (C/10) to very high rates (4C or 8C) and operating temperature from ambient to low levels. Figure 13 shows galvanostatic curves obtained at room temperature. As can be seen, polarisation increased with increasing applied current. At 8C, a difference of almost 1 V existed between the reduction and oxidation processes. This behaviour is certainly due to the poor ionic conductivity of the bulk material which limits the diffusion of the lithium ions into the very core of the particles. However, the specific capacities referred to the cathode obtained at the different cycling rates exhibit interesting values (Figure 14). Thus, the capacity at C/10 was 160 mAh·g^{-1}, that at C 150 mAh·g^{-1}, and those obtained under more severe conditions such as 4C and 8C were 125 mAh·g^{-1} and 110 mAh·g^{-1}, respectively. Moreover these capacities seem to be stable upon cycling since the slope of the C rate curve (capacity *vs* cycles) was -0.008 % per cycle. If we extrapolate this last result, we can expect our electrochemical system to retain 80 % of its initial specific capacity (*i.e.* 120 mAh·g^{-1}) after charge–discharge 2500 cycles (each in 1 hour). Such an excellent behaviour was confirmed with a real long-time cycling battery (Figure 15), which exhibited less than 5 % of capacity fading after 800 cycles at C/5. In addition, operating at lower temperatures had little effect on the specific capacity; thus, 77 % of the initial capacity was recovered at –20°C when using a charge–discharge rate of C/10 (Figure 16).

Optimizing the carbon coating of the particles to increase the electronic conductivity and reducing crystallite size to overcome the weak ionic conductivity are no doubt two major keys to making lithium iron phosphate a powerful positive electrode material.

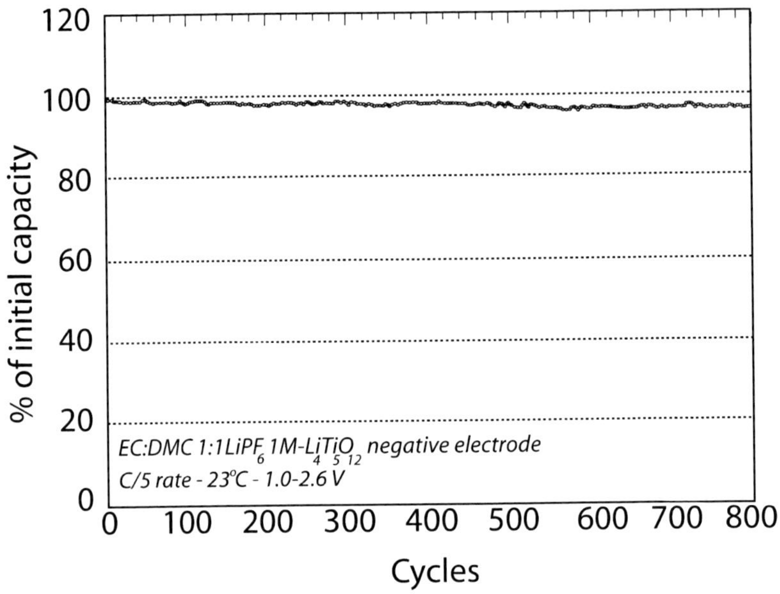

Figure 15. Capacity (referred to the cathodic material) of the $Li_4Ti_5O_{12}$/C-$LiFePO_4$ cell over 800 cycles of charge/discharge at C/5.

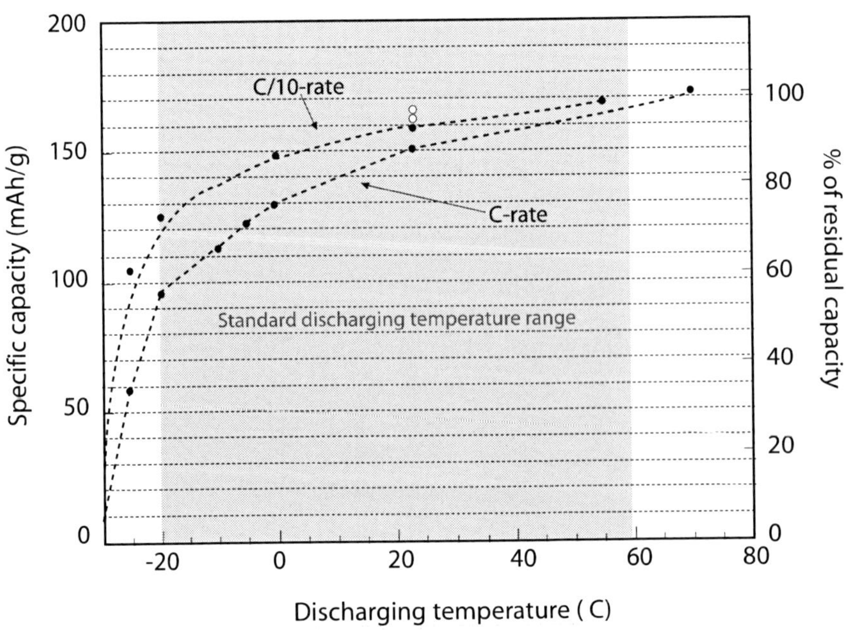

Figure 16. Capacity (referred to the cathodic material) of the $Li_4Ti_5O_{12}$/C-$LiFePO_4$ cell in function of the operating temperature.

α-$LiFeO_2$ Based Nanobelts

Lithium cobalt oxide, $LiCoO_2$, is the most widely used positive electrode material in commercial Li-ion batteries at present [2] by virtue to its high reversible capacity (130–150 mAh g^{-1}), long cycle life (300–500 cycles) and easy preparation. The cell charging reaction is:

$$LiCo^{III}O_2 \rightarrow x\,Li^+ + x\,e^- + Li_{1-x}Co^{IV}O_2 \quad (4)$$

A voltage plateau at 3.8 V is obtained and the cell delivers a capacity close to 150 mAh·g^{-1}[93] during the extraction of ½ mole of lithium per mole of oxide. However, Co compounds are toxic and expensive. Moreover, $LiCoO_2$-based cells are subject to safety problems associated to the instability of the delithiated phase, which contains Co(IV); this is a strong oxidant which can give a highly exothermic reaction upon contact with the electrolyte solvent [16]. Various strategies have been proposed and tested to avoid some of the previous drawbacks, including replacing cobalt with another transition metal [94–96] or using of a protective coating consisting of some inert matrix such as an oxide (ZrO_2 [16], Al_2O_3 [97], SiOx [98]) or phosphate ($AlPO_4$ [16]). One interesting alternative is the use of $LiFeO_2$ given the low cost and environmental friendliness of iron. $LiFeO_2$ crystallizes in a Na-Cl type structure where in which Li and Fe atoms occupy the octahedral sites in a cubic close packing of oxygen atoms. Four polymorphs have been identified from cation arrangement [99,100]. In the α-$NaFeO_2$ type structure, alternate layers of trigonally distorted MO_6 and LiO_6 octahedra share edges. In the α-$LiFeO_2$ structure, which crystallizes in the cubic system, Li^+ and Fe^{3+} randomly occupy the octahedral sites; in the $LiMnO_2$-type structure, however, (corrugated layered structure) the oxygen anion array is distorted and alternating zigzag layers of Li^+ and Fe^{3+} cations result in a reduced symmetry in the orthorrombic system. Finally, in the γ-$LiFeO_2$ (a goethite structure), metal ions are ordered and the resulting symmetry is tetragonal. Recently, a tunnel structure bearing some similarities to hollandite α-MnO_2 was reported [101]. This compound contains FeO_6 octahedra creating tunnels with oxygen in the center. Lithium ions surround the oxygens and strongly coordinate to other oxygens in the framework structure.

The hypothetical reaction taking place at the electrode during the charge process in a $LiFeO_2$ based cell is:

$$LiFe^{III}O_2 \rightarrow x\,Li^+ + x\,e^- + Li_{1-x}Fe_{1-x}{}^{III}\,Fe_x{}^{IV}O_2 \quad (5)$$

With X = 1, this reaction provides a capacity of 283 mAh·g^{-1}. The four materials have been reported to exhibit disparate electroactivity. Thus, Kanno [102] reported a maximum value of X = 0.1 for the α-$NaFeO_2$-type structure. $LiFeO_2$ with a corrugated structure obtained from LiOH and γ-FeOOH using a conventional ceramic method [103] was reported to yield 150 mAh·g^{-1} (X = 0.53). However, the capacity decayed over the next few cycles by effect of the transformation of the layered structure into the $LiFe_5O_8$ spinel. This value was higher than the first reported for the same material by Kanno et al. [102] (lithium removed X = 0.4); also the lithium ferrite becomes an amorphous phase upon cycling. Goethite-structure $LiFeO_2$ has been reported to provide similar capacities [104]. The tunnel structure [101], with a maximum lithium removal value, X, close to 0.7, was found to exhibit higher electroactivity. However, there is some controversy concerning the reaction behind the electrochemical activity as Mössbauer spectra for the charged electrode suggest the release of oxygen from the structural framework rather than reaction (5). Finally, there is little literature on the electrochemistry of the α-$LiFeO_2$ polymorph. A preliminary x value of 0.2 was reported for a low temperature form [105]. Later, an x value close to 0.3 was reported for the compound in the form of nanorods [106]. Higher reactivity was observed in a $Li_{4/3}Ti_{2/3}O_2$–$LiFeO_2$ solid solution prepared by Tabuchi et al. [107] in Fe/Fe+Ti ratios between 0.25 and 0.75. In any case, the previous three studies revealed a low reversibility in reaction (5).

We adapted the method reported by Wang et al. [106] by using water, LiOH and α-FeOOH instead of LiOH·H_2O and β-FeOOH as precursors, respectively, in order to obtain lithium ferrite as nanorods. Briefly, $LiNO_3$, LiOH and goethite α-FeOOH were mixed in a 2:2:1 mole proportion and ground for 30 min. The resulting slurry was heated to 250 °C at 3° C·min^{-1} in 3h. The brown product obtained was thoroughly washed with distilled water in order to remove excess lithium compounds, then centrifugated and dried at 60 °C for 2h.

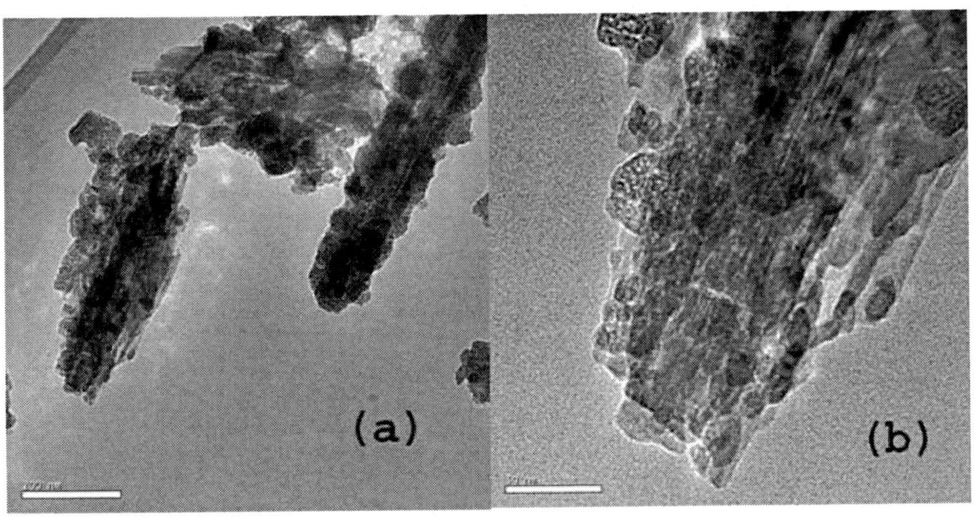

Figure 17. HRTEM images of hematite/lithium ferrite nanocomposite. White bar corresponds to 200 nm (left) and 50 nm (right).

The XRD pattern for the solid exhibited several peaks that were ascribed to α-$LiFeO_2$ and four others below 35°. A similar set was observed, albeit not examined, for the α-$LiFeO_2$ system prepared by Sakurai [105]. Lithium ferrite crystallizes in the *Fm3m* group, with a = 4.155(2) Å, which is quite consistent with the results reported by several authors [105–107]. The TEM images of the sample in Figure 17 reveal that goethite retains its belt shape. The belts consist of tiny highly porous nanoparticles a few tenths of a nanometer in size.

Figure 18 shows the Mössbauer spectra recorded at 295 and at 89.6 K. The doublet is indicative of a paramagnetic state. The hyperfine parameters for the major iron site (IS=0.323±0.003 mm·s^{-1} and QS=0.602±0.002 mm·s^{-1}) are reasonably consistent with the results of a comprehensive study conducted by Cox et al. [108]. Moreover, a small amount (10% in weight ratio judging by the areas under the resonance curve) of a magnetically ordered iron site is present that can be assigned to the impurity detected in the XRD pattern.

The corresponding hyperfine field is very small, which suggests that species other than Fe_2O_3 were present in our samples. The spectrum obtained at 89.6 K revealed complete magnetic ordering leading to a sextet. However, the sextet exhibits asymmetric broadening, which is suggestive of hyperfine field distribution associated to a distribution of environments around the iron atom. The spectra also show the presence of the magnetically ordered impurity.

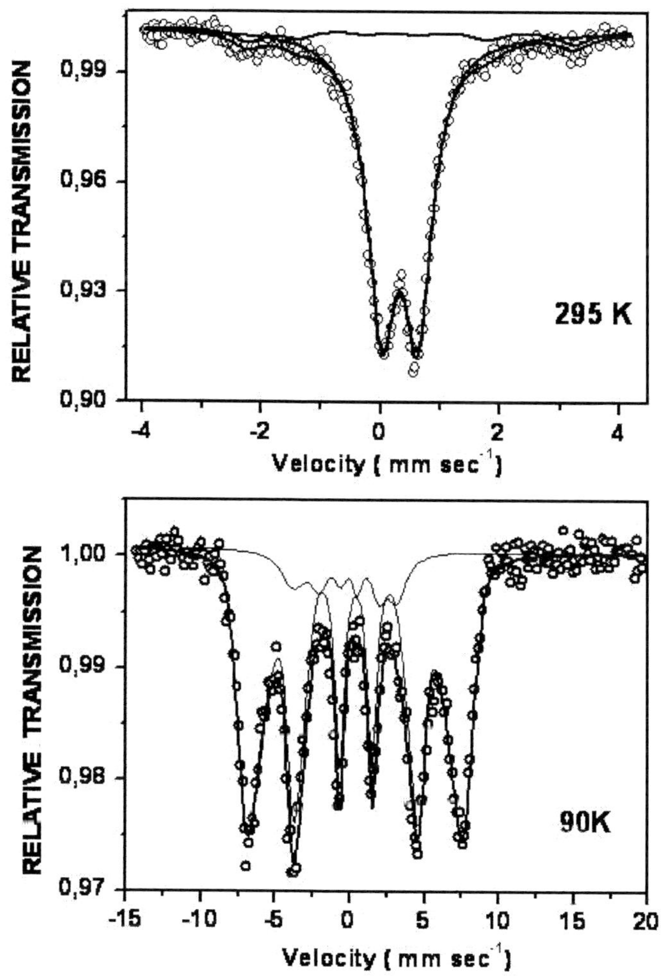

Figure 18. Mössbauer spectra recorded for the α-$LiFeO_2$ nanobelts at different temperatures.

In order to study the electrochemical properties of this nanocomposite, the electrode was prepared from a mixture of active material, carbon black and Teflon in a 75:17:8 weight proportion. The cell configuration was identical with that for the ferrites studied as negative electrodes, but the electrolyte solvent (EC:DMC, 1:1 %v) was different owing to the increased voltages required. The electrochemical tests were carried out over the 1.5–4.5 voltage range and under two different regimes (*viz.* C/4 and C/2.5, which are referred to the content in lithium ferrite, C representing 1 Li^+ ion exchanged in 1 h).

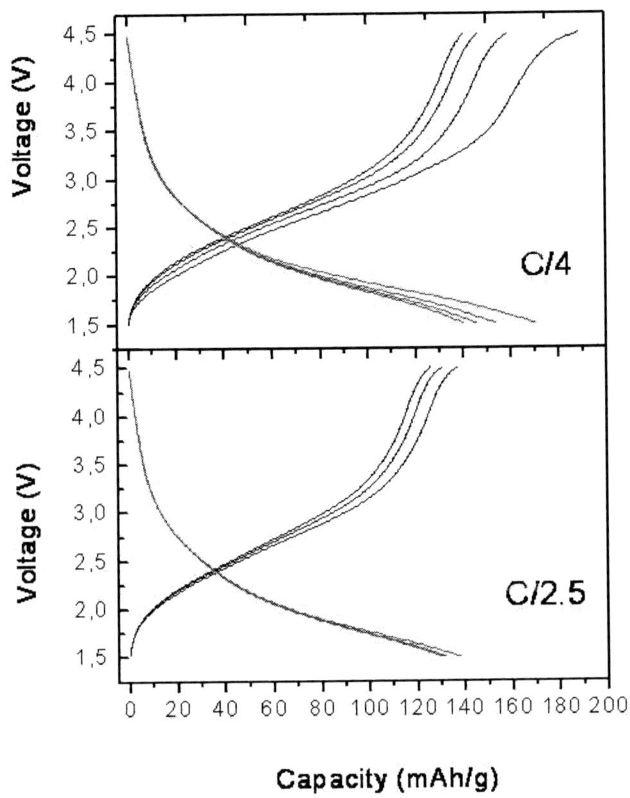

Figure 19. First galvanostatic cycles of the cells based on α-$LiFeO_2$ nanobelts under different current regimes.

Figure 19 shows the galvanostatic curves for the four first cycles. On charging the cell, the voltage exhibited an abrupt increase from 2.6 (OCV) to 3.9 V; this accounts barely for 0.1 lithium ion removed from the structure. From this value, a significant slope change was observed and 0.5 Li ions were removed up to 4.5 V (the upper limit recorded). This is a common behavior for other $LiFeO_2$ with different structures and the amount of lithium removed depends on several factors that of electrolyte and the cell configuration used. In fact, tests on swagelock and coin cells provide disparate results. The best were obtained by using coin cells, which was thus the configuration adopted for subsequent cycles. There was a pronounced drop in voltage at the beginning of the discharge process. At 3.0 V, the slope changed with the presence of a pseudo-plateau that extended to *ca.* 1.7 V. Like the first discharge curve, subsequent charge/discharge curves were s-shaped, similarly to iron oxides with different structures such as α-$LiFeO_2$ [105,106], corrugated layer $LiFeO_2$ [102-104], goethite-type $LiFeO_2$ [104], spinel-type $LiFe_5O_8$ [109] or even FeOOH nanorods [110]. In this respect, Sakurai et al. [104,103] noted that unusual Fe^{IV} ions generated during charging may play an important role in the development of voltage hysteresis in these systems. However, this model must be inappropriate for the latter compound, the electrochemical activity of which is independent of the presence of this oxidation state.

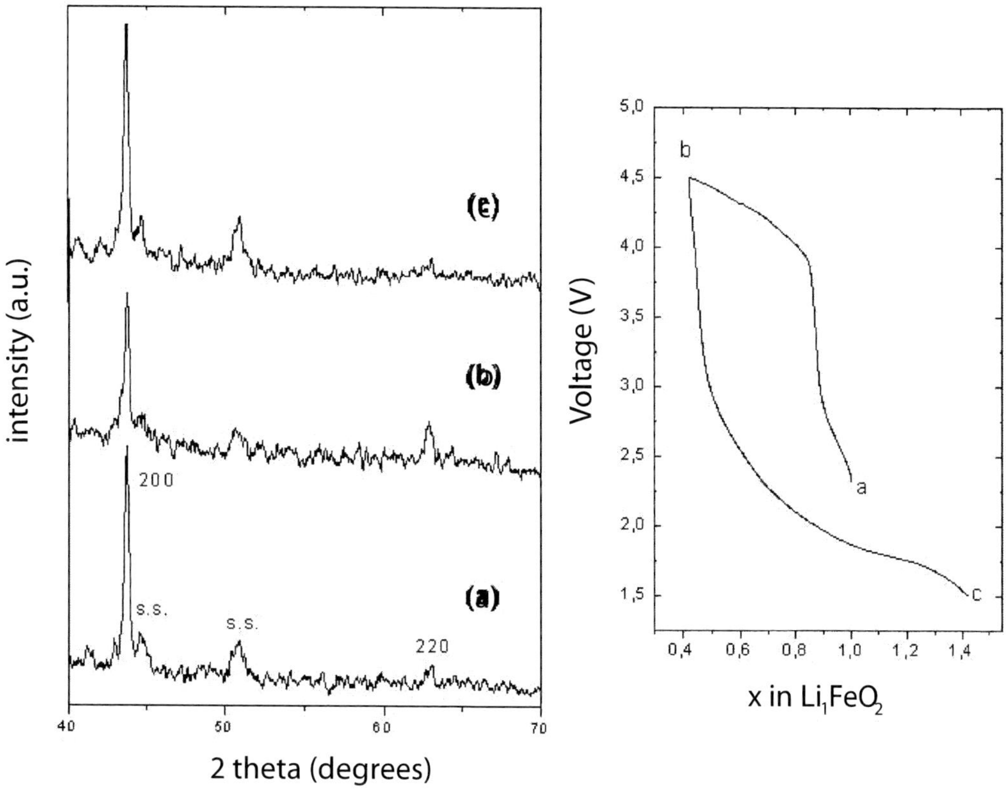

Figure 20. XRD patterns (left) of the α-$LiFeO_2$ electrodes at the different states represented in the charge/discharge (right).

Figure 20 shows the *ex-situ* XRD patterns for the reaction products obtained in the charged (4.5 V) and discharged (1.5 V) states. The electrochemical process occurring at the electrode degrades the crystallinity of the components to an extent that the peaks for the impurities become barely distinguishable. Interestingly, the peaks for lithium ferrite retained their same position, as previously reported by Sakurai et al. [105] However, we found the $I_{(220)}/I_{(200)}$ ratio to be higher during charge (0.37) than at OCV (0.17). Moreover, the intensity ratio recovered its initial value, 0.17, during the discharge. These results are suggestive of a structural rearrangement upon lithium removal. Similar results were reported for a $Li_{4/3}Ti_{2/3}O_2$–$LiFeO_2$ solid solution prepared by Tabuchi et al. [107]. In this work, the authors proposed that Fe^{IV} atoms formed during reaction (5) are displaced from octahedral *4a* sites to tetrahedral *8c* positions, and also that lithium ions must use the *8c* sites as a conduction pathway, similarly to Na^+ and Ag^+ in the cubic rock-salt structures NaCl and AgCl [111]. Therefore, iron (IV) must hinder lithium diffusion during charge, which accounts for the disparate profiles of the charge–discharge curves. However, other authors believe that no clear free spaces exist for lithium diffusion as the metal ions are randomly arranged in the α-$LiFeO_2$ structure [105]. A small plateau at 4.2V was observed during the second charge (Figure 20), the width of which decreased on cycling until completely vanishing at the fourth cycle. Then, there was substantial profile retention in the charge curves.

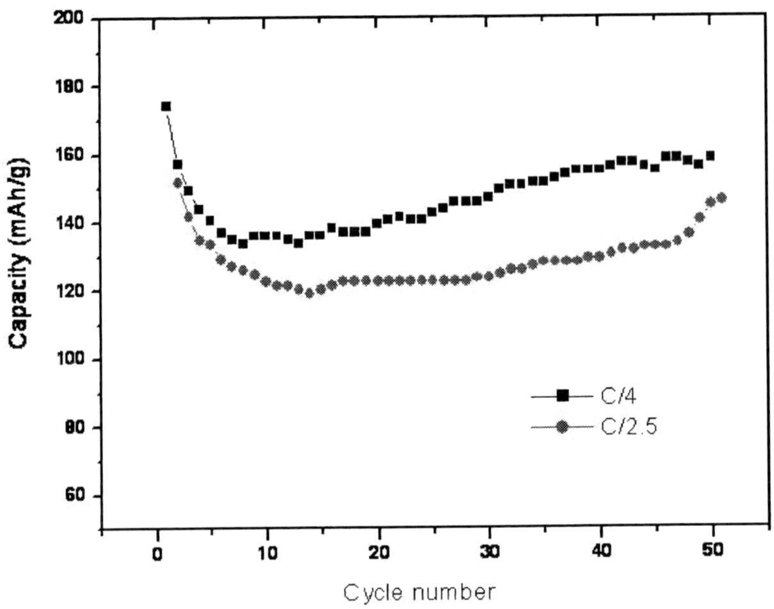

Figure 21. Capacity evolution as a function of cycle number for cells based on α-$LiFeO_2$ under different regimes.

Figure 21 shows the variation of the capacity as a function of the number of cycles at C/4 and C/2.5. The cells, which were cycled between 4.5–1.5 V, exhibited a similar trend, namely a capacity drop in the first cycles, followed by a gradual increase on further cycling. Thus, the capacity increased to *ca.* 150 and 130 mAh·g^{-1}, at the 9th (C/4) and 14th (C/2.5) cycles, respectively. These values are much higher than those reported for other electroactive α-$LiFeO_2$ forms [102,105,106] (particularly those for the nanorods based cells, which provide only 80 mAh·g^{-1} at the 50th cycle [106]). The nanorods were 80 nm in diameter and 900 nm in length on average. Our nanocomposite consists of linear arranged individual nanocrystals sized of 50 nm on average. Therefore, we believe that the origin for the discrepancy between the electrochemical response of the nanorods prepared by Wang et al. [106] and our nanocomposites can be the smaller size of the latter. Moreover, the impurity phase detected by XRD and Mössbauer techniques, may play an additional role that is unclear at the moment. Thus, our nanocomposites possess good properties as positive electrodes for low voltage batteries. Their nanometric particle size has a favourable effect on the electrochemical performance. An increased surface area in the nanobelts may facilitate deintercalation and intercalation of lithium ions.

Nanospinels

Spinels have received increasing attention as anode materials for lithium batteries ever the discovery of the reversible reaction between transition metal binary oxides and lithium by Tarascon et al. [4,7]. For a typical formula AB_2O_4, where A and B are a divalent and trivalent ion, respectively, electrochemical reaction should be of the type:

$$AB_2O_4 + 8\ Li \rightarrow A + 2\ B + 4\ Li_2O \quad (6)$$

This reaction model has been validated for various ferrites and cobaltites (B = Fe, Co) [45,60-66,112-115]. The large amount of lithium involved in the reaction (6) results in very high capacities *e.g.* 3 times that of graphite, which is the most common negative electrode material. This ability and the low voltages where the reaction takes place are valuable properties for negative electrode applications. From an electrochemical point of view, and taking into account the capacity delivered during the first discharge and the reversibility of the reaction, some authors have correlated the spinel activity with that observed for an intimate mixture of two binary oxides AO + B_2O_3 [116]. Thus, the good properties found for CoO [4,7,55] led to the study of the cobalt spinel, Co_3O_4, the performance of which was found to be remarkably good [60–66]. There is, however, some exceptions, the most outstanding being magnetite Fe_3O_4 [117], the electrochemical properties of which in lithium cells are poorer than those of $Fe^{II}O$. Besides cobaltite, electrochemical studies have focused on spinels containing cobalt [112-114], nickel [112] or iron [112-114] in their composition.

The general method used to prepare spinels is ceramic; depending on the precursors and the transition metals involved, temperatures as high as 1000 °C can be required in order to obtain a phase free of unreacted binary metal oxides. Temperature is a critical factor influencing composition and other physico–chemical parameters of the end products. For example, the lithium and oxygen contents of spinels of formula $Li_xMn_2O_4$ [118,119] depends on the temperature used to obtain the material. Also, changes in particle size and morphology have been observed in several spinels when synthesis temperature was increased [118–122]. High temperatures promote particles growth and increase crystallinity. In order to obtain crystalline, pure, nanosized materials, some authors have prepared spinels via sol–gel [119,123,124] or hydrothermal methods [125,126] that use much lower temperatures to obtain pure phases.

Table V. Physicochemical parameters of the ferrites synthesized by a hydrothermal treatment.

Sample	*z* (Å)	*Rp, Rwp (%)*	M/Fe (XPS)
$MnFe_2O_4$	8.479	18.6, 13.5	0.54
$CoFe_2O_4$	8.388	5.34, 6.38	0.54

The use of surfactants as templates can influence the final particle size and the morphology of the compounds [127,128]. Template agents used for this purpose include cetyltrimethylammonium bromide (CTAB), sodium dodecylsulfate and polymers such as poly(ethylene glycol). The first compound is an ammonium quaternary salt with a long carbon chain that is generally used to prepare mesoporous materials of controlled pore size. Templates are also used to avoid particle growth. Thus, CTAB has been used to synthesize nanosized particles of cobalt ferrite. However, the electrochemical properties of these materials as negative electrodes in lithium batteries have not been examined [126].

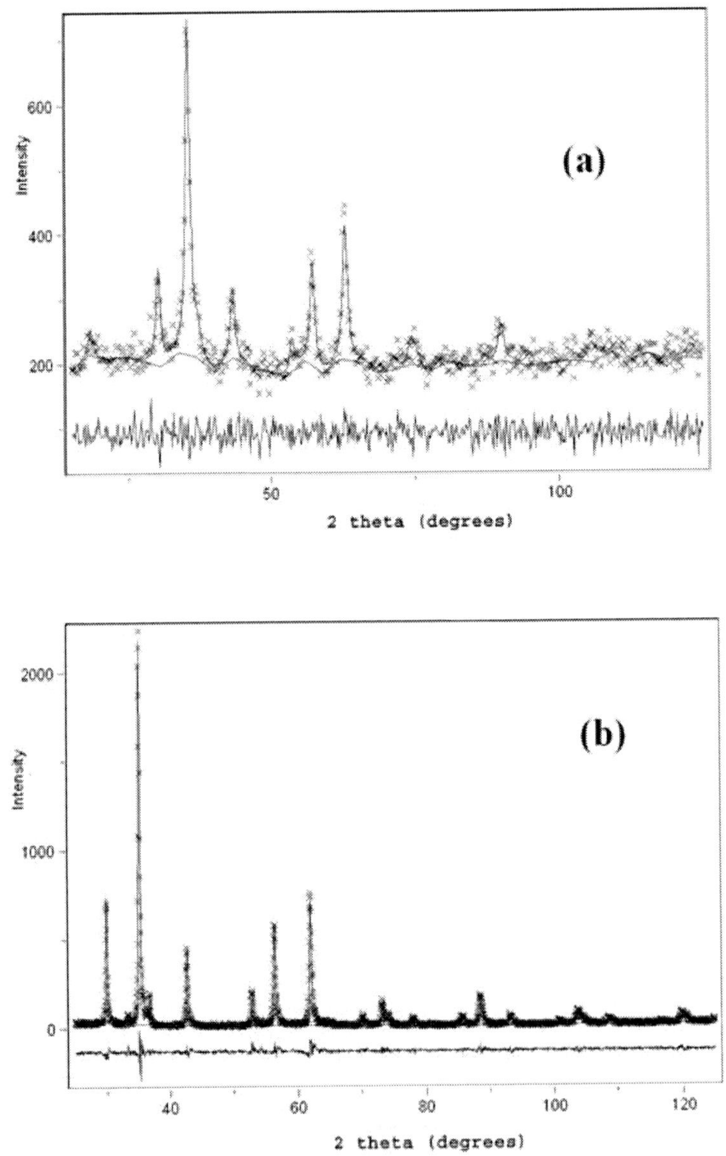

Figure 22. XRD patterns for nanosized $LiFePO_4$. Experimental (dotted), calculated (line) and the difference (bottom)

$CoFe_2O_4$ was prepared by using a hydrothermal method reported elsewhere [126]. Briefly, $FeCl_3{\cdot}6H_2O$ and $CoCl_2{\cdot}6H_2O$ were added to an aqueous solution containing CTAB. The suspension formed after adjusting the pH to 11 with sodium hydroxide and ultrasonic treatment, was transferred to an autoclave. The hydrothermal synthesis was carried out at 130 °C for 15 h. The black precipitate thus obtained was filtered and extensively washed with water. No CTAB was required to prepare nanosized $MnFe_2O_4$ [127]. $MnCl_2{\cdot}4H_2O$ and $FeCl_3{\cdot}6H_2O$ were dissolved in water and sodium hydroxide to the solution in order to rise the pH and facilitate the precipitation of a brown solid which was hydrothermally treated at 180

°C for 12 h. The final product was filtered and washed several times with water and ethanol. Prior to analysis, the cobalt and manganese ferrites were dried at 60 °C for 2 h.

XRD patterns were recorded over the 15–125 2θ range, using 8 s per step. The XRD patterns obtained are shown in Figure 22. The patterns only contain peaks corresponding to the spinel structure (inverse-type) and are consistent with others previously reported for these ferrites [126,127]. The peaks for $MnFe_2O_4$, which are stronger and sharper, reveal an increased crystallinity relative to $CoFe_2O_4$ by effect of the higher temperature used in its synthesis. XRD data were refined using the Rietveld method as implemented in the GSAS software suite [86]. The results are shown in Figure 23, and the calculated parameter values and cell dimensions in Table V. These values are consistent with those reported by Wang [127] and Ferreira [126,129], and reflect the good quality of the spinels.

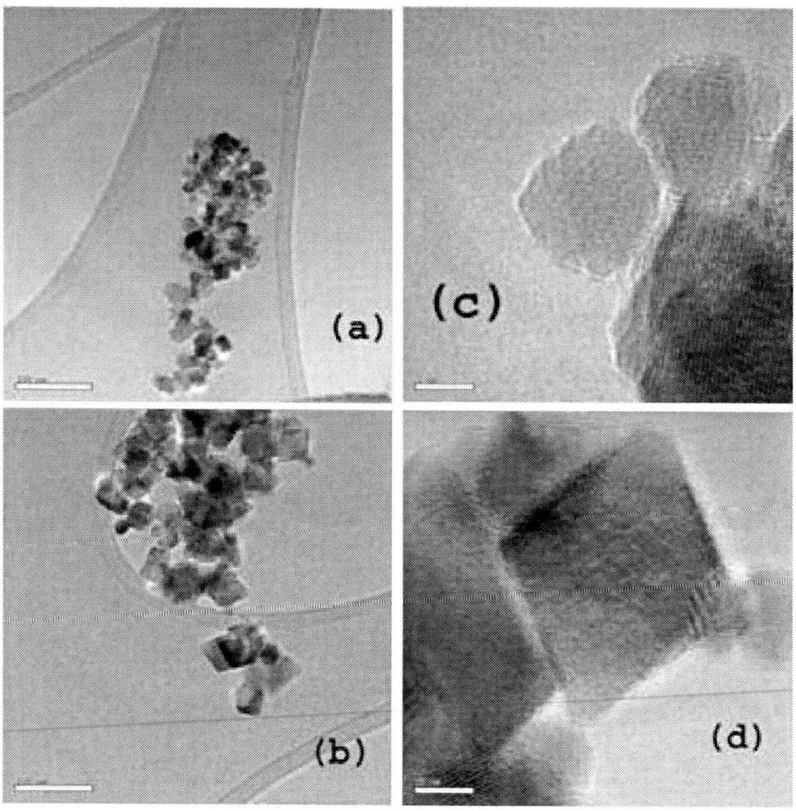

Figure 23. HRTEM images of (a), (c) $CoFe_2O_4$ and (b), (d) $MnFe_2O_4$ prepared in this work. Bar corresponds to (a),(b) 100 nm, (c) 5 nm and (d) 10 nm.

Figure 23 shows TEM images of the spinel particles. The Mn ferrite particles exhibit a pseudopolyhedral morphology typical of a spinel. By contrast, the particle shape of the Co ferrite is more rounded and has ill-defined edges. A similar morphology was found for the cobalt ferrite synthesized by Olsson et al. [121]. These differences are consistent with the disparate crystallinity of the materials as inferred from the XRD patterns and must be related to the thermal treatment rather than to the presence of CTAB which is otherwise used to improve crystallinity. Particle size was quite uniform in both cases, with an average value of

30 nm and 20 nm for the manganese and cobalt ferrite, respectively. Fringes 2.55 nm wide were observed at a higher resolution (Figs. 23c and d). This value is consistent with the spacing of the (311) planes All these features confirm that the hydrothermal procedure is an effective choice for obtaining highly homogenous, pure, crystalline nanoparticles.

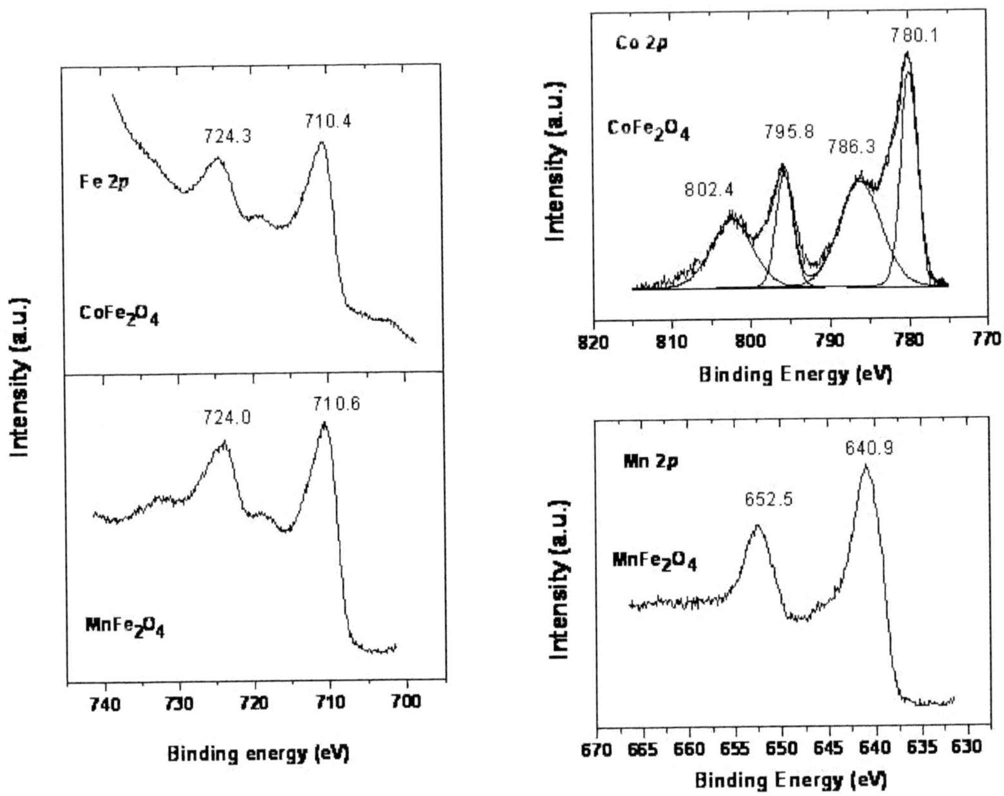

Figure 24. XPS spectra of Co *2p*, Mn *2p* and Fe 2*p* for the nanoferrites prepared in this work.

Supplementary information about the oxidation states of the elements in the spinels was obtained from the high resolution XPS spectra for Co *2p*, Mn *2p* and Fe *2p* in Figure 24. The M/Fe atomic ratios at the surface level, Table V, are consistent with the expected stoichiometry, based on which, Mn and Co must be divalent and Fe trivalent. These oxidation states were confirmed by the binding energies calculated from the spectra. The complex profile for the Co 2p spectrum (Figure 24) was fitted to two components. The first component was assigned Co^{+2} species generating peaks at 780.1 and 795.8 eV. The doublet separation (DS), 15.7 eV, virtually coincides with those for other compounds containing divalent cobalt (15.6 eV) [130]. The second component was resolved into two peaks with higher binding energy peaks (786.3 and 802.38 eV) and assigned to cobalt shake-up satellites [130]. The Mn *2p* profile was somewhat more simple, with peaks at 640.9 and 652.5 eV; these are consitent with reported values for MnO [130] and markedly lower than those reported for other Mn spinels such as $LiNi_{0.5}Mn_{1.5}O_4$ (BE $2p_{3/2}$ 642.4 eV), where the element is present in a higher oxidation state [23]. Figure 24 also shows the Fe 2*p* profiles. The spectrum is quite similar for the two spinels, which suggests that Fe atoms are in a similar chemical environment. The two peaks at 710.5 eV and 724.3 eV correspond to Fe $2p_{3/2}$ and $2p_{1/2}$ core levels, respectively. A shake-up satellite can also be seen at 719 eV. These characteristics are consistent with the presence of Fe^{3+} as the main component [130].

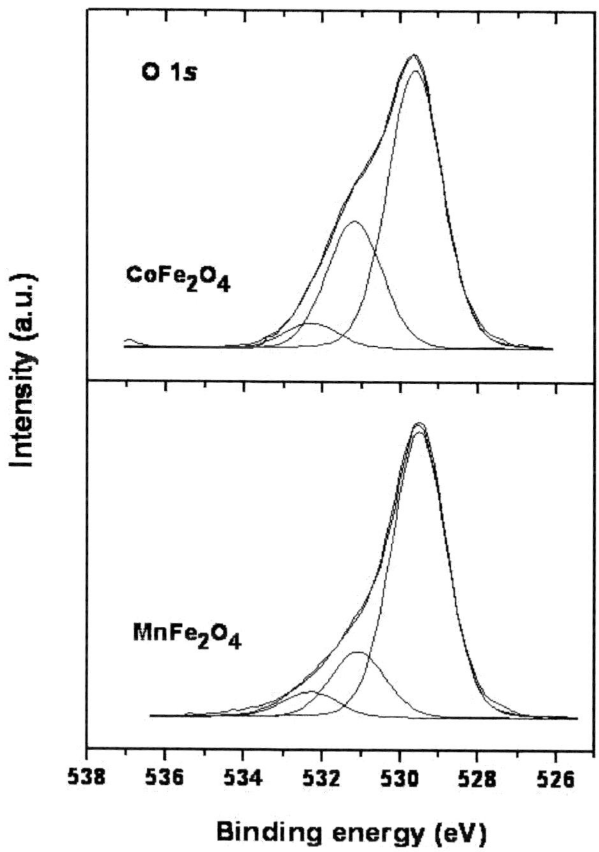

Figure 25. O *1s* XPS spectra in the nanoferrites prepared in this work.

The O 1s spectrum exhibited a complex profile (Figure 25) with a major component centred at 529.6–529.8 eV that was assigned to M–O bonds. The components at higher binding energies, of lower intensity, are typically associated to either OH^- groups, O^{2-} or the multiplicity of physisorbed and chemisorbed water on and into the surface [131]. These latter signals were stronger for the cobalt ferrite. Its lower crystallinity and smaller particle may have resulted in greater ease of hydration. The two ferrites also contained carbon species that gave peaks at 284.8 (adventitious carbon), and 286.2 and 288.5 eV (carboxyl groups) in the C1*s* spectra. The peaks for the Co ferrite can be assigned to the presence of traces of the organic compound. Figure 26 shows the FTIR spectrum recorded over the 4000–400 cm^{-1} range. The broad band between 3000 and 3700 cm^{-1}, and the peak at 1627 cm^{-1} can be assigned to water. The peak at 1051 cm^{-1} testifies to the presence of carboxyl groups. The presence of CTAB is confirmed by the weak peaks at 2923 and 2855 cm^{-1}, which correspond to asymmetric and symmetric stretching vibrations of C–CH_2 bonds in the methylene chains. These peaks were not observed in the $MnFe_2O_4$ system and disappeared on calcining the $CoFe_2O_4$ spinel at 800 °C.

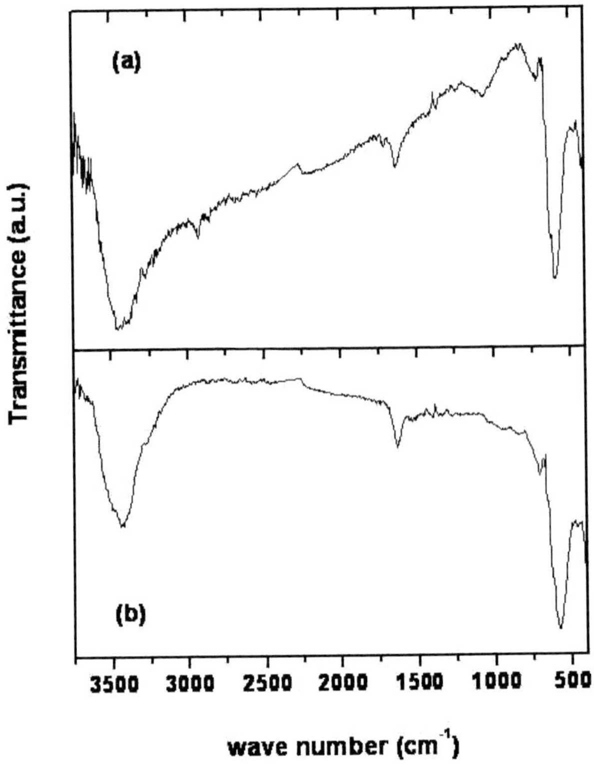

Figure 26. FTIR spectra of (a) as prepared $CoFe_2O_4$, (b) this sample heated at 800°C.

The electrode was prepared from a mixture of the spinel with carbon black and Teflon in a 85:10:5 weight proportion. Galvanostatic cycling was done by using coin cells with lithium foil as reference and counter electrode, and 1M $LiPF_6$ in EC, DEC as electrolyte. Galvanostatic discharge was done at 1.6C, which is a fast regime. Cells were monitored with a battery testing equipment supplied by Arbin.

Figure 27 shows the first galvanostatic discharge/charge curves. There is some similarity in the shape of the first discharge curve, particularly as regards the presence of an extended plateau around 0.7 V. Above this potential, the voltage drop occurs in at least two steps defined by very short pseudoplateaux. Therefore, the curve shape, and hence the electrochemical reaction with lithium, are mainly governed by Fe^{3+}, which is present in both spinels. Below 0.7, the voltage decreased steadily. If one assumes that, below 0.5 V, the main electrochemical reaction is the electrolyte decomposition [7,55,60–66,112–114], the capacity delivered by the spinels can be calculated to be 750 $mAh{\cdot}g^{-1}$ for the Co and 1050 $mAh{\cdot}g^{-1}$ for the Mn ferrite, respectively. These capacities are equivalent to 6.6 and 9.2 mole of lithium per mole of compound, respectively. These values differ from the 8 Li atoms calculated for the reaction

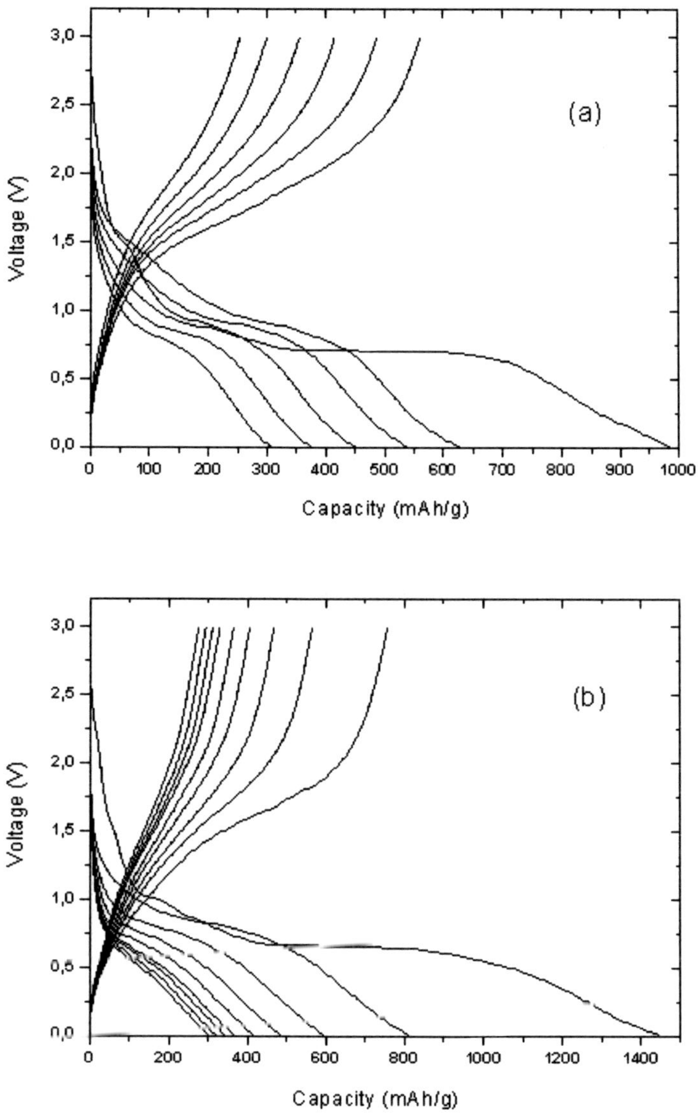

Figure 27. First discharge/charge curves for cells based on (a) $CoFe_2O_4$, (b) $MnFe_2O_4$ under a C/5 regime.

$$MFe_2O_4 + 8\ Li \rightarrow M + 2Fe + 4\ Li_2O\ (M{=}Mn,Co) \tag{7}$$

As stated above, these spinels (particularly $CoFe_2O_4$) have received special attention for use in lithium cells. At least two recent papers have been devoted to the electrochemical properties of $CoFe_2O_4$ nanocrystalline thin films for lithium ion batteries that were prepared in two different manners [112,114]. Although the deposits were identical in nature, the discharge curves exhibited significant differences in both shape and delivered capacity, the latter shifting from 4 to 8.4 Li atoms per mole of compound. This difference simply reflects that differences in variables such as the experimental conditions used to perform the electrochemical measurement can influence the reactivity towards lithium. The origin of the increased reactivity of the Mn ferrite is unclear. In fact, the only report available on manganese oxides [55,116] indicates that Mn(II) or Mn(III) are scarcely reduced to Mn(0). This contradicts the results of Hara et al. [132], who claimed that Mn(II) in a $Mn_{0.6}Mo_{0.8}V_{1.2}O_6$ compound with a brannerite structure was fully reduced under a C/5 regime.

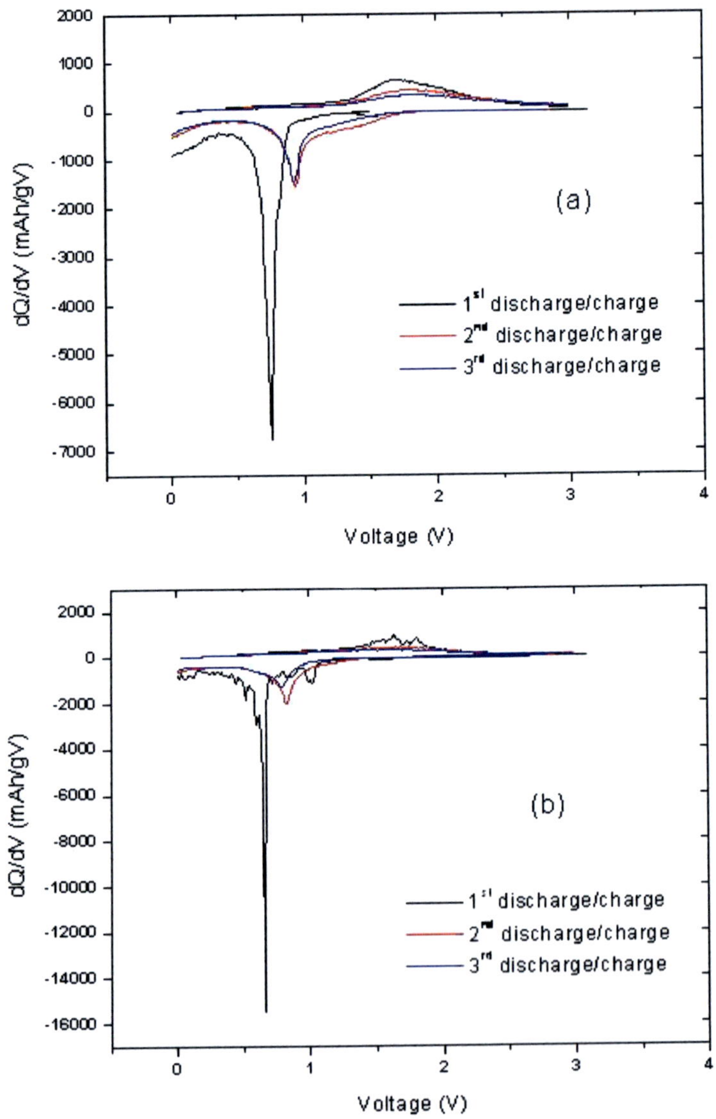

Figure 28. First derivative of the discharge/charge curves of cell based on (a) $CoFe_2O_4$, (b) $MnFe_2O_4$.

Figure 28 shows the differential capacity plots obtained from the galvanostatic curves. The most salient feature of the first discharge is a strong peak at *ca.* 0.8 V resulting from the above-described extended plateau. The other peaks are considerably weaker and may account for the different steps in the galvanostatic curve.

On charging, the metal nanoparticles formed are believed to be reoxidized according to the following reaction

$$M + Li_2O \rightarrow MO + 2Li^+ + 2e^- \quad (8)$$

$$Fe + 3\,Li_2O \rightarrow Fe_2O_3 + 6\,Li^+ + 6\,e^- \quad (9)$$

Neither reconstruction of the spinel structure nor reversibility of reaction (7) has yet been demonstrated. This contradicts the reported reversibility of the reaction $Co_3O_4 + 8\,Li \rightarrow 3Co + 4\,Li_2O$ [45,60–64,66], but some authors claim that reoxydation of cobalt particles stops at CoO [65]. Experimentally, we observe a high polarization of the cells that resulted in the removal of an amount of Li much smaller than that calculated for the discharge process. This is also reflected in the differential capacity plots, which exhibit a weak, broad signal. These data clearly expose a significant capacity loss from the first to the second cycle. As stated above, secondary reactions can take place during the discharge process. This is well documented for this type of electrode and must chiefly affect the low voltage region. Capacities above 400 mAh·g^{-1} are calculated for the 0.8–0.0 V voltage range that can be assigned to the electrolyte solvent reduction. The product of the electrolyte reduction is a solid electrolyte interface (SEI) that hinders electronic diffusion in the electrode. This film surrounds the electrode particles and acts as an electronic barrier, thus decreasing the performance of the electrochemical reactions. However, several authors suggested that the barrier electrochemical dissolution may undergo by effect of charging [7,45,55–57,61–63] and generate additional capacity increasing that resulting from reactions (8) and (9). Therefore, the above-described charge profile might include a contribution associated to dissolution of the SEI.

The discharge profiles changed dramatically after the first cycle (Figure 27). The most salient feature of the second and subsequent discharge profiles was the presence of a pseudo-plateau at 0.9 V the width of which was found to depend on the pristine spinel. The corresponding first derivatives are shown in Figure 28. The $MnFe_2O_4$ based electrode exhibits a weak peak at 0.83V, whereas the $CoFe_2O_4$ electrode exhibits a broad peak centred at 1.45 V and a somewhat stronger one at 0.94 V. Furthermore, the second charge profiles indicate that reoxidation of M and iron particles is difficult (especially for the Mn–Fe–O system), so the capacity recovered during the charge is lower than in the first cycle.

The cycling properties of the electrodes were examined under a 1.6C regime and are shown in Fig. 29. The variation of the discharge capacities with the number of cycles reflects the high interest of these materials as electrodes for lithium ion batteries. Thus, the capacities delivered in the first discharge were almost four times greater than that of graphite. However, the capacity continuously faded on cycling, in part probably as a result of the presence of a significant amount of SEI when the lower voltage used was 0.0 V. In any case, every cell provided capacities close to 200 mAh·g^{-1} after 25 cycles.

Subsequent research was focused on the cobalt ferrite, the discharge profile of which exhibited a lower capacity by effect of the electrolyte reduction. In order to improve the performance of this material, we tried to circumvent the major drawback associated to the presence of the SEI layer. Taking into account that SEI is generally assumed to form at low working voltages and its development to be promoted by the nanosized nature of electrode materials, we projected an improvement in performance by using three different approaches, namely: (i) rising the lower limit voltage to 0.5 V, (ii) adding a stabilizing agent to the electrolyte solvent in order to hinder its reduction and (iii) increasing particle size by heating under air at 800 °C.

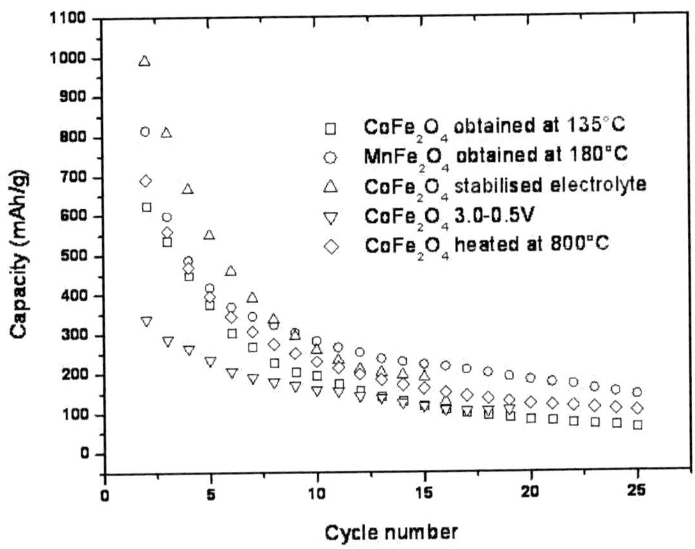

Figure 29. Capacity variation as a function of the cycle number for several cells based on cobalt and manganese ferrites.

Approach (iii) may be very interesting as it may expose the influence of particle size and morphology on cell performance. The TEM images of Fig. 30 clearly reveal that the particles undergo sintering and that this results in an increase in size that triplicates that for the pristine particles. Also, particles retain a high crystallinity as revealed by the regular fringes of Fig. 30 b, which are associated to the (311) planes. The discharge/charge profile for this material was similar to that obtained for pristine $CoFe_2O_4$; however, the capacity values of the cell were somewhat higher. In any case, the cycling properties (Fig. 29) continued to suffer. This continuous capacity decrease is not surprising since the textural properties of the reduced particles after the first discharge can be quite similar irrespective of the initial crystallinity or size of the particles. However, there is an interesting observation: the influence of particle size on the spinel reactivity. This result contradicts the generally accepted statement that firing reduced particles surface areas and the reactivity as a result. However, there are some reports on the beneficial influence of an initial crystallized state of the materials on their electrochemical performance [56,61,62]. In any case both cobalt ferrites behave similarly after a few cycles.

A stabilizing agent was added to the electrolyte solvent and the resulting cell cycled (Fig. 29). The beneficial effect of the additive was obvious in the first cycles: the cell delivered 400 and 300 $mAh \cdot g^{-1}$, respectively, more than the pristine material in thc 2nd and 3rd cycles. However, no effect was observed after ten cycles. On the other hand, limiting the lower voltage to 0.5 V (Fig. 29) resulted in better capacity retention at the expense of lower capacity values. In fact, only 300 $mAh \cdot g^{-1}$ were recovered after the second cycle. Both studies clearly underline the influence of the SEI formation on the cycling efficiency of these cells. Finally, we explored the response of the cobalt ferrite cell when exposed to a higher temperature (*e.g.* 55 °C). Some authors claim that raising the temperature favours the nanoparticle-driven SEI growth process. If dissolution of the SEI is reversible, then the use of temperatures above room level (25 °C) can be useful with a view to improving performance in ferrite based cells.

However, we have earlier shown that this is not the case with our nanoferrites. Thus, in cells tested at 55 °C exhibited substantial capacity fading after the first cycle.

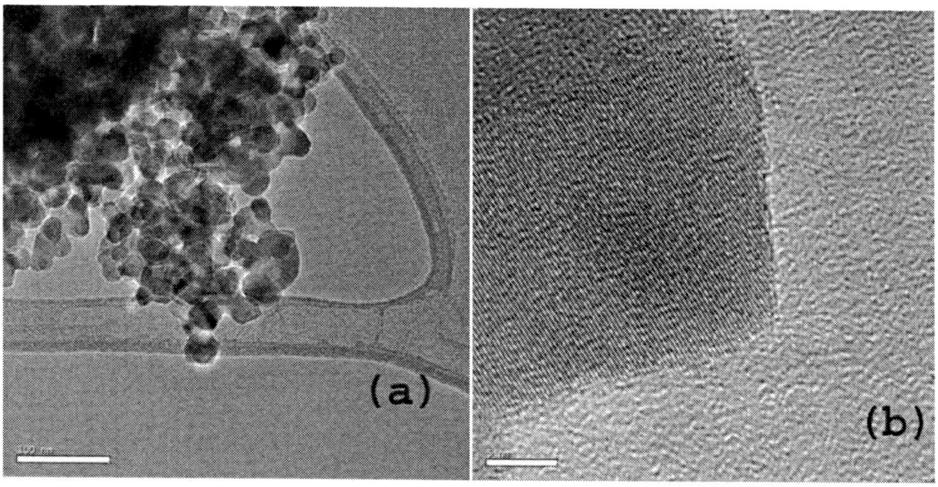

Figure 30. HRTEM images of $CoFe_2O_4$ obtained at 800°C.

In conclusion, the electrochemical performance of nanoferrites is extremely dependent on SEI growth. Despite the high first capacity value provided during the discharge, irreversible dissolution of the SEI results in continuous capacity fading on cycling. Other factors including particles shape, porosity, dimensionality, the presence of defects, particle orientation and certain cell components [4,60,61,66] are also influential on the capacities delivered in the first few cycles.

Conclusions

The advantages of nanomaterials relative to microsized materials can be summarized as follows:

1. A decreased path length for the transport of electrons and lithium ions, which results in faster kinetics of lithium insertion/deinsertion and hence in increased battery power.
2. Higher electrode/electrolyte contact areas, which lead to increased reactivity of the material towards lithium and are beneficial to cell capacity.
3. Better electrode integrity retention upon cycling by effect of better product accommodation (particularly for anode-based materials).
4. Promotion of new reactions.

However, there still exist a large number of other problems to be solved. Thus, electrodes of nanosized materials provide a much larger interface between the solid electrode and the liquid electrolyte. In general, increased interface areas are advantageous as regards the overall current that can be drawn relative to small interfaces. Thus, increased surface areas may have an effect on the amount of lithium that can be reversibly inserted into the electrodes. Also related to the problem of irreversibility is the formation of a Solid-Electrolyte-Interface (SEI), which results in an irreversible capacity and in decreased transfer of lithium ions and electrons across the electrolyte/electrode interface. It remains be determined whether the irreversible capacity loss due to SEI formation is related to the composition and structure of the electrode materials as well as to particle size. Also, electrical and thermal conductivities are two extremely important issues if rapid recharging is the goal. It is of paramount importance that each small particle involved in the intercalation/deintercalation reaction should maintain good electrical contact. Ocassionally, this situation is problematic (*e.g.* when a volume change occurs during the insertion reaction). This conductivity can limit the amount of nanomaterial that is active in the electrochemical process. Much effort must be made with in order to avoid these potential problems with a view to developing new applications for high stored energy Li ion batteries with high recharge rates, as well as to improving devices currently operating on batteries. Detailed studies on different materials in various sizes may provide effective solutions to these problems.

Acknowledgements

This work was supported by CICyT (MAT2005-03069) and Junta de Andalucía (Group FQM 175). The authors acknowledge the help of Prof. R. H. Herber and Dr. I. Nowik (Racah Institute of Physics, The Hebrew University of Jerusalem, Israel), for recording and discussing the Mössbauer spectra; Dr. A. Caballero-Amores for HRTEM micrographs of $LiFePO_4$; and Dr. M. Cruz-Yusta and B. Sc. J. C. Arrebola-Haro for the Rietveld refinements. JSP is also grateful to Junta de Andalucía (Spain) for inclusion in its Researcher Return Program.

References

[1] Scrosati, B. *Nature* 1995, 373, 557-558.

[2] van Schalkwijk,W.A.; Scrosati,B. (Eds.), *Advances in Lithium Ion Batteries*, Kluwer Academic Publishers: New York, 2002.

[3] Yang, J.; Winter, M.;Besenhard, J.O. *Solid State Ionics* 1996, 90, 281-287.

[4] Poizot, P.;Laruelle,S.;Grugeon,S.; Dupont, L.; Tarascon, J.M. *Nature* 2000, 407, 496-499.

[5] Tarascon, J.M.;Armand, M. *Nature* 2001, 414, 359-367.

[6] Graetz,J.;Ahn,C.C.;Yazami,R.;Fultz,B. *Electrochem. Solid State Lett.* 2003, 6, A194-A197.

[7] Poizot,P.;Laruelle,S.;Grugeon,S.;Dupont,L.;Tarascon,J.M. *J. Power Sourc.* 2001, 97-98, 235-239.

[8] Larcher,D.;Masquelier,C.;Bonnin,D.;Chabre,Y.;Masson,V.;Leriche,J.B.;Tarascon, J.M. *J. Electrochem. Soc.* 2003, 150, A133-A139; Larcher, D.;Bonnin,D.;Cortes,R.; Rivals,L.;Personnaz,L.;Tarascon,J.M. *J. Electrochem. Soc.* 2003, 150, A1643-A1650.

[9] Chen, J.;Xu,L.;Li,W.;Gou,X. *Adv. Mater.* 2005, 17, 582-586.

[10] Sides, C.R. ; Li, N. ; Patrissi, C.J. ; Scrosati, B. ; Martin, C.R. *MRS Bull.* 2002, 27, 604-607.
[11] Amatucci,G.G.;Badway,F.;Singhal,A.;Beaudoin,B.; Skandan,G.;Bowmer,T.;Plitz,I.; Periera,N.;Chapman,T.;Jawoski, R. *J. Electrochem. Soc.* 2001, 148, A940-A950.
[12] Huang,S.Y.;Kavan,L.;Exnar,I.;Grätzel,M. *J. Electrochem. Soc.* 1995, 142, L142-L144.
[13] Schoonman,J. *Solid State Ionics* 2000, 135, 5-19; Schoonman, J. *Solid State Ionics* 2003, 157, 1-4.
[14] Singhal,A.;Skandan,G.;Amatucci,G.;Badway,F.;Ye,N.;Manthiram,A.;Ye,H.;Xu,J.J. *J. Power Sourc.* 2004, 129, 38-44.
[15] Jamnik, J.; Maier, J. *Phys. Chem. Chem. Phys.* 2003, 5, 5215-5220.
[16] Cho,J.;Kim,Y.J.;Kim,T.J.;Park,B. *Angew. Chem. Int. Ed.* 2001, 40, 3367-3369.
[17] Chen,H.; Qiu,X.;Zhu,W.;Hagenmuller,P. *Electrochem. Comm.* 2002, 4, 488-491.
[18] Balaya,P.;Li,H.;Kienle,L.;Maier,J. *Adv. Func. Mater.* 2003, 13(8), 621-625; Balaya, P.; Bhattacharyya, A.J.;Jamnik,J.;Zhukovskii,Y.F.;Kotomin,E.A.;Maier,J. *J. Power Sourc.* 2006, 159, 171-178.
[19] Padhi,A.K.;Nanjundaswany,K.S.;Masquelier,C.;Okada,S.;Goodenough,J.B. *J. Electrochem. Soc.* 1997, 144, 1609-1613.
[20] Franger, S.;Le Cras,F.;Bourbon,C.;Rouault,H. *J. Power Sourc.* 2003, 252, 119-121; Franger,S.;Le Cras,F.;Bourbon,C.;Rouault,H. *Electrochem. Solid-State Lett.* 2002, 5, A231-A233; Franger,S.;Bourbon,C.;LeCras,F. *J. Electrochem. Soc.* 2004, 151, A1024-A1027.
[21] Panero,S.;Scrosati,B.;Wachtler,M.;Croce,F. *J. Power Sourc.* 2004, 129, 90-95.
[22] Fu,L.J.;Liu,H.;Li,C.;Wu,Y.P.;Rahm,E.;Holze,R.;Wu,H.Q. *Prog. Mater. Sci.* 2005, 50, 881-928.
[23] Caballero,A.;Cruz,M.;Hernán,L.;Melero,M.;Morales,J.;Rodríguez-Castellón,E. *J. Power Sourc.* 2005, 150, 192-201.
[24] Sides,C.R.;Martin,C.R. *Adv. Mater.* 2005, 17, 125-128.
[25] Patrissi,C.J.;Martin,C. R. *J. Electrochem. Soc.* 1999, 146, 3176-3180.
[26] Sun,D.;Kwon,C.W.;Baure,G.; Richamn,E.;MacLean,J.;Dunn,B.;Tolbert,S.H. *Adv. Funct. Mater.* 2004, 14(12), 1197-1204.
[27] Wei,M.;Sugihara,H.;Honma,I.;Ichikara,M.;Zhou,H. *Adv. Mater.* 2005, 17, 2964-2969.
[28] Shen,C.M.;Zhang,X.G.;Zhou,Y.K.;Li,H.L. *Mater. Chem. Phys.* 2003, 78, 437-441.
[29] Franger,S.;Le Cras,F.;Bourbon,C.;Benoit,C.;Soudan,P.;Santos-Peña,J. in: S.G. Pandalai (Ed.), *Recent Research Development in Electrochemistry,* Transworld Research Network Ed.: Kerala (India), 2005, pp.225-256.
[30] Chen,J.;Tao,Z.L.;Li,S.L. *Angew. Chem. Int. Ed.* 2003, 42, 2147-2153.
[31] Dominko,R.;Arcon,D.;Mzrel,A.;Zorko,A.; Cevc,P .;Venturini,P.;Goberscek,M.; Remskar,M.;Mihaiolovic,D. *Adv. Mater.* 2002, 14, 1531-1534.
[32] Wang,G.X.;Bewlay,S.;Yao,J.;Liu,H.K.;Dou,S.X. *Electrochem. Solid-State Lett.* 2004, 7, A321-A323.
[33] Natarajan,C.;Setoguchi,K.;Nogami,G. *Electrochim. Acta* 1998, 43, 3371-3374.
[34] Armstrong,A.R.;Armstrong,G.;Canales,J.;García,R.;Bruce,P.G. *Adv. Mater.* 2005, 17, 862-865.
[35] Wilson,A. M.;Way, B.M.;Dahn, J. R. *J. Appl. Phys.* 1995, 77, 2363-2369.
[36] Li, H.;Huang, X.;Chen, L.;Wu,Z.;Liang,Y. *Electrochem. Solid-State Lett.* 1999, 2, 547-549.

[37] Li,H.;Huang,H.;Chen,L.;Zhou,G.;Zhang,Z.;Yu,D.;Mo,Y.J.;Pei,N. *Solid State Ionics* 2000, 135,181-191.
[38] Holzapfel,M.;Buqa,H.;Hardwick,L.J.;Hahn,M.;Würsig,A.;Scheifele,W.;Novák,P.; Kötz,R.;Veit,C.;Petrat,F.M. *Electrochim. Acta* 2006, 52, 973-978.
[39] Yang,J.;Wang,B.F.;Wang,K.;Liu,Y.;Xie,J.Y.;Wen,Z.S. *Electrochem. Solid-State Lett.* 2003, 6, A154-A156.
[40] Lee,H.Y.;Lee,S.-M. *Electrochem. Comm.* 2004, 6, 465-469.
[41] Holzapfel, M.;Buqa,H.;Scheifele,W.;Novák,P.;Petrat,F.-M. *Chem. Comm.* 2005, 1566-1568.
[42] Wang,C.;Appleby,A.J.;Little,F.E. *J. Power Sourc.* 2001, 93(1-2), 174-185.
[43] Chandra-Bose, A.; Kalpana,D.;Thangadurai,P.;Ramasamy, S. *J. Power Sourc.* 2002, 107, 138-141.
[44] Santos-Peña,J.;Brousse,T.;Schleich,D.M. *Solid State Ionics* 2000, 135, 87-93; Balan,L.; Ghanbaja, J.;Willmann,P.;Billaud,D. *Carbon* 2005, 43, 2311-2316.
[45] Pralong,V.;Leriche,J.B.;Beaudoin,B.;Naudin, E. ;Morcrette,M.;Tarascon,J.M. *Solid State Ionics* 2004, 166, 295-305.
[46] Ren,J.;He,X.;Pu,W.;Jiang,C.;Wan,C. *Electrochim. Acta* 2006, 52, 1538-1541.
[47] Xie,J.;Zhao,X.B.;Cao,G.S.;Su,S.F. *J. Electrochem. Soc.* 2005, 152, A601-A606.
[48] Crosnier,O.;Brousse,T.;Devaux,X.;Fragnaud,P.;Schleich,D.M. *J. Power Sourc.* 2001, 2, 169-174.
[49] Wolfenstine,J.;Campos,S.;Foster,D.;Read,J.;Bel,W.K. *J. Power Sourc.* 2002, 109, 230-233.
[50] Che,G.;Lakshmi,B.B.;Fisher,E.R.;Martin,C.R. *Nature* 1998, 93, 346-349.
[51] Frackowiak,E.;Gaucher,H.;Bonnamy,S.;Béguin,F. *Carbon* 1999, 37, 61-69.
[52] Shimoda,H.;Gao,B.;Tang,X.P.;Kleinhammes,A.;Flemming,L.;Wu,Y.;Zhou,O. *Phys. Rev. Lett.* 2002, 88, 155021-155024.
[53] Xu,J.J;Jain,G. *Electrochem. Solid-State Lett.* 2003, 6, A190-A193.
[54] Chen,J.;Xu,L.;Li,W.;Gou, X. *Adv. Mater.* 2005, 17, 582-586.
[55] Poizot,P.;Laruelle,S.;Grugeon,S.;Tarascon,J.M. *J. Electrochem. Soc.* 2002, 149, A1212-A1217.
[56] Grugeon,S.;Laruelle,S.;Herrera-Urbina, R. ; Dupont, L.; Poizot, P.; Tarascon, J.M. *J. Electrochem. Soc.* 2001, 148, A285-A292.
[57] Debart,A.;Dupont, L.;Poizot,P.;Leriche,J.B.;Tarascon,J.M. *J. Electrochem. Soc.* 2001, 148, A1266-A1274.
[58] Morales,J.;Sánchez,L.;Martín,F.;Ramos-Barrado,J.R. *Thin Solid Films* 2005, 474, 133-140.
[59] Morales,J.;Sánchez,L.;Bijani,S.;Gabás,M.;Martínez,L.;Ramos-Barrado,J. *Electrochem. Solid-State Lett.* 2005, 8, A159-A162.
[60] Wang,G.X.;Chen,Y.;Konstantinov,K.;Lindsay,M.;Liu,H.K.;Dou,S.X. *J. Power Sourc.* 2002, 109, 142-147; Wang, G.X.; Chen, Y.;Konstantinov, K.;Yao, J.;Ahn, J.H.;Liu, H.K.;Dou,S.X. *J. Alloy Compd.* 2002, 340, L5-L10.
[61] Larcher,D.:Sudant,G.;Leriche,J-B.;Chabre,Y.;Tarascon,J.M. *J. Electrochem. Soc.* 2002. 149, A234-A241.
[62] Badway,F.;Plitz,I.;Grugeon,S.;Laruelle,S.;Dollé,M.;Gozdz,A.S.;Tarascon,J.M. *Electrochem. Solid-State Lett.* 2002, 5, A115-A118.

[63] Laruelle,S.;Grugeon,S.;Poizot,P.;Dollé,M.;Dupont,L.;Tarascon,J.M. *J. Electrochem. Soc.* 2002, 149, A627-A634.
[64] Yuan, Z.;Huang,F.;Fena,C.;Sun,J.;Zhou,Y. *Mater. Chem. Phys.* 2003, 79,1-4.
[65] Fu,Z.W.;Wang,Y.;Zhang,Y.;Qin,Q-Z. *Solid State Ionics* 2004, 170,105-109.
[66] Li,W.Y.;Xu,L.N.;Chen,J. *Adv. Funct. Mater.* 2005, 15, 851-857.
[67] Andersson,A.S.;Kalska,B.;Häggström,L.;Thomas,J.O. *Solid State Ionics* 2000, 130, 41-52.
[68] Burba,C.M.;Frech,R. *J. Electrochem. Soc.* 2004, 151, A1032-A1038.
[69] Srinivasan,V.;Newman,J. *J. Electrochem. Soc.* 2004, 15, A1517-A1529; Srinivasan,V.;Newman,J. *Electrochem. Solid-State Lett.* 2006, 9, A110-A114.
[70] Ravet,N.;Chouinard,Y.;Magnan,J.F.;Berver,S.;Gauthier,M.;Armand,M. *J. Power Sourc.* 2001, 97-98, 503-507.
[71] Huang,H.;Yin,S.C.;Nazar,L.F. *Electrochem. Solid-State Lett.* 2001, 4, A170-A172.
[72] Myung,S.T.;Komaba,S.;Hirosaki,N.;Yashiro,H.;Kumagai,N. *Electrochim. Acta* 2004, 29, 4213-4222.
[73] Chung,H.T.;Jang,S.K.;Ryu,H.W.;Shim,K.B. *Solid State Comm.* 2004, 131,549-554.
[74] Belharouak,I.;Johnson,C.;Amine,K. *Electrochem. Comm.* 2005, 7, 983-988.
[75] Amine,K.;Liu,J.;Belharouak,I. *Electrochem. Comm.* 2005, 7, 667-673.
[76] Sides,C.R.;Croce,F.;Young,V.Y.;Martin,C.R.;Scrosati,B. *Electrochem. Solid-State Lett.* 2005, 8, A484-A487.
[77] Liao,X.Z.;Ma,Z.F.;He,Y.S.;Zhang,X.M.;Wang, L.;Jiang,Y. *J. Electrochem. Soc.* 2005, 152, A1969-A1973.
[78] Croce,F.;D'Epifanio,A.;Hassoun, J.;Deptula,A.;Olczac,T.;Scrosati,B. *Electrochem. Solid-State Lett.* 2002, 5, A47-A50.
[79] Caballero,A.;Cruz-Yusta,M.;Morales, J.;Santos-Peña, J.;Rodríguez-Castellón,E. *Eur. J. Inorg. Chem.* 2006, 9,1758-1764.
[80] Wang,G.X.;Yang,L.;Chen,Y.;Wang,J. Z.;Bewlay, S.;Liu,H. K. *Electrochim Acta* 2005, 50, 4649-4654.
[81] Yamada,A.;Chung,S.C.;Hinokuma,K. *J. Electrochem. Soc.* 2001, 148, A224-A229.
[82] Prosini,P.P.;Carewska,M.;Scaccia,S.;Wisniewski,P.;Pasquali,M. *Electrochim. Acta* 2003, 48, 4205-41211.
[83] Hsu,K.F.;Tsay,S.Y.;Hwang,B.J. *J. Mater. Chem.* 2004,14, 2690-2695.
[84] Ye,X.R.;Jia,D.Z.;Yu,J.Q.;Xin,X.Q.;Xue,Z. *Adv.Mater.* 1999, 11, 941-942.
[85] Haber,J.A.;Gunda,N.V.; Buhro,W.E. *J. Aerosol Sci.* 1998, 29, 637-645.
[86] Larson,A.C.;Von Dreele,R.B. *Los Alamos National Lab. Rep.* No. LA-UR-86, 1994, 748.
[87] Dokko,K.;Shiraishi,K.;Kanamura, K. *J. Electrochem. Soc.* 2005, 152, A2199-2202.
[88] Cano,E.;Torres,C.L.;Bastidas,J. M. *Mater. Corros.* 2001, 52, 667-676.
[89] Lu,J.;Tang,J.;Zhang,Z.;Shen,W. *Mater. Res. Bull.* 2005, 40, 2039-2046.
[90] Jones,D.J.;Aptel,G.;Brandhorst,M.;Jacquin,M.;Jiménez,J.J.;López,A.J.;Torres,P.M.;Pi wonski,I.;Castellón,E.R.;Zajac,J.;Rozière,J. *J. Mater. Chem.* 2000,10, 1957-1963.
[91] Subramanya Herle,P.;Ellis,B.;Coombs,N.;Nazar,L.F. *Nature Mater.* 2004, 2, 147-152.
[92] Arrebola,J.A.;Caballero,A.;Hernán,L.;Morales,J.;Rodriguez-Castellón,E. *Electrochem. Solid State Lett.* 2005, 8, A303-A307.
[93] Cho, J.;Kim,G. *Electrochem. Solid-State Lett.* 1999, 2, A253-A255.

[94] Caballero,A.;Hernán,L.;Morales,J.;Rodríguez-Castellón,E.; Santos-Peña,J. *J. Power Sourc.* 2004, 128, 286-291.

[95] Kobayashi,H.;Shigemura,H.;Tabuchi,M.;Sakaebe,H.;Ado,K.;Kageyama,H.;Hirano,A.; Kanno,R.;Wakita,M.;Morimoto,S.;Nasu,S. *J. Electrochem. Soc.* 2000, 147, A960-A969.

[96] Ceder,G.;Chiang,Y.M.;Sadoway,D.R.;Aydinol,M.K.;Jang,Y.I.;Huang,B. *Nature* 1998, 392, 694-696.

[97] Cho,J.;Kim,Y.J.;Park,B. *Chem. Mater.* 2000, 12, 3788-3791.

[98] Omanda,H.;Brousse,T.;Schleich,D.M. *J. Electrochem. Soc.* 2004, 51, A922-A929.

[99] Barriga,C.;Barrón,V.;Gancedo,R.;Gracia,M.;Morales,J.;Tirado,J.L.;Torrent,J. *J. Solid State Chem.* 1988, 77, 132-140.

[100] Tabuchi,M.;Tsutsui,S.;Masquelier,C.;Kanno,R.;Ado,K.;Matsubara,I.;Nasu,S.; Kageyama,H. *J. Solid State Chem.* 1998, 140, 159-167.

[101] Matsumura,T.;Kanno,R.;Inaba,Y.;Kawamoto,Y.;Takano,M. *J. Electrochem. Soc.* 2002, 149, A1509-A1513.

[102] Kanno,R.;Shirane,T.;Inaba,Y.;Kawamoto,Y. *J. Power Sourc.* 1997, 68, 145-152.

[103] Lee,Y.S.;Yoon,C.S.;Sun,Y.K.;Kobayakawa,K.;Sato,Y. *Electrochem. Comm.* 2002, 4, 727-731;Lee,Y.S.;Sato,S.;Tabuchi,M.;Yoon,C.S.;Sun,Y.K.;Kobayakawa,K.;Sato,Y. *Electrochem. Comm.* 2003, 5, 549-554; Lee,Y.S.;Sato,S.;Sun,Y.K.; Kobayakawa,K.; Sato,Y. *J. Power Sourc.* 2003, 119-121, 285-289.

[104] Sakurai,Y.;Arai,H.;Okada,S.;Yamaki,J. *J. Power Sourc.* 1997, 68, 711-715.

[105] Sakurai, Y.; Arai, H.; Yamaki, J. *Solid State Ionics* 1998, 113-115, 29-34.

[106] Wang,X.;Gao,L.;Zhou,F.;Zhang,Z.;Ji,M.;Tang,C.;Shen,T.;Zheng,H. *J. Crystal Growth* 2004, 265, 220-223.

[107] Tabuchi,M.;Nakashima,A.;Shigemura,H.;Ado,K.;Kobayashi,H.;Sakaebe,H.; Tatsumi,K.;Kageyama,H.;Nakamura,T.;Kanno,R. *J. Mater. Chem.* 2003, 13, 1747-1757.

[108] Cox,D.E.;Shirane,G.;Flinn,P.A.;Ruby,S.L.;Takei,W.J. *Phys. Rev.* 1963, 132, 1547-1553.

[109] Kim,J.;Manthiram,A. *J. Electrochem. Soc.* 1999, 146, 4371-4374.

[110] Wang,X.;Chen,X.;Gao,L.;Zheng,H.;Ji,M.;Tang,C.;Shen,T.;Zhang,Z. *J. Mater. Chem.* 2004, 14, 905-907.

[111] West,A.R. *Basic Solid State Chemistry, second edition*, John Wiley & Sons,(Eds.): New York, 1999, pp. 324-332.

[112] NuLi,Y.N.;Qin,Q.Z. *J. Power Sourc.* 2005, 142, 292-297.

[113] NuLi,Y.N.;Chu,Y.Q.;Qin,Q.Z. *J. Electrochem. Soc.* 2004, 151, A1077-A1083.

[114] Chu,Y.Q.;Fu,Z.W.;Qin,Q.Z. *Electrochim. Acta* 2004, 49, 4915-4921.

[115] Ai,C.;Yin,M.;Wang,C.;Sun,J. *J. Mater. Sci.* 2004, 39,1077-1079.

[116] Liu,X.;Yasuda,H.;Yamachi,M. *J. Power Sourc.* 2005, 146, 510-515.

[117] Taberna,P.L.;Mitra,S.;Poizot,P.;Simon,P.;Tarascon,J.M. *Nature Mater.* 2006, 5, 567-573.

[118] Takada,T.;Hayakawa,H.;Akiba,E. *J. Solid State Chem.* 1995, 115, 420-426.

[119] Hernán,L.;Morales,J.;Sánchez,L.;Santos,J. *Solid State Ionics* 1997, 104, 205-213.

[120] Tarascon,J.M.;Guyomard,D. *J. Electrochem. Soc.* 1991, 138, A2864-A2868.

[121] Olsson,R.T.;Salazar-Alvarez,G.;Hedenqvist,M.S.;Gedde,U.W.;Lindberg,F.; Savage,S.J. *Chem. Mater.* 2005, 17, 5109-5118.

[122] Bensebaa,F.;Zavaliche,F.;L'Ecuyer,P.;Cochrane,R.W.;Veres,T. *J. Colloid Interf. Sci.* 2004, 277, 104-110.

[123] Peñarroya-Mentruit,M.;Turnes-Palomino,G.;Otero-Areán,C. *Trends Inorg. Chem.* 2001, 7, 1-13.

[124] Valdés-Solís,T.; Marbán,G.;Fuertes,A.B. *Chem. Mater.* 2005, 17, 1919-1922.

[125] Rozman,M.;Drofenik,M. *J. Am. Ceram. Soc.* 1995, 78, 2449-2455.

[126] Ji,G.B.;Tang,S.L.;Ren,S.K.;Zhang,F.M.;Gu,B.X.;Du,Y.W. *J. Crystal Growth* 2004, 279,156-161.

[127] Wang,J.;Chen,Q.;Hou,B.;Peng,Z. *Eur. J. Inorg. Chem.* 2004, 6, 1165-1168.

[128] Zhang,Z.;Rondinone,A.J.;Ma,J.-X.;Shen,J.;Dai,S. *Adv. Mater.* 2005, 17, 1415-1419.

[129] Ferreira,T.A.S.;Waserneborgh,J.C.;Mendonça,M.H.R.M.;Nunes,M.R.;Costa,F.M. *Solid State Sci.* 2003, 5, 383-392.

[130] Allen,G.C.;Harris,S.J.;Jutson,J.A. *Appl. Surf. Sci.* 1989, 37, 111-134.

[131] Choudhury,T.;Saied,S.O.;Sullivan,J.L.;Abbot,A.M. *J. Phys. D: Appl. Phys.* 1989, 22, 1185-1195.

[132] Hara,D.;Ikuta,H.;Uchimoto,Y.;Wakihara,M. *J. Mater. Chem.* 2002, 12, 2507-2512.

In: Diffusion and Reactivity of Solids
Editor: James Y. Murdoch, pp. 209-226

ISBN: 978-1-60021-890-3

Chapter 5

STRUCTURE AND DIFFUSE SCATTERING OF SUPERIONIC CONDUCTOR CUI

Takashi Sakuma**, *Xianglian and Khairul Basar
Institute of Applied Beam Science, Ibaraki University, Mito 310-8512, Japan

Abstract

The structure and diffuse scattering of CuI that has high ionic conductivity at high temperature have been studied by X-ray diffraction, anomalous X-ray scattering and neutron diffraction methods. The expression of the diffuse scattering intensity including the correlation effects among the thermal displacements of atoms was shown and applied to the analysis of diffuse scattering of γ-, β- and α-CuI. The calculated energy dependence of the intensities of Bragg lines based on the ordered arrangement of Cu atoms could explain the characteristics of the observed scattering intensities of γ-CuI by anomalous X-ray scattering measurement. The model which includes the ordered arrangements of Cu atoms could explain the observed neutron diffuse scattering intensities of γ-CuI at 8 and 290 K. From the structural model with trigonal system the intensities of X-ray and neutron diffuse scattering was estimated based on the disordered arrangement of Cu atoms in β-CuI. Numerical calculations of the diffuse background of α-CuI have been made based on the short range order of copper atoms and the correlation effects among the thermal displacements of atoms. The cubic system of the space group Fm3m with the disordered arrangement of copper atoms could explain the diffuse scattering of α-CuI. The low-energy excitation in CuI by neutron inelastic scattering method was discussed. The temperature dependence of the damping factor of the excitation would be related to that of the ionic conductivity.

Keywords: diffuse scattering, short-range-order, disorder, thermal vibration, correlation effect, superionic conductor, CuI, X-ray diffraction, anomalous X-ray scattering (AXS), neutron diffraction

---

* E-mail address: sakuma@mx.ibaraki.ac.jp. Tel.: +81-29-228-8357; Fax: +81-29-228-8357. (Corresponding author.)

1. Introduction

The diffuse scattering includes information about a short-range-order in a disordered arrangement (static disorder) and correlation effects among thermal displacements of atoms (thermal disorder) in crystals [1-11]. Anomalously strong and oscillatory diffuse scattering had been observed from α-AgI type superionic conductors by X-ray and neutron diffraction experiments. The oscillatory forms of the diffuse scattering were explained by the correlation effects between thermal displacements of nearest neighboring atoms.

The Rietveld method is essential in the study of characterizing polycrystalline materials [12-14]. The profile-shape functions for Bragg lines have been extensively investigated by many researchers. However, the studies of background functions for diffuse scattering are not enough. Legendre polynomials which have no physical meanings have been used for the background function. Recently, the background function with the correlation effects among the thermal displacements of atoms was applied to the analysis of α-AgI type superionic conductors.

The highest temperature phases having a famous bcc structure of α-AgI type superionic conductors had been widely studied by the usual X-ray diffraction, neutron diffraction and the Extended X-ray Absorption Fine Structures (EXAFS) measurements [15-17]. Numerical calculations of the diffuse background were also performed based on the fcc crystal structure of α-Cu_2Se including the correlation effects between the thermal displacements of atoms. The expected intensities of the diffuse scattering were calculated based on the disordered distribution of Cu atoms and Debye-Waller temperature parameters which were obtained from the analysis of Bragg lines of α-Cu_2Se by X-ray diffraction measurement. This result agreed with the observed intensities by anomalous X-ray scattering (AXS) measurement at the Cu K-absorption edge. The crystal structure model of α-Cu_2Se was also supported by the analysis of diffuse scattering.

Copper iodide exhibits phase transitions at 369°C (γ–β) and 407°C (β–α). The high temperature α-phase is well known as having a high ionic conductivity of about 10^{-1} Scm^{-1} [18-21]. CuITe is one of the copper halide-chalcogen compounds which have ionic conductivity of about 10^{-5} Scm^{-1} at room temperature, which is much greater than that of CuI [22-24]. The crystal structure of $(Cs_{1-y}Rb_y)Cu_4Cl_3I_2$ which is synthesized from CuI is iso-structural with that of $RbAg_4I_5$. $(Cs_{1-y}Rb_y)Cu_4Cl_3I_2$ has a high ionic conductivity at room temperature [25,26]. The crystal structures of CuI have been studied by X-ray and neutron diffraction methods. The crystal structure of γ-CuI had been studied by the ordered arrangement and the disordered arrangement of Cu atoms [27-30]. The structure of β-CuI had been reported as wurtzite structure first. However, the forbidden line for the wurtzite structure was observed by neutron diffraction method. Considering systematic absence of reflections, various structural models of β-CuI based on ordered or disordered arrangement of Cu were examined. Both structural models with the ordered arrangement (ZnS type) and the disordered arrangement (CaF_2 type) could explain the relative intensities of Bragg lines of α-CuI. Crystal structure of α-CuI was first investigated by X-ray diffraction measurement and determined to be a cubic ZnS type. Later, the structure of α-CuI was suggested to be a CaF_2 type by neutron diffraction experiment. As the temperature increases, the anharmonicity of atoms is larger in CuI [31,32]. The presence of low-energy excitations in CuI was investigated by neutron inelastic scattering measurements.

The energy dependence of the intensities of Bragg lines is investigated by the AXS measurements to confirm the crystal structure of γ-CuI. The diffuse scattering intensity of superionic conductors is closely connected with the crystal structures, short–range-order of disordered arrangement of atoms and correlation effects among thermal displacements of atoms. In this paper the diffuse scattering intensities of CuI are discussed in connection with the crystal structures of γ-, β- and α-phase.

2. Diffuse Scattering and Correlation Effects among Thermal Displacements

The background intensity consists of coherent diffuse scattering and incoherent scattering. The incoherent scattering for X-ray scattering measurement is from Compton scattering and that for neutron scattering measurement from spin and isotope effects. The diffuse scattering intensity by X-ray and neutron diffraction measurement is given as

$$I_{\mathrm{D}} = k \sum_{n} \sum_{n'} \exp[i\mathbf{Q}.(\mathbf{R}_n - \mathbf{R}_{n'})] \langle \Delta F_n \Delta F_{n'}^{*} \rangle \,, \tag{1}$$

where k is a function depending on the experimental conditions [4,33]. The structure factor F includes the scattering factor f and atomic position $\mathbf{r}$.

$$F = \sum_{j} f_j \exp[-i\mathbf{Q}.\mathbf{r}_j] \tag{2}$$

ΔF_n is defined as the deviation of the structure factor at the nth site from the mean structure factor, namely

$$F_n = \langle F \rangle + \Delta F_n \,. \tag{3}$$

Therefore we could obtain the information of Δf and $\Delta \mathbf{r}$ from the analysis of diffuse scattering. $\Delta \mathbf{r}_s$ is the displacement from the mean position caused by thermal vibration. At high temperature many crystalline superionic conductors have the averaged structure in which the number of available atomic sites is greater than that of atoms. In this case the contribution of Δf to diffuse scattering is important. From the value of $\langle \Delta f_s \Delta f_{s'} \rangle$ and $\langle \Delta \mathbf{r}_s \Delta \mathbf{r}_{s'} \rangle$, the static correlation among atoms (short-range order) and the thermal correlation among atoms (thermal correlation effect) are obtained, respectively. The thermal average is obtained by cumulant expansion;

$$\left\langle \exp\left[-i\mathbf{Q}.\left(\Delta\mathbf{r}_{s(\mathrm{i})}-\Delta\mathbf{r}_{s'(\mathrm{j})}\right)\right]\right\rangle \cong \exp\left[\left\{-i\mathbf{Q}.\left\langle\left(\Delta\mathbf{r}_{s(\mathrm{i})}-\Delta\mathbf{r}_{s'(\mathrm{j})}\right)\right\rangle\right\}+\frac{\mathbf{Q}^2}{2}\left\{\left\langle\left(\Delta\mathbf{r}_{s(\mathrm{i})}-\Delta\mathbf{r}_{s'(\mathrm{j})}\right)^2\right\rangle-\left\langle\left(\Delta\mathbf{r}_{s(\mathrm{i})}-\Delta\mathbf{r}_{s'(\mathrm{j})}\right)\right\rangle^2\right\}\right]$$

$$=\exp\left[-\frac{\mathbf{Q}^2}{2}\left(\left\langle\Delta\mathbf{r}^2_{s(\mathrm{i})}\right\rangle+\left\langle\Delta\mathbf{r}^2_{s'(\mathrm{j})}\right\rangle\right)\left(1-2\frac{\left\langle\Delta\mathbf{r}_{s(\mathrm{i})}\Delta\mathbf{r}_{s'(\mathrm{j})}\right\rangle}{\left\langle\Delta\mathbf{r}^2_{s(\mathrm{i})}\right\rangle+\left\langle\Delta\mathbf{r}^2_{s'(\mathrm{j})}\right\rangle}\right)\right]$$

$$=\exp\left[-\left(M_\mathrm{i}+M_\mathrm{j}\right)\left(1-\lambda_{r_{s(\mathrm{i})s'(\mathrm{j})}}\right)\right], \tag{4}$$

where the correlations among the thermal displacements of atoms $\lambda_{r_{s(\mathrm{i})s'(\mathrm{j})}}$ is defined as;

$$\lambda_{r_{s(\mathrm{i})s'(\mathrm{j})}}=2\left\langle\Delta\mathbf{r}_{s(\mathrm{i})}\Delta\mathbf{r}_{s'(\mathrm{j})}\right\rangle\Big/\left(\left\langle\left(\Delta\mathbf{r}_{s(\mathrm{i})}\right)^2\right\rangle+\left\langle\left(\Delta\mathbf{r}_{s'(\mathrm{j})}\right)^2\right\rangle\right). \tag{5}$$

The diffuse scattering intensity I_D from a powder sample including the correlations among the thermal displacements of atoms (λ) and the probability functions (α and β) is expressed as follows;

$$\begin{aligned} I_\mathrm{D} &= kN_0\sum_\mathrm{i} u_\mathrm{i} f_\mathrm{i} f_\mathrm{i}^*\left\{1-\exp\left(-2M_\mathrm{i}\right)\right\} \\ &+ kN_0\sum_\mathrm{i}\sum_\mathrm{j}\sum_{s'(\mathrm{j})}{}' u_\mathrm{i} f_\mathrm{i} f_\mathrm{j}^*\left[\alpha_{r_{s(\mathrm{i})s'(\mathrm{j})}}\left[\exp\left\{-\left(M_\mathrm{i}+M_\mathrm{j}\right)\left(1-\lambda_{r_{s(\mathrm{i})s'(\mathrm{j})}}\right)\right\}-\exp\left\{-\left(M_\mathrm{i}+M_\mathrm{j}\right)\right\}\right]\right. \\ &\left.+\left(1-p_\mathrm{i}\right)\left(\alpha_{r_{s(\mathrm{i})s'(\mathrm{j})}}-\beta_{r_{s(\mathrm{i})s'(\mathrm{j})}}\right)\exp\left\{-\left(M_\mathrm{i}+M_\mathrm{j}\right)\right\}\right] Z_{r_{s(\mathrm{i})s'(\mathrm{j})}} S_{r_{s(\mathrm{i})s'(\mathrm{j})}}, \end{aligned} \tag{6}$$

where Z_r is the number of sites belonging to the s'th j type neighbor around a sth i type site. S_r is equal to $\sin(Qr)/Qr$. In the case of X-ray diffraction k includes the term of polarization effect. The probability p_i of finding the atom in any site is equal to the ratio of the number of i atoms to the number of the sites in the crystal. The Debye-Waller factor $\exp(-M_\mathrm{i})$ is equal to $\exp\{-B_\mathrm{i}(\sin\theta/\lambda)^2\}$. The probability function α_r gives the probability of finding an atom at a site apart by a distance r from a site occupied by an atom, and β_r the probability of finding an atom at a site apart by r from a vacant site. The oscillatory form is expressed as $\sin(Qr)/Qr$, where Q is equal to $4\pi\sin\theta/\lambda$. The values of the correlation among the thermal displacements of s(i) and s'(j) atoms is 0 in the case of no correlation among atomic displacements, and in the case of perfect correlation λ is $2\sqrt{B_\mathrm{i}B_\mathrm{j}}/\left(B_\mathrm{i}+B_\mathrm{j}\right)$. The prime added to the summation symbol in equation (6) means to omit the term of $r_{s(\mathrm{i})s'(\mathrm{j})}=0$. There are relationships among the short range order parameters;

$$\begin{aligned} \alpha_{r_{s(\mathrm{j})s'(\mathrm{i})}} &= p_\mathrm{i}\alpha_{r_{s(\mathrm{i})s'(\mathrm{j})}}\big/p_\mathrm{j}, \\ \alpha_{r_{s(\mathrm{i})s'(\mathrm{j})}}-\beta_{r_{s(\mathrm{i})s'(\mathrm{j})}} &= \left(\alpha_{r_{s(\mathrm{i})s'(\mathrm{j})}}-p_\mathrm{j}\right)\big/\left(1-p_\mathrm{i}\right). \end{aligned} \tag{7}$$

There are restrictions among the probability functions $\alpha_{r_{s(j)s'(i)}}$;

$$\sum_{j}\sum_{s'(j)}{}' Z_{r_{s(i)s'(j)}}\left(\alpha_{r_{s(i)s'(j)}} - \beta_{r_{s(i)s'(j)}}\right) = -1 . \tag{8}$$

The theoretical expression was applied to the analysis of the diffuse background intensities of superionic conductor CuI.

3. Crystal Structure and Diffuse Scattering of γ-CuI

According to the diffraction study, the room temperature γ-phase has a ZnS structure with the space group $F\bar{4}3m$. The lattice constant is a = 6.059 Å at room temperature. In regard to the distribution of Cu atoms, both the structural models with the ordered arrangement and with the disordered arrangement could explain the intensities of the Bragg lines of CuI at room temperature. Two structural models of γ-CuI have been reported; one has the ordered arrangement of Cu atoms (ZnS type) and the other the disordered arrangement of Cu atoms. The atomic positions for the ordered arrangement of Cu atoms are as follows;

$$\left(0,0,0;\ 0,\tfrac{1}{2},\tfrac{1}{2};\ \tfrac{1}{2},0,\tfrac{1}{2};\ \tfrac{1}{2},\tfrac{1}{2},0\right)+$$

I	4(a)	0,0,0.
Cu	4(c)	$\tfrac{1}{4},\tfrac{1}{4},\tfrac{1}{4}$.

Four Cu atoms occupy 4(c) sites. The atomic positions of γ-CuI for the disordered arrangement of Cu atoms are as follows;

$$\left(0,0,0;\ 0,\tfrac{1}{2},\tfrac{1}{2};\ \tfrac{1}{2},0,\tfrac{1}{2};\ \tfrac{1}{2},\tfrac{1}{2},0\right)+$$

I	4(a)	0,0,0.
Cu	16(e)	$x,x,x;\ x,\bar{x},\bar{x};\ \bar{x},x,\bar{x};\ \bar{x},\bar{x},x$.

Four Cu atoms are statistically distributed over the 16(e) sites around 4(c) sites. It is required to investigate the distribution of Cu atoms in CuI at room temperature. The anomalous X-ray scattering method (AXS) is one of the powerful tools for structural characterization of solid electrolytes. AXS method had been applied to the structural study of the superionic phase of Cu_2Se and succeeded in explaining the observed Bragg lines and diffuse scattering.

The arrangement of Cu atoms in CuI at room temperature was examined by the AXS measurement. The intensities of the Bragg lines for powder CuI were observed using several incident X-ray wavelengths. The X-ray measurements were carried out by the double axis goniometer with an intrinsic pure germanium solid state detector and a multi-channel analyzer system (Rigaku RINT 2000). In the measurement a rotating molybdenum anode X-ray tube having a fine focus was used as the X-ray source. The incident X-ray energies of

8.05, 8.30, 8.70 keV monochromated with the 220 reflection of germanium, 17.48, 21.00, 24.00, 27.50 keV with the 440 reflection of germanium and 32.50, 32.90 keV with the 660 reflection of germanium were used. Molybdenum tube was operated with electron current 350 mA and voltage 40 kV. X-ray intensity data were collected for 100 s per step at 0.05° intervals over the 2θ range of 5 to 100° by a step-scan mode at room temperature. The energy dependence in X-ray scattering intensities for 111, 200, 220, 311, 400, 331, 511, 440, 531, 600, 620, 533, 444, 640, 551, 642, 222, 420 and 622 lines was studied. Although the operating condition of electron current and voltage for molybdenum tube is same, the incident X-ray intensity to the sample varies with the X-ray energy. To compensate the difference of the incident flux of X-ray beam, the observed intensities of Bragg lines were divided by the intensity of 422 line for each measurement. The results of the X-ray scattering intensity of 420 line divided by the intensity of 422 line are shown by broken lines in Fig.1.

Table 1. The values of real (*f′*) and the imaginary part (*f″*) of the anomalous dispersion terms for Cu and I atoms.

	Energy of incident X-rays (keV)								
	8.05	**8.30**	**8.70**	**17.48**	**21.00**	**24.00**	**27.50**	**32.50**	**32.90**
f'_{Cu}	-2.028	-2.306	-3.131	0.264	0.275	0.251	0.215	0.163	0.159
f''_{Cu}	0.589	0.557	0.511	1.266	0.911	0.714	0.554	0.404	0.395
f'_{I}	-0.562	-0.469	-0.358	-0.719	-1.030	-1.308	-1.712	-3.465	-4.272
f''_{I}	6.831	6.496	6.010	1.812	1.307	1.028	0.802	0.591	0.577

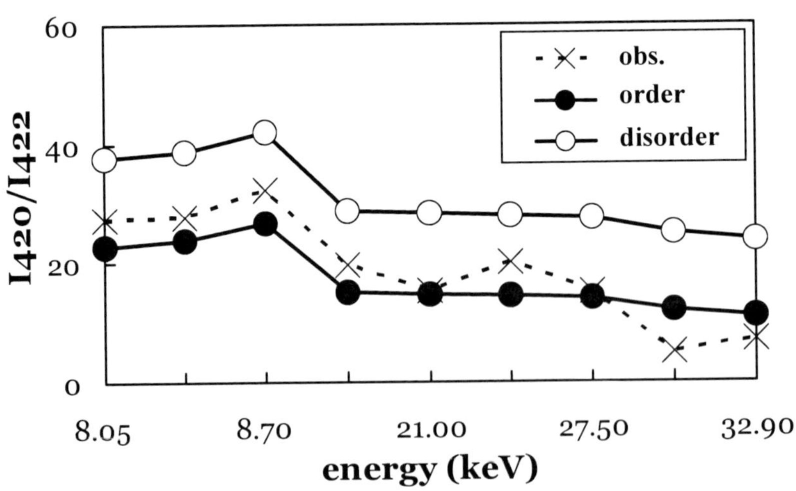

Figure 1. Energy dependence of the ratio of I_{420} to I_{422} for γ-CuI at room temperature by AXS measurement.

The scattering intensity indicates the distinct energy dependence, arising from the so-called anomalous dispersion phenomena. The atomic scattering factor f is expressed as $f = f_0 + f' + if''$ for the AXS. The real part (f') and the imaginary part (f'') of the anomalous dispersion terms for Cu and I atoms used in the present data analysis are listed in Table 1. The calculation of the scattering intensity of Bragg lines for CuI at room temperature at several X-ray energies was based on the structural models of Bührer and Hälg [28]. The calculated scattering intensities for 420 line divided by the intensities of 422 line with the ordered and disordered models are shown in Fig.1. The differences in the calculated intensities between ordered and disordered models for 111, 200, 220, 311, 400, 331, 511, 440, 531, 600, 620, 533, 444, 551 and 642 lines are small. The energy dependence in the observed intensities of these lines is explained well by both the ordered and disordered arrangements of Cu atoms. Contrary to these, the calculated intensities of 222, 420 and 622 lines with ordered model differ very much from those with disordered model. It is found from Fig. 1 that the calculated energy dependence of the intensities of Bragg lines with the ordered Cu atoms in CuI could explain the characteristics of the observed scattering intensities. This result would show that the arrangement of ordered Cu atoms is reasonable in γ-CuI at room temperature. The anomalous X-ray scattering method is the powerful tool for a research in the atomic distribution in superionic conductors. It would be interesting to extend this technique to the structural study of the high-temperature phase of superionic conductors.

Neutron scattering measurements were performed from a powder CuI in a cryostat at 8 K and 290 K. The incident neutron energy of 41.2 meV (λ = 1.41 Å) was used. Figs. 2 and 3 show the results of the double-axis neutron diffraction measurement of CuI at 8 K and 290 K, respectively. The existence of oscillatory background intensity was confirmed by the measurement of CuI at 290 K. The oscillatory characteristic in the diffuse scattering of CuI at 8 K is not clear. The first, second and third peak of the oscillatory diffuse scattering appear at $2\theta \sim 36°$, 67° and 104°, respectively.

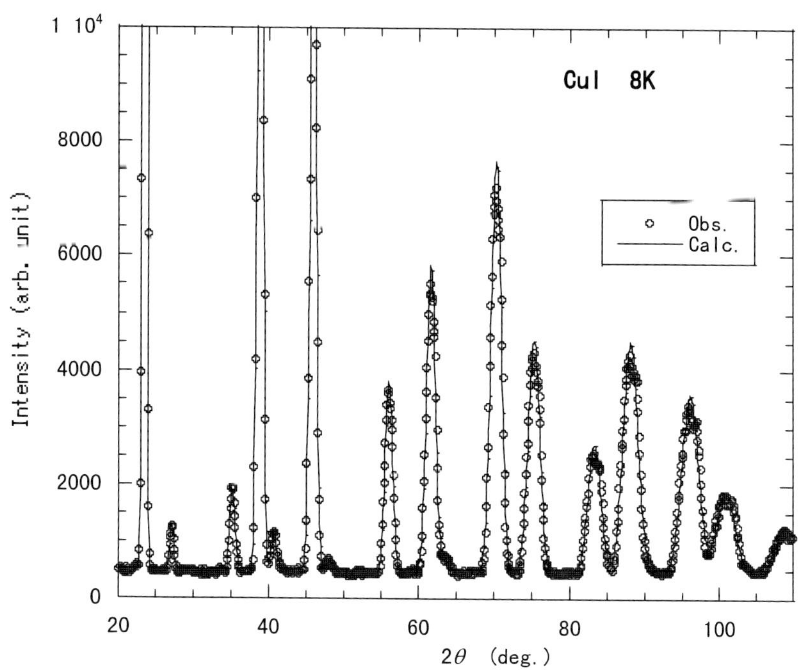

Figure 2. Observed and calculated neutron powder diffraction intensity of γ-CuI at 8 K. The incident wave length is 1.41 Å.

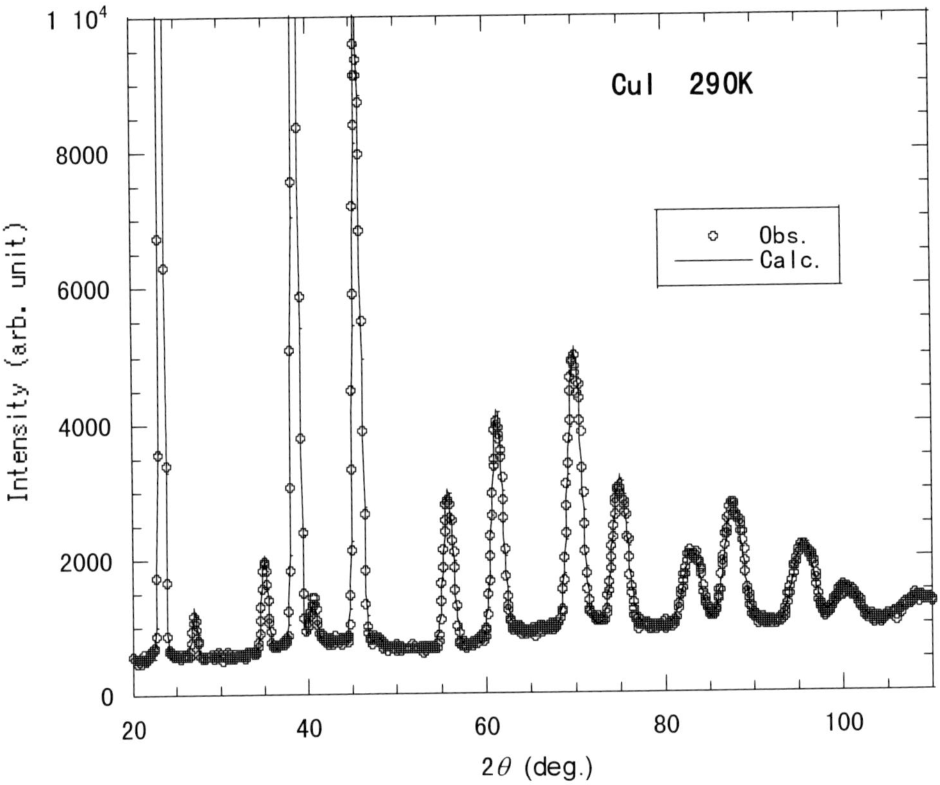

Figure 3. Observed and calculated neutron powder diffraction intensity of γ-CuI at 290 K. The incident wave length is 1.41 Å.

Rietveld refinements of the neutron scattering intensities of CuI with RIETAN-94 have been carried out with the modified background function (equation (6)) including the correlation effects among the thermal displacements of atoms. The structure of γ-CuI belongs to the cubic system with the space group $F\bar{4}3m$. The ordered arrangement of Cu atoms was used in the calculation. Copper and iodine atoms occupy 4 (c) and 4 (a) sites, respectively. The derived thermal parameters at 8 K in Rietveld refinements are $B_I \sim 0.1$ Å^2 and $B_{Cu} \sim 0.1$ Å^2. The values of the thermal parameters at 290 K are relatively large; $B_I \sim 1.4$ Å^2 and $B_{Cu} \sim 1.8$ Å^2. The observed and calculated neutron diffraction intensities at 8 and 290 K are shown in Figs. 2 and 3, respectively. The value of the correlation effects between the thermal displacements of the nearest-neighboring atoms at 8 and 290 K is 0.75 in the calculation. The values of the correlation effects except the nearest-neighboring atoms are 0. The number of nearest neighboring sites is equal to 4 in the zinc blende type structure. The large values of the Debye-Waller temperature parameters at 290 K contribute to the thermal diffuse scattering in the observed background intensity. The background function including the correlation effects among the thermal displacements of atoms gives somewhat lower R factors in the Rietveld refinements at 290 K than those with background function by Legendre polynomials which are usually used in the Rietveld refinements. The background function including the correlation effects between the thermal displacements of atoms is available in the analysis of Rietveld refinements for the angle-dispersive diffraction dada above room temperature. Recently, the structure of γ-AgI was reexamined by synchrotron X-ray powder diffraction method [34]. The ordered copper model could explain the observed synchrotron X-ray diffraction pattern.

4. Crystal Structure and Diffuse Scattering of β–CuI

The crystal structure of β-CuI had been investigated by many researchers. From the early X-ray diffraction study it had been reported that the structure of β-CuI had wurtzite structure $P\bar{6}_3mc$. However, the measurement didn't cover the low scattering angle where Bragg line could appear. Later the 001 line for the space group $P\bar{6}_3mc$ was observed in the region. This means that β-CuI is not of the wurtzite structure. There is a condition limiting possible reflections l=2n for hhl in the wurtzite structure.

X-ray diffraction pattern of the β-phase of CuI indicates that the Bragg lines are indexed by hexagonal (trigonal) system with the lattice constants a = 4.279 Å, c = 7.168 Å at 400°C. There are two CuI units in the unit cell. Considering systematic absence of reflections, various structural models based on ordered and disordered arrangement of Cu were examined. As the results it was found that the structure of β-CuI belonged to the trigonal system with the space group P3m1 [35]. The structure has the ordered arrangement of Cu atoms. The atomic positions are as follows;

I	1(a)	0,	0,	0.
	1(b)	$\frac{1}{3}$,	$\frac{2}{3}$,	$\frac{1}{2}$.
Cu	1(a)	0,	0,	z.
	1(b)	$\frac{1}{3}$,	$\frac{2}{3}$,	z.

The Inter-atomic distances and the coordination number with z_a=0.636 and z_b=0.896 in β-CuI are given in Table 2. The nearest neighboring atoms of Cu atoms are arranged in distorted tetrahedra of I atoms and those of the I atoms distorted tetrahedra of Cu atoms. The temperature parameters B of Cu and I are 4.1 $Å^2$ and 15.1 $Å^2$, respectively. The temperature parameters are considerably large especially for Cu atoms. Cu atoms would possess large anharmonicity of the thermal vibration in the β-phase. Because of the low site symmetry $3m$ for Cu(a) and Cu(b) sites many anharmonicity parameters are necessary to describe the effective potential field with terms up to even the third order. Detailed experimental studies using the single crystal of CuI would provide further information on the thermal behavior of the ions. The easiest channel of cation movement is the path connecting Cu(a) and Cu(b) sites. It is well known that the low temperature phase of superionic conductors is often ZnS or wurtzite. The γ-phase of CuI has ZnS structure. The β-phase of CuI has a structure similar to the distorted wurtzite type. The other structural model for β-CuI with disordered arrangements of Cu atoms has been reported. The space group of the disordered model is $P\bar{3}m1$, with I in 2(d) sites at $(\frac{1}{3},\frac{2}{3},z)$ and $(\frac{2}{3},\frac{1}{3},z)$ with z = 0.242. The Cu atoms are predominantly on 2(d) sites at z = 0.621 and other on 2(d) sites with z = 0.878 [36,37].

Table 2. Inter-atomic distances and the coordination numbers *Z* in β-CuI (space group P3m1).

		Z	Distance (Å)
Cu(a)	- I(a)	1	2.612
	- I(b)	3	2.655
	- Cu(b)	3	3.095
Cu(b)	- I(a)	3	2.581
	- I(b)	1	2.837
	- Cu(a)	3	3.095
I(a)	- Cu(b)	3	2.581
	- Cu(a)	1	2.612
I(b)	- Cu(a)	3	2.655
	- Cu(b)	1	2.837

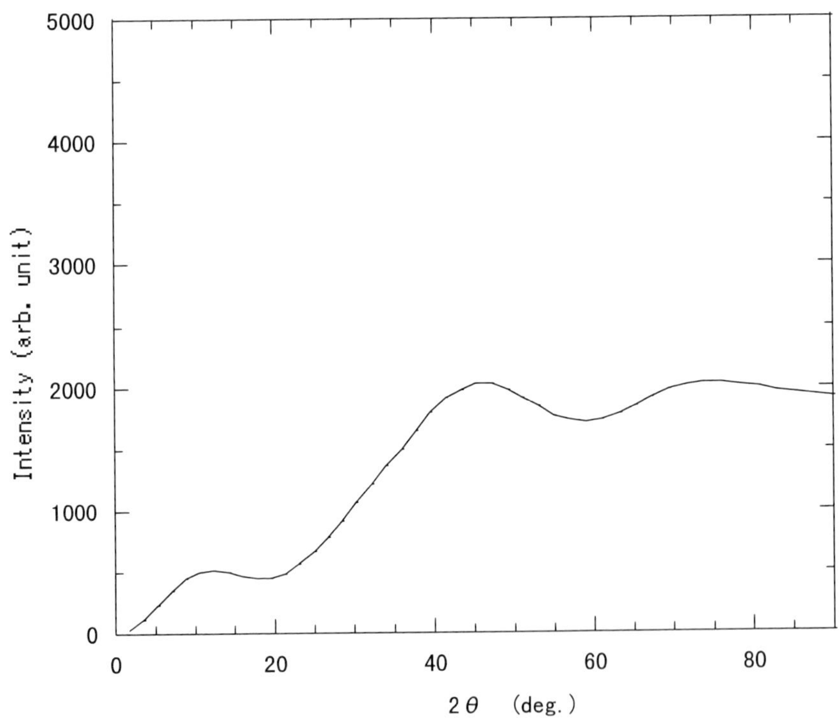

Figure 4. Expected diffuse scattering intensity of β-CuI at 400°C by X-ray diffraction experiment (λ = 1.54 Å).

The diffuse scattering intensity of β-CuI has not been reported. We calculate the expected diffuse scattering intensities by X-ray and neutron diffraction experiments based on the above crystal structure of trigonal system P3m1. The inter-atomic distances and coordination numbers of the first nearest Cu-I pairs for the space group P3m1 are almost same as those of $P\bar{3}m1$. The calculated X-ray (λ = 1.54 Å) and neutron (λ = 1.41 Å) diffuse scattering intensities of β-CuI at 400°C are shown in Figs. 4 and 5, respectively. The number of neighboring sites and inter-atomic distances in Table 2 are used in the calculation. The values of the correlation effects between the thermal displacements of the neighboring atoms ($r < 3.0$) are 0.7 in the calculation. The values of the correlation effects except the nearest-neighboring atoms ($3.0 < r$) are 0. The large values of the Debye-Waller temperature parameters contribute to the thermal diffuse scattering in the observed background intensity at 400°C. In Fig. 4 the oscillatory peaks in the calculated diffuse scattering appear around $2\theta \sim 43$ ($Q \sim 3.0$) and 75° ($Q \sim 5.1$) for X-ray diffraction measurement. In the case for neutron diffraction measurement the peaks appear around $2\theta \sim 40$ ($Q \sim 3.0$) and 70° ($Q \sim 5.0$) in Fig. 5. These peak positions would be related to the correlation effects between the thermal displacements of nearest neighboring copper and iodine atoms. As the inter-atomic distances of first nearest neighboring atoms are almost same in γ-, β- and α-CuI and α-AgI type superionic conductors, the positions of peaks in diffuse scattering intensities appear around same Q positions.

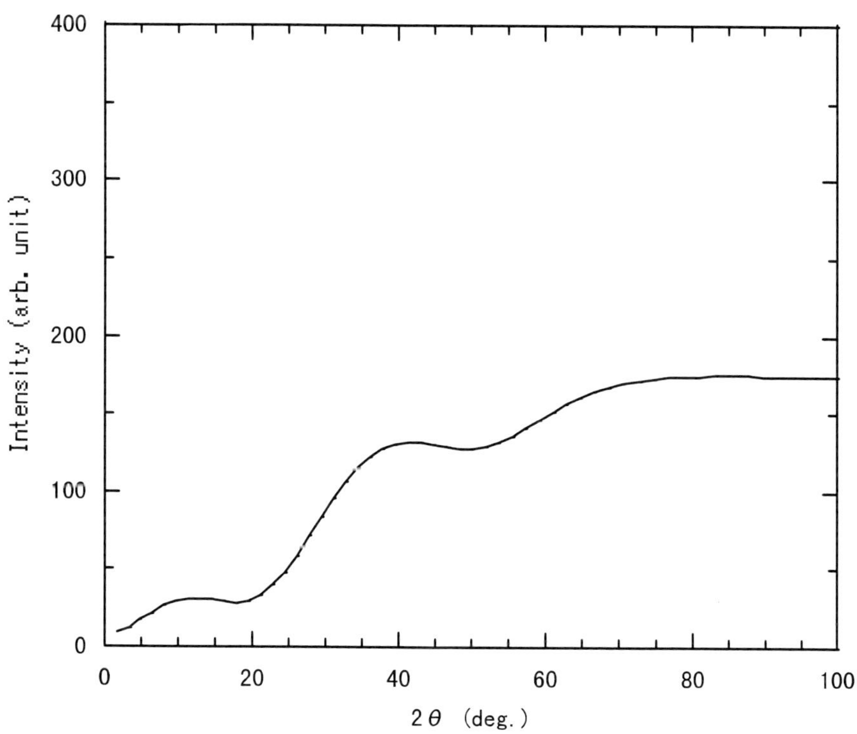

Figure 5. Expected diffuse scattering intensity of β-CuI at 400°C by neutron diffraction experiment (λ = 1.41 Å).

5. Crystal Structure and Diffuse Scattering of α-CuI

Neutron diffraction measurements were performed from a powder CuI in an electric furnace at 475°C. The incident neutron energy of 41.2 meV (λ = 1.41 Å) was used. Fig. 6 shows the result of the diffraction measurement of CuI. Several sharp Bragg lines and a large oscillatory diffuse scattering were observed.

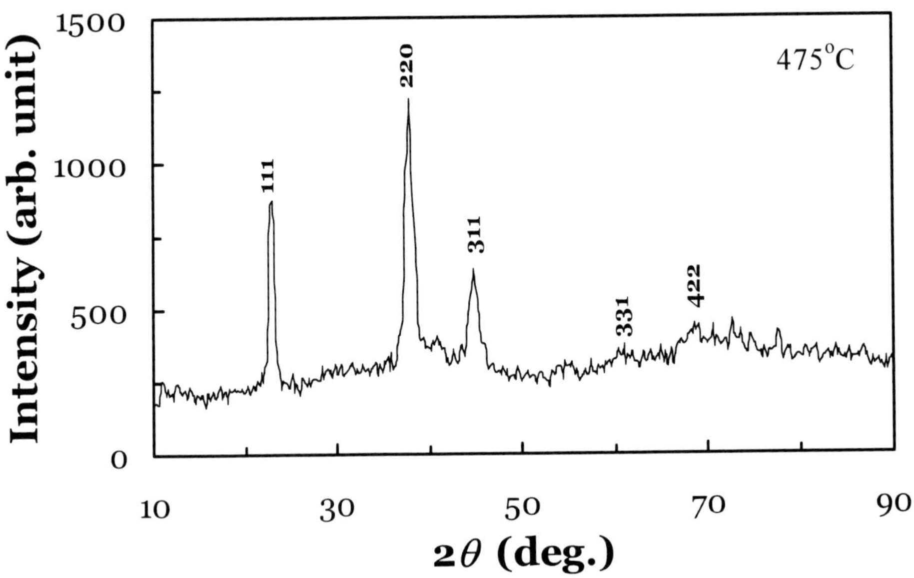

Figure 6. Observed neutron diffraction intensity for α-CuI at 475°C (λ = 1.41 Å).

Rietveld refinements of the neutron diffraction data of CuI have been carried out with RIETAN-94. Two structural models have been applied; one has an ordered arrangement of Cu atoms with the space group $F\bar{4}3m$ and the other a disordered arrangement of Cu atoms with the space group $Fm3m$. The atomic positions of α-CuI with the space group $F\bar{4}3m$ (ZnS type) are as follows;

$$\left(0,0,0;\ 0,\tfrac{1}{2},\tfrac{1}{2};\ \tfrac{1}{2},0,\tfrac{1}{2};\ \tfrac{1}{2},\tfrac{1}{2},0\right)+$$

I	4(a)	0,0,0.
Cu	4(c)	$\tfrac{1}{4},\tfrac{1}{4},\tfrac{1}{4}$.

In this case Cu atoms show ordered arrangement. The atomic positions of α-CuI with the space group $Fm3m$ (CaF_2 type) are as follows;

$$\left(0,0,0;\ 0,\tfrac{1}{2},\tfrac{1}{2};\ \tfrac{1}{2},0,\tfrac{1}{2};\ \tfrac{1}{2},\tfrac{1}{2},0\right)+$$

I	4(a)	0,0,0.
Cu	8(c)	$\tfrac{1}{4},\tfrac{1}{4},\tfrac{1}{4};\ \tfrac{3}{4},\tfrac{3}{4},\tfrac{3}{4}$.

Four Cu atoms are statistically distributed over the 8(c) sites.

The lattice constant, Debye-Waller temperature parameters of Cu and I atoms and the reliability factor with the space group $F\bar{4}3m$ by the analysis of the intensities of Bragg lines are a = 6.126 Å, B_I = 3.1 Å^2, B_{Cu} = 3.5 Å^2, R_I = 13.9 %, respectively. On the other hand, the obtained lattice constant, Debye-Waller temperature parameters and the reliability factor with the space group Fm3m are a = 6.126 Å, B_I =2.9 Å^2, B_{Cu} = 11.8 Å^2, R_I = 4.8 %, respectively. Although both two models could explain the relative intensities of observed Bragg lines, the reliability factors show that the structural model with the space group Fm3m (CaF_2 type) would be adapted for the crystal structure of α-CuI.

The diffuse scattering intensities of α-CuI were calculated based on the ordered and disordered arrangement of Cu atoms. In the case of the ordered arrangement of Cu atoms with the space group $F\bar{4}3m$, the value of α_r -β_r in equation (6) is equal to 0, and α_r and p_{Cu} equal to 1. The calculated diffuse background intensity of CuI at 475°C with the space group $F\bar{4}3m$ is shown in Fig.7. The Debye-Waller temperature parameters (B_{Cu} = 3.5 Å^2, B_I = 3.1 Å^2) that were obtained from the analysis of Bragg lines were used in equation (6). The values of the correlations terms $\lambda_{r_{s(i)s'(j)}}$ are:

$$\lambda_{r_{s(Cu)s'(I)}} = 0.7 \text{ at } r < 3.0 \text{ Å},$$
$$\lambda_{r_{s(i)s'(i)}} = 0 \text{ at } r > 3.0 \text{ Å}. \qquad (9)$$

The correlation effects between the thermal displacements of nearest neighboring copper and iodine atoms are strong.

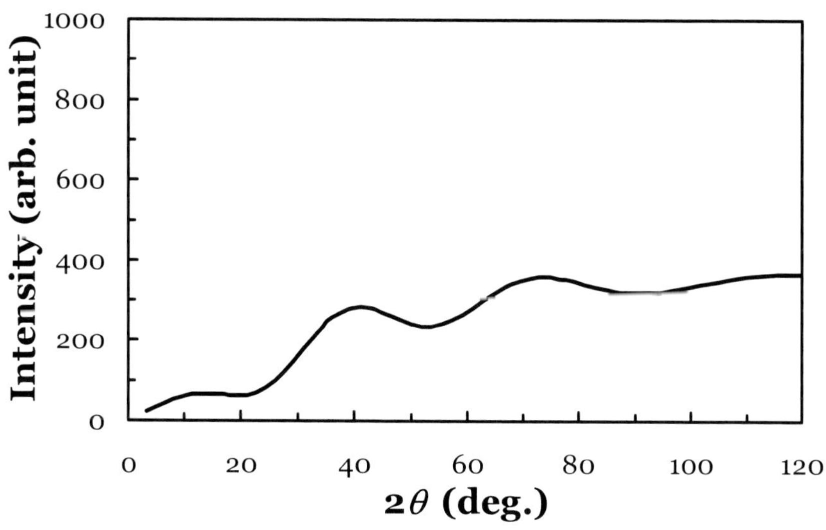

Figure 7. Calculated diffuse neutron background intensity of α-CuI based on the ordered arrangement of Cu atoms with the space group $F\bar{4}3m$ (ZnS type).

The calculation of diffuse scattering intensity was also performed with the space group Fm3m with the same values of the correlation effects $\lambda_{r_{s(i)s'(j)}}$ in equation (9). In the space group Fm3m copper atoms have disordered arrangement. The values of probability function $\alpha_{r=3.063}$ ranging from 0.2 to 0.7 were used. 3.063 Å corresponds to the inter-atomic distance between nearest neighboring Cu atoms. The other values of α_r in the calculation are:

$$\alpha_r = \left(17 - 12\alpha_{r=3.063}\right)/24 \text{ at } r = 4.331 \text{ Å},$$
$$\alpha_r = p_{Cu} \text{ at } r > 4.4 \text{ Å}. \tag{10}$$

p_{Cu} is equal to 4/8. The Debye-Waller temperature parameters (B_{Cu} = 11.8 Å^2, B_I = 2.9 Å^2) that were obtained from the analysis of Bragg lines were used. The obtained results are shown in Fig. 8.

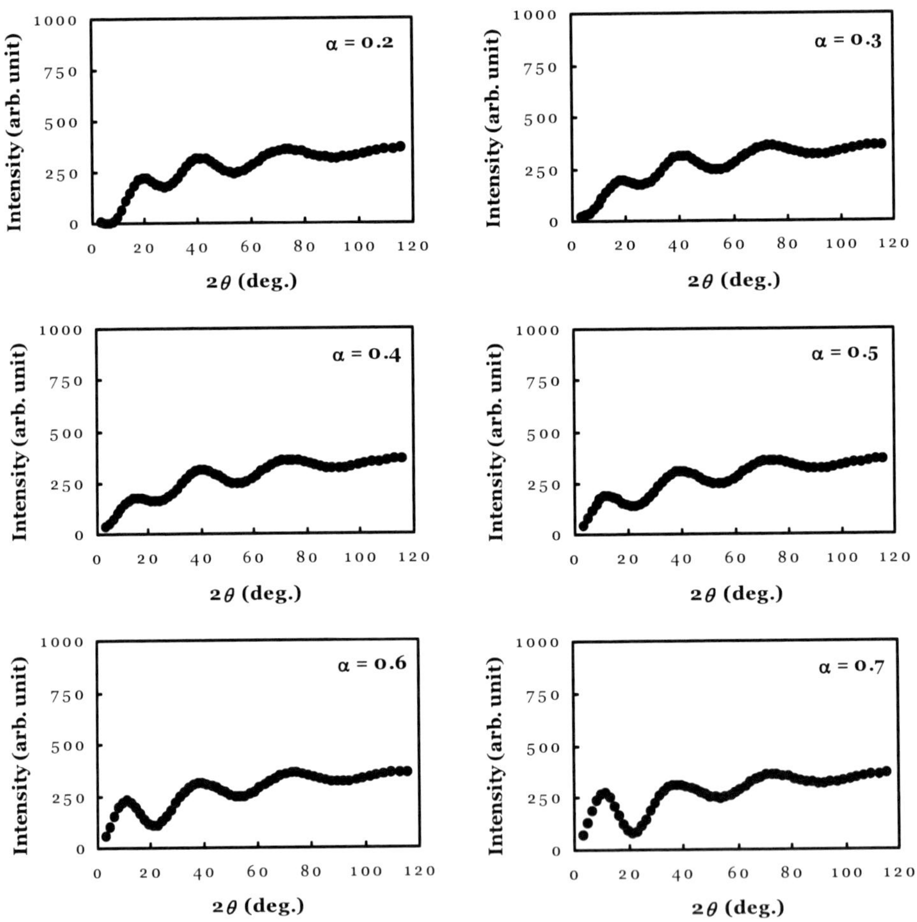

Figure 8. Calculated neutron diffuse background intensity of α-CuI based on the disordered arrangement of Cu atoms with the space group Fm3m (CaF_2 type). α gives the probability of finding Cu atom at a site apart by a distance r = 3.063 Å from a site occupied by Cu atom.

The peaks of the oscillatory diffuse scattering appear at $2\theta \sim 40°$ and 70° in Figs. 7 and 8. The similar oscillatory peaks of the diffuse scattering were obtained in other superionic conductors. These positions of the peaks were explained by the correlation effects among the thermal displacements of nearest neighboring silver and iodine atoms in α-AgI type superionic conductors. In α-CuI the observed oscillatory peaks in the diffuse scattering at $2\theta \sim 40$ and 70° would be related to the correlation effects between the thermal displacements of nearest neighboring copper and iodine atoms. The difference of the diffuse scattering between two structural models appears below $2\theta\sim 30°$ in Figs. 7 and 8. The intensity of the diffuse scattering below $2\theta\sim 30°$ is very weak in Fig.7. However, the intensity below $2\theta\sim 30°$ is relatively strong and have a maximum peak around $2\theta\sim 10°$ in Fig.8. The peak below $2\theta\sim 30°$ in Fig. 8 is explained by the disordered arrangement of Cu atoms. The maximum peak at $2\theta\sim 12°$ ($Q\sim 0.93$ Å^{-1}) for $\alpha_r\sim 0.5$ in Fig. 8 would correspond to the formerly reported hump in the diffuse scattering of α-CuI [38]. This hump would be related to the short range order of the disordered arrangement of Cu atoms. It is found that the agreement between observed and calculated diffuse scattering intensities is obtained by the disordered arrangements model (Fm3m). To obtain a good agreement between the observed and calculated Bragg intensities in α-CuI, other structural models having disordered arrangement of Cu atoms would be available, for example 16 (e) sites with the space group $F\bar{4}3m$ and 32(f) sites with the space group Fm3m [39]. The structural models have to explain the peculiar diffuse scattering intensities of CuI.

$\Delta E = 0$ scan method (neutron elastic scattering measurement) is effective to separate the contribution from thermal vibration and static disorder to diffuse scattering [40]. In the strict sense there are contributions from over-damped thermal diffuse scattering (phonon mode) besides the elastic static diffuse scattering to the observed intensity by $\Delta E = 0$ scan method. A part of the acoustic branch near the zone center where Bragg peaks exist is also included within the limitation of the experimental energy resolution. This method was applied to the analysis of diffuse scattering from the single crystal of α-AgI. The calculation of the diffuse scattering intensity of AgI with $\Delta E = 0$ scan method was carried out based on the disordered distribution of two Ag atoms into 48(j) sites of space group Im3m . The qualitative feature of the observed intensity is explained by the calculation. This method would be effective to analyze the diffuse scattering intensities of superionic conductors.

By the synchrotron X-ray powder diffraction method the model which includes the disordered arrangement of Cu atoms over 8(c) sites was supported in the α-phase. A large spatial distribution of copper ions along the <111> directions around the 8(c) sites and diffusion pathway of mobiles copper ions along the <100> direction have been reported [31]. From the analysis of EXAFS spectrum of CuI, however, the likely conduction path between sites was suggested to be in the <111> directions. The nature of ionic motions in CuI had been studies with the molecular dynamics technique [41]. It is found that <100> jumps between tetrahedral sites are more frequent than <111> jumps [42]. To resolve the dynamic behavior of Cu atoms we need to perform a neutron inelastic scattering or other energy transfer measurements. The dynamic properties of mobile Cu ions in the superionic conductor CuI are studied by *ab initio* molecular-dynamics simulations. The covalent bonding around the Cu ions weakens when they diffuse in the octahedron cage. The ionicities of the Cu ions at the octahedral sites are larger than those at the tetrahedral sites [43].

6. Low-Energy Excitation in CuI

The dynamic scattering function $S(Q, \omega)$ is obtained from the inelastic neutron scattering spectra. The components of the density of state are approximated by the low-lying local vibrational mode at the lower-energy side and the phonon modes mainly due to the acoustic branch at the higher-energy side. From the model function we could obtain the frequency of the local vibrational mode ω and the half width of the local mode *Γ*.

A number of experiments have been made for various kinds of crystalline superionic conductors containing conducting ions. Low-energy dispersionless excitations which are 2.0-3.0 meV in the case of Ag ion conducting superionic conductors have been found. It is assumed that the excitation is due to an isolated vibrational mode of AgI_4 structural unit existing in the AgI crystal. The superionic conduction also appears in various kinds of amorphous or glassy electrolytes. If the short-range order for Ag atoms in the typical superionic glasses AgI-$AgPO_3$ was very alike with the crystalline material, the low-energy excitation of about 2.0 ~ 3.0 meV would be observed. The low energy excitation was observed near 2.0 ~ 3.0 meV in the composite glass by the inelastic neutron scattering measurements [44-46]. The excitation would be due to an isolated vibrational mode of AgI_4 structural unit in the glass. This may be an experimental evidence of the similarity of the crystals and glasses concerning the local vibrational mode arising from conducting ions.

The inelastic neutron scattering spectrum of CuI was measured at several temperatures by the use of the TOF spectrometer. The Q range covered by the spectrometer is 0.2 – 2.6 A^{-1} and the energy resolution was about 200 μeV (FWHM). The measurements were performed at seven Q positions where Bragg lines don't appear. A low-lying dispersionless excitation near 3.4 meV was observed in the inelastic scattering spectra of a powder sample of CuI over a wide range of Q at low temperature. As the temperature is increased the position of the excitation peak shifts to lower value and the intensities of the inelastic scattering spectra in the energy range from 1.0 to 3.0 meV increase. Considering the damping effect *Γ* into the analysis, it is found that the value of low-energy excitation is almost same over the temperature. The obtained value of low-energy excitation is 3.4 meV. The low-energy excitation of about 3.4 meV is common to other copper ion conductors. The temperature dependence of the damping factor *Γ* would be related to that of the ionic conductivity. The damping factor shows the degree of anharmonicity of thermal vibration, which is proportional to the inverse of the life time of the mode. In the case of CuI, the measured Debye-Waller temperature parameters increase rapidly above 200°C. As the thermal vibration becomes large, a mobile ion can easily diffuse over the barrier of the activation energy.

The inelastic neutron scattering experiment showed that the low-energy dispersionless excitation in Cu_2Se at room temperature was about 3.4 meV. The low-energy dispersionless excitations of 3.4 meV were also obtained in $Cu_{1.8}S$ and $Cu_{1.8}Se$. The low-energy dispersionless excitation 3.4 meV would be caused by the local vibration of Cu ions. Assuming that the excitation energies of Ag is 2.6 meV, those for Cu and Na are obtained as 3.4 meV and 5.7 meV, respectively, from the relation of excitation energy and cation mass. These calculated values almost coincide with the observed values in superionic conductors. A future work of the low-energy excitation in nanocrystals is expected in addition to crystals and glasses.

Conclusion

The high ionic conductivity of the high-temperature phase of CuI is related to the disordered arrangements of atoms and large Debye-Waller temperature parameters of ions. The crystal structures of γ-, β- and α-CuI belong to cubic system (ZnS type), trigonal system and cubic system (CaF_2 type), respectively. Many structural models with ordered and disordered arrangements of Cu atoms could explain the relative intensities of Bragg lines. AXS measurement and diffuse scattering measurement are effective to conclude the crystal structure of superionic conductors.

Acknowledgments

This research was partially supported by the Ministry of Education, Science, Sports and Culture, Grant-in-Aid for Scientific Research on Priority Areas, 17041001, 2006, Ibaraki Prefecture and Renkei Jigyou at Ibaraki University. Authors would like to thank to Prof. H. Takahashi, Dr. M. Arai and Dr. Y. Ishii for their useful discussions. Authors express their thanks to Mr. T. Satou for helping the anomalous X-ray scattering measurement.

References

[1] Sakuma, T. *B. Electrochem.* 1995, 11, 57-80.

[2] Sakuma, T.; Basar, K.; Shimoyama, T.; Hosaka, D.; Xianglian; Arai, M. In *Physics of Solid State Ionics*; Sakuma, T.; Takahashi, H.;Eds.; Static and dynamic structure in solid state ionics; Research Signpost, 2006, 323-346.

[3] Hoshino, S.; Fujishita, H.; Sakuma: T. *Phys. Rev.* 1982, *B*25, 2010-2011.

[4] Sakuma, T. *J. Phys. Soc. Jpn.* 1993, 62, 4150-4151.

[5] Sakuma, T.; Hoshino, S. *J. Phys. Soc. Jpn.* 1993, 62, 2048-2050.

[6] Sakuma, T.; Thomas, J. O. *J. Phys. Soc. Jpn.* 1993, 62, 3127-3134.

[7] Sakuma, T.; Aoyama, T.; Takahashi, H.; Shimojo, Y.; Morii, Y. *Physica* 1995, *B* 213&214, 399-401.

[8] Nield, V. M.; Keen, D. A.; Hyes, W.; McGreevy, R. L. *Solid State Ionics* 1993, 66, 247-258.

[9] Basar, K.; Shimoyama, T.; Hosaka, D.; Xianglian; Sakuma T.; Arai, M. *J. Thermal Anal. Cal.* 2005, 81, 507-510.

[10] Arai, M.; Shimoyama, T.; Sakuma, T.; Takahashi, H.; Ishii, Y. *Solid State Ionics* 2005, 176, 2477-2480.

[11] Sakuma, T.; Shimoyama, T.; Basar, K.; Xianglian; Takahashi, H.; Arai, M.; Ishii, Y. *Solid State Ionics* 2005, 176, 2689-2693.

[12] Arai, M.; Sakuma, T. *J. Phys. Soc. Jpn.* 2001, 70, 144-147.

[13] Sakuma, T.; Nakamura, Y.; Hirota, M.; Murakami, A.; Ishii, Y. *Solid State Ionics* 2000, 127, 295-300.

[14] Kim Y. I.; Izumi F. *J. Ceram. Soc. Jpn.* 1994, 102, 401-404.

[15] Sakuma, T.; Sugiyama, K.; Matsubara, E.; Waseda, Y. *Materials Transactions, JIM* 1989, 30, 365-369.

[16] Sugiyama, K.; Waseda, Y. *Materials Transactions, JIM* 1989, 30, 235-241.

[17] Sakuma, T.; Sugiyama, K.; Matsubara, E.; Waseda, Y. *Materials Transactions, JIM* 1989, 30, 365-369.

[18] Wagner, J. B.; Wagner, C. *J. Chem. Phys.* 1957, 26, 1597-1601

[19] Matsui,T.; Wagner, J. B. *J. Electrochem. Soc.* 1977, 124, 300-305
[20] Boyce, J. B; T.; Hayes, M.; Mikkelsen, J. C. *Phys. Rev.* 1981, *B*23, 2876-2896.
[21] Boyce, J. B.T.; Hayes, M.; Mikkelsen, J. C; Stutius, W. *Solid State Commu.* 1981, 33, 183-189.
[22] Arai, M.; Sakuma, T.; Atake, T.; Kawaji, H. *J. Thermal Analysis and Calorimetry* 2002, 69, 905-908.
[23] Sakuma, T.; Kaneko, T.; Takahashi, H.; Honma, K. *J. Phys. Soc. Jpn.* 1991, 60, 1136-1137.
[24] Beeken, R. B.; Dean, J. E.; Jetzer, W. L.; Lee, D. S.; Sakuma, T. *Solid State Ionics* 1992, 58, 189-191.
[25] Geller, S.; Ray, A. K.; Sakuma, T. *Solid State Ionics* 1983, 9, 1227-1232.
[26] Geller, S.; Sakuma, T. *Solid State Chem.* 1983, 50, 256-260.
[27] Miyake, S.; Hoshino, S.; Takenaka, T. *J. Phys. Soc. Jpn.* 1952, 7, 19-24.
[28] Bührer, W.; Hälg, W. *Electrochimica Acta* 1977, 22, 701-704.
[29] Matsubara, T. *J. Phys. Soc. Jpn.* 1975, 38, 1076-1079.
[30] Krug, J.; Sieg, L. *Z. Naturforsch.* 1952, 7*a*, 369-371.
[31] Matsubara, T. *Prog. Theor. Phys.* 1975, 53, 1210-1211.
[32] Yoshiasa, A; Okube, M.; Kamishima, O.; Arima, H.; Okudera, H.; Terada, Y.; Nakatsuka, A. *Solis State Ionics* 2005, 176, 2487-2491.
[33] Sakuma, T. *J. Phys. Soc. Jpn.* 1992, 61, 4041-4048.
[34] Yashima, M.; Xu. Q.; Yoshiasa, A.; Wada, S. *J. Mater. Chem.* 2006, 16, 4393-4396.
[35] Sakuma, T. *J. Phys. Soc. Jpn.* 1988, 57, 565-569.
[36] Keen, D. A.; Hull, S. *J. Phys. Condens. Matter* 1994, 6, 1637-1644.
[37] Keen, D. A.; Hull, S. *J. Phys. Condens. Matter* 1995, 7, 5793-5804.
[38] Chahid, A.; McGreevy, R. L. *J. Phys. Condens. Matter* 1998, 10, 2597-2609.
[39] Burns, G.; Alben, R.; Dacol, F. H.; Shafer, M. W. *Phys. Rev.* 1979, *B*15, 638-647.
[40] Hoshino, S.; Sakuma, T.; Fujishita; H.; Shibata, K. *J. Phys. Soc. Jpn.* 1983, 52, 1261-1269.
[41] Vashishta, P.; Rahman. A. In *Fast Ion Transport in Solids*; Vashishta, Mndy, Shenoy; Eds.; Nature of Ionic Motions in AgI and CuI; Elsevier North Holland, 1979, 535-540.
[42] Boyce, J. B.; Hayes, T. M. In *Fast Ion Transport in Solids*; Vashishta, Mndy, Shenoy; Eds.; EXAFS investigation of superionic conduction; Elsevier North Holland, 1979, 535-540.
[43] Shimojo, F.; Aniya, M. *J. Phys. Soc. Jpn.* 2003, 72, 2702-2705.
[44] Sakuma, T.; Shibata, K.; Hoshino, S. *Solid State Ionics* 1992, 53-56, 1278-1281.
[45] Takahashi, H.; Hiki, Y.; Sakuma, T.; Funahashi, S. *Solid State Ionics* 1992, 53-56, 1164-1167.
[46] Sakuma, T.; Shibata, K. *J. Phys. Soc. Jpn.* 1989, 58, 3061-3064.

In: Diffusion and Reactivity of Solids
Editor: James Y. Murdoch, pp. 227-241

ISBN: 978-1-60021-890-3

Chapter 6

OXYGEN DIFFUSION IN $YBA_2CU_3O_{7-X}$ AND ITS POTENTIAL APPLICATIONS

Xing Hu[1*]***, Delin Yang***[1] ***and Jie Hu***[1,2]
[1] School of Physical Engineering and Material Physics Laboratory, Zhengzhou University, Zhengzhou 450052, PR China
[2] Henan Textile College, Zhengzhou 450007, PR China

Abstract

Oxygen diffusion properties of high temperature superconductor material $YBa_2Cu_3O_{7-x}$ (YBCO) was studied by thermogravimetry (TG), oxygen static adsorption, oxygen permeability and resistance measurement. The non-isothermal TG experiment in air shows that the mass of sample exhibits periodic variation with temperature increase and decrease. The isothermal kinetic TG experiment indicates that the oxygen in-diffusion is faster than out-diffusion. The TG experiments with different heating rates indicates that between 500°~800°C the oxygen desorption activation energy has some relations with the oxygen stoichiometry of the material. The activation energy increases obviously with temperature in the range of 500°~650°C, from 184kJ/mol to 290kJ/mol. But the energy increases smoothly from 293kJ/mol to 315kJ/mol when temperature changing from 650°~800°C. The influences of oxygen partial pressure and temperature on saturated oxygen adsorption of the material were also evaluated by the static oxygen adsorption experiments. The application of YBCO membrane in the process of partial oxidation of methane (POM) to syngas was also investigated. Methane conversion, CO and H_2 selectivity can reach almost 100%, 95%, and 86% respectively at 900°C. However, the stability of YBCO in reducing atmosphere is questionable because of the reduction of copper from the YBCO membrane.

Key words: Oxygen diffusion; $YBa_2Cu_3O_{7-x}$; Oxygen permeation membranes

* E-mail address: xhu@zzu.edu.cn,; Tel: 8637167767671; Fax: 8637167766629.(Corresponding author.)

1. Introduction

High temperature superconducting material $YBa_2Cu_2O_{3-x}$ (YBCO) has a layered perovskite orthorhombic (P_{mmm} space group) structure with a mixed copper valence, and therefore, variable oxygen stoichiometry. The Superconducting properties is related closely with the oxygen content in YBCO and oxygen diffusion in YBCO have been widely studied. The results of oxygen diffusion for different experiment methods and different temperature ranges are not in agreement or even contradicted [1-10], therefore, there is no convincible and acceptable data. The reason may be partially due to the fact that some of the methods used may not be suitable for giving reliable results. There seems also to be disagreement about the exact mechanism of diffusion. Both the interstitial mechanism[1,11] and vacancy mechanism[4] have been proposed to account for the observed oxygen diffusion in YBCO. Furthermore, some researcher reported that the activation energy for oxygen diffusion or diffusion coefficient is very anisotropic and is different along the c-axis from that in the a-b plane. But the anisotropic data are also disparate. Maier et al.[12] reported that the activation energy is about 3 times higher along the c-axis of YBCO than it is in the a-b plane. However Tsukui et al.[13] obtained that in the orthorhombic phase the activation energies in the a-b plane and along the c-axis are almost the same, but the diffusion coefficients in the a-b plane are larger than that in the c-axis direction by more than three orders of magnitude. At higher temperatures in the tetragonal phase, the activation energy in the c-axis direction is about 3 times larger than that in the a-b plane, but the diffusion coefficients of the c-axis direction and of the a-b plane become closer each other with the increase of temperature. The other reported results[14-17] of the oxygen diffusion coefficient in YBCO scatter also by several orders of magnitude.

In fact, the oxygen diffusion in YBCO is a complex process and has relation with stoichiometry, temperature, and oxygen partial pressure of the circumstance. The stoichiometry itself correlates with oxygen pressure and temperature. The results of oxygen diffusion experiments are method correlation. In this article we report our experiment results of oxygen diffusion of YBCO and its application as oxygen permeation membrane.

2. The Static Oxygen Adsorption of YBCO

The influence of temperature and oxygen partial pressure on oxygen adsorption of YBCO was measured with a set of static adsorption equipments shown in Fig. 1[18]. In this experiment, small pellets about 50 g of YBCO were put into a stainless steel cell. The temperature of the cell was heated to 950°C and a mechanical pump pumped the oxygen released by the samples. The cell was kept at vacuum and cooled down to a measured temperature, and then a certain amount of oxygen was introduced into the cell. Since the samples were in an oxygen deficient state they will adsorb oxygen and reach balance with a certain oxygen partial pressure. From the oxygen partial pressure at beginning and balance time and the volume of the cell, the amount of oxygen adsorbed by the samples can be obtained. Several isothermal lines were measured for different temperatures.

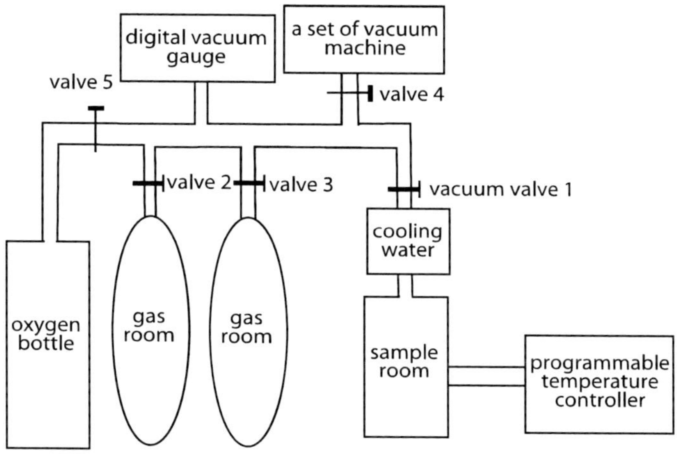

Figure 1. The sketch of the statistic adsorption experiment.

From the oxygen partial pressure at beginning and balance of adsorption and the volume of the cell the amount, the oxygen adsorbed by the samples can be obtained. The relationship between oxygen vacancy C_v, x and sample weight change (%) is

$$x = weight\ change\ (\%) / \mathbf{2.4} \qquad C_v = x / V_{mol} \tag{1}$$

where V_{mol} is mole volume of YBCO which is $107.45cm^3/mol$. The sample mass change of 2.4% corresponds to x change of 1, correspondingly about $16cm^3$ oxygen will be released or absorbed by 1g sample. We will use x to indicate the oxygen vacancy concentration in this paper.

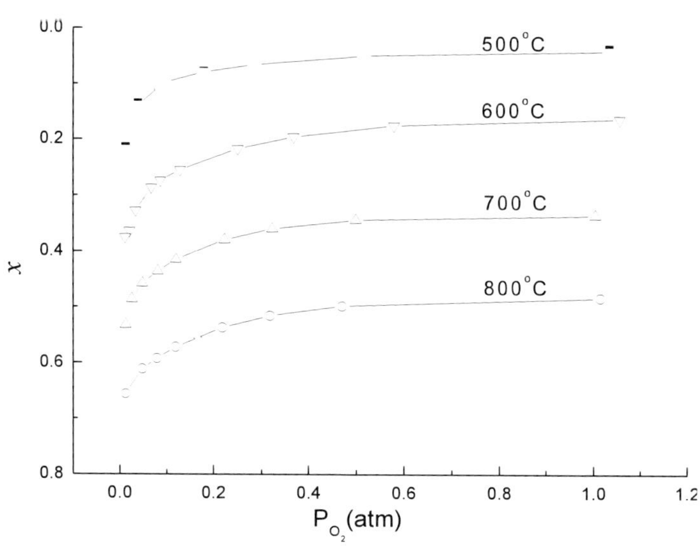

Figure 2. Dependence of equilibrium oxygen vacancy concentration x on oxygen partial pressure at different temperature. The solid lines are obtained by Eq.(1).

The dependence of equilibrium oxygen vacancy concentration x on oxygen partial pressure at different temperature is shown in Fig.2. From Fig.2, one can see that: (1) as for different isothermal lines the amounts of saturated adsorption decrease gradually with temperature increasing. Thus x increases with temperature increasing at same oxygen partial pressure. It indicates that the absorption of oxygen and desorption reaches a homeostasis at that temperature, oxygen content at homeostasis comes down with rising temperature, and more oxygen vacancy will be generated at a higher temperature. (2) the amounts of saturated adsorption increase gradually with oxygen pressure increasing along any isothermal lines. The increase rates are very rapidly below about 20kPa and become slowly above 20kPa. With the oxygen pressure increase the amount of saturated adsorption tends to a constant at a certain temperature. Fig.3 gives a plot of x vs $\log(P_{O_2})$ which shows a linear behavior approximately and is consisted with the result of Kishio et al obtained by thermogravimetric method[1]. The experimental result indicates that at comparative low oxygen partial pressure (20kPa, for example) the amount the absorption of oxygen into YBCO reaches almost its maximum at this temperature.

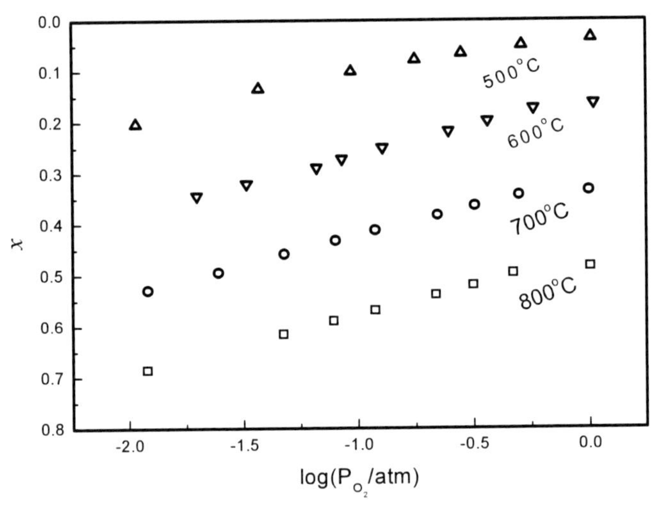

Figure 3. Plots of x versus $\log(P_{O_2})$.

3. Transient Thermogravimetric Study

Two kinds of thermogravimetric (TG) experiment were carried out with a thermal analyzer (SETARAM Labsys™). The isothermal kinetic experiments or the transient thermogravimetric experiments were done on YBCO powder with the average particle size 1.67μm and specific surface 3890cm^2/g. The powder was heated with a certain rising temperature speed to a certain temperature and held at this temperature in flowing nitrogen atmosphere. When the weight of the sample did not change any more the atmosphere was switched to oxygen. After the sample adsorbed oxygen and reached equilibrium the oxygen atmosphere was switched to nitrogen once again.

Another TG experiment intent to obtain the oxygen desorption activation energy of YBCO was performed with heating rate γ=5, 10, 15, and 20K/min respectively. To ensure the equal initial oxygen content of the sample each experiment was made up of two cycles. The data for calculating activation energies came from the second cycle. The detail of above TG experiments can be found in our previous work [18~20].

Figure 4 shows the mass variation with temperature that indicates the change of oxygen content of the sample with temperature. We can see that the mass loss begins obviously at about 400°C and arrives at a value of 1.2% of its original mass (corresponding to 0.5 oxygen atom released per cell) at 800°C. Oxygen content can recover completely to its original value when temperature descends to 400°C. The mass changes with temperature variation show a very good repetition.

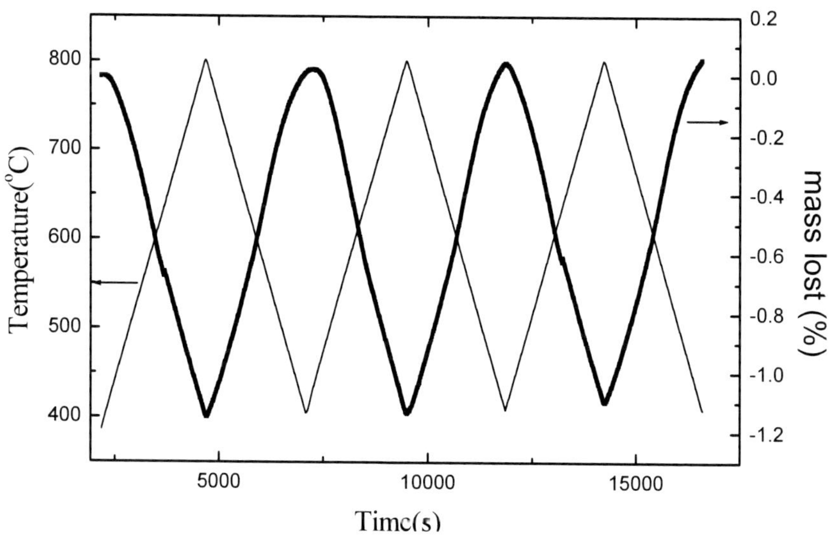

Figure 4. The relationship between sample mass variation and temperaturc.

Fig. 5 shows the 850°C isothermal oxygen in-diffusion and out-diffusion. Such result at other temperatures is similar with Fig. 5. From Fig. 5, we can see that when nitrogen is changed to oxygen atmosphere the weight of the sample increases very quickly and reaches balance in a very short time. However when oxygen switches to nitrogen, the weight decreases slowly and needs a quite long time to reach its balance value. This indicates that the rate of oxygen absorption is remarkably faster than the rate of oxygen desorption, and the results consisted with many other authors' discussions[4,10], they assumed that the mechanisms of oxygen in-diffusion and out-diffusion were different.

Many authors have investigated the oxygen kinetics of $YBa_2Cu_3O_{7-x}$ at high temperatures since the discovery of YBCO[4,6,8,10,21-24] and discrepancy still exists between the results of different authors. Tu[25] et al stated that the out-diffusion of oxygen from YBCO is independent of x and its rate is surface-reaction limited with activation energy of 1.7eV, while the in-diffusion of oxygen has a strong dependence on x. They had proposed a defect mechanism for the anisotropic diffusion of oxygen in the CuO plane to explain their experimental results. Kishio[1] et al studied the chemical diffusion of oxygen in $YBa_2Cu_3O_{7-x}$ in temperature range from 550°C to 850°C under oxygen pressure from 1 to 10^{-2} atm by thermogravimetry. They found that the chemical diffusion coefficient of oxygen is strongly sensitive to the oxygen composition. The diffusion coefficient sharply increases with the decrease of x. They thought that the oxygen diffusion should be viewed to proceed via an interstitial-like mechanism. Jin[26] et al reported that oxygen out-diffusion of porous samples is slower than its in-diffusion, while the rates of oxidation and reduction in powders and dense samples are equal. By a solid-state potentiostatic step technique Gür[7] et al found that the oxygen chemical diffusion coefficient was $5\times10^{-8}cm^2/s$ at 800°C. They reported that the out-diffusion rates were not affected by the oxygen stoichiometry and indicated definitely that YBCO is a mixed-conductor at elevated temperatures and exhibits magnificent p-type electrical conductivity in which both holes and oxygen ions are mobile and contribute to the total charge–transport.

As a matter of fact, the diffusion of oxygen in YBCO is controlled by temperature, oxygen content of the sample, and oxygen partial pressure around the sample. The isothermal oxygen in-diffusion and out-diffusion experiment indicates that oxygen in-diffusion is very fast for an oxygen deficient sample in oxygen atmosphere, while the out-diffusion rate is relatively slow for an oxygen deficient sample in lower oxygen pressure (in nitrogen).

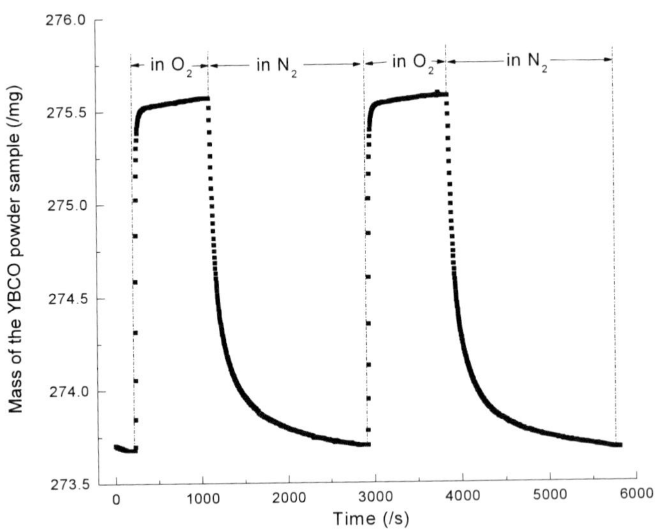

Figure 5. The mass variation of YBCO with atmospheres at 800°C.

Oxygen transport in the small YBCO grains involves surface reactions on the grain surface and diffusion of the oxygen ions in the grain bulk phase. However, if the grain size is small enough the surface reaction will become the main factor controlling the oxygen absorption and desorption[27]. In our transient thermogravimetric measurement, the average size of YBCO powder is less than 2μm. Therefore, we can assume safely that the surface reactions are the rate limiting steps for oxygen transport in the YBCO grains and the oxygen vacancy in the grain can be assumed a uniform concentration profiles. So that we can use the model proposed by Zeng and Lin to obtain the lumped surface reaction rate constant of transient oxygen transport into (or out of) the solid grains undergoing a change in surrounding oxygen partial pressure. The detail of calculation can be found elsewhere[20]. The result is shown in table 1. It can be seen that surface reaction rate constant for the oxygen absorption

k_a is much larger than that of the oxygen desorption k_d. As a result the sample shows a rapid weight gain in the oxygen absorption period and a slow weight loss in the oxygen desorption period.

Table 1. Values of k_a, and k_d, at different temperature

	500°C	600°C	700°C	800°C
k_a (10^{-6}cm.s^{-1})	0.836	1.037	1.378	1.561
k_d (10^{-6}cm.s^{-1})	0.111	0.123	0.203	0.25

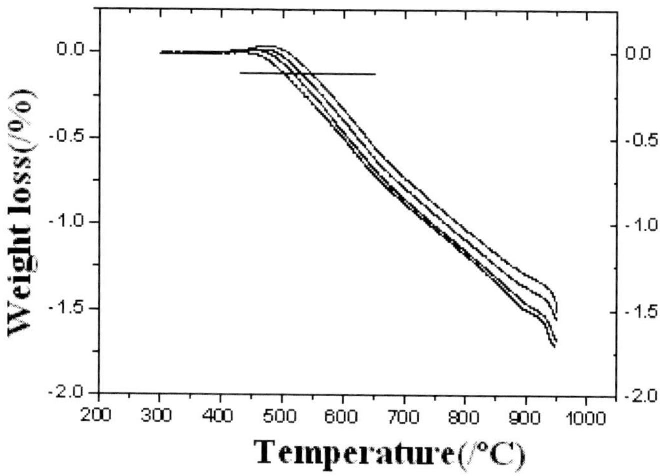

Figure 6. Weight loss versus temperature of YBCO at different heating rates (from left to right: 5, 10, 15 and 20°C/min) in static air circumstance. The horizon line cross the four curves denotes the same mass loss.

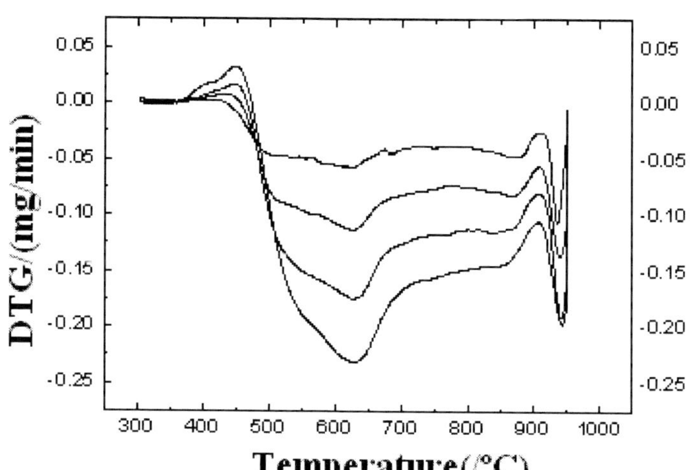

Figure 7. Weight loss rate versus temperature for different heating rates (from top to bottom: 5, 10, 15 and 20°C/min).

Fig. 6 shows the TG curves with heating rate of 5, 10, 15 and 20°C/min in static air circumstance respectively and Fig. 7 shows DTG curves. From Fig. 6, we can see that there is a clear change of the TG curves slope from about 650°C with a little difference for different heating rate. The slope change can be more clearly seen from the DTG curves. The mass loss rates have a rapidly increase and reach a maximum at about 650°C. The transition from orthorhombic phase to tetragonal phase with oxygen content about 6.5 may occur at this temperature. After the phase transition the mass loss rates have a rapid decrease until 700°C and then the rates are into a range of smooth variation. A second rapid decrease of the mass loss rate takes place at 860°C followed by a speedy increase of the rate. This maybe means that the oxygen in the basal plane almost exhausted and oxygen in other sites began to diffuse out.

The oxygen desorption activation energy of YBCO can be calculated by assuming that the desorption of oxygen from the YBCO obeys the Arrhenius equation, the detail is given in Ref. [19]. The curve of the calculated activation energy versus temperature is shown in Fig. 8. The interesting feature of the oxygen desorption activation energy of YBCO is that we can divide the dependence of E on T into two ranges approximately. In the first range of 500~650°C, the activation energy increases obviously with temperature elevation from 184kJ/mol (1.9eV) to 290kJ/mol (3.01eV). However, E varies very smoothly in the second range of 650~800°C and increases from 293kJ/mol (3.04eV) to 315 kJ/mol (3.27eV). This behavior of the activation energy of YBCO may be caused by the phase difference in these two temperature ranges. When T is below about 650°C, the sample is in its orthorhombic phase, while when T is above 650°C, the specimen is in its tetragonal phase. The detailed discussion can be found in [19].

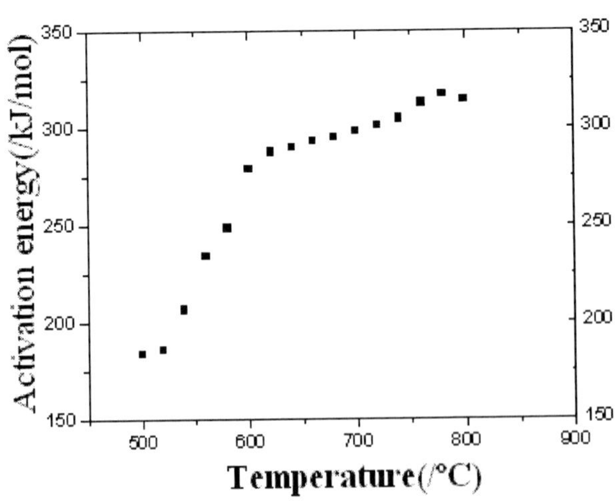

Figure 8. The dependence of activation energy of YBCO on temperature.

4. The Oxygen Permeability in YBCO

Oxygen permeability experiments were carried out in a vertical high-temperature gas permeation system, as shown in Fig. 9. The YBCO membrane disks with different thickness were respectively sealed to alumina tubes using a kind of high-temperature glue. One side of the sealed membrane was exposed to air and the other side to flowing high purity helium. The composition of the effluent helium stream was analyzed with an gas chromatograph.

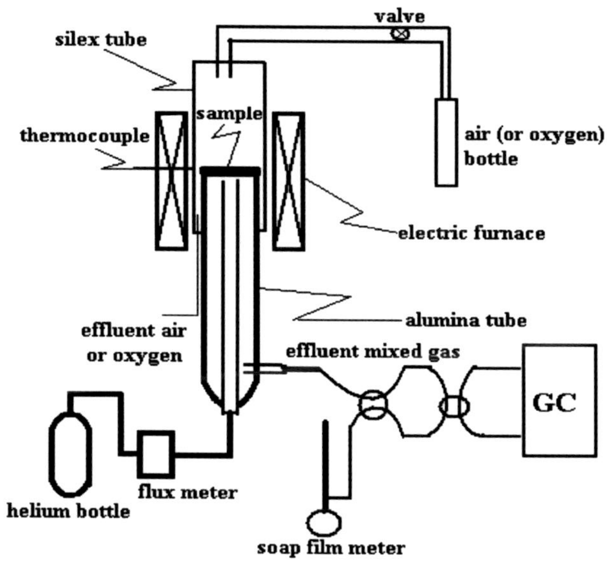

Figure 9. High temperature gas permeation measuring system.

The oxygen permeation rate is shown in Fig. 10. From the figure we can see that thinner membranes yield higher oxygen flux. An oxygen flux of 3.36×10^{-7} mol·cm^{-2}·s^{-1} was observed for the 1.00mm thick membrane at 900°C, while 1.51×10^{-7} mol·cm^{-2}·s^{-1} for the 2.23mm thick membrane at the same temperature. The oxygen flux increases with temperature increase as one expected since oxygen permeation is a thermally activated process. When the atmosphere at the feed side changed from air to pure oxygen we found that the oxygen permeation flux increases only slightly. This is because oxygen content of YBCO in air is almost as the same as in oxygen as Fig.5 shown. Therefore, the oxygen permeation flux through the YBCO membranes didn't increase much.

The oxygen permeation flux density J_{O_2} can be theoretically calculated by the formula (in our research, no nonaxial transport of oxygen, taking G=1) [32]:

$$J_{O_2} = \frac{\sigma_{amb}}{4FL}(E-\eta) \tag{10}$$

where L is membrane thickness, η the driving force consumed by surface oxygen exchange, σ_{anb} ambipolar conductivity, respectively. And E is the driving force for oxygen permeation expressed by

$$E=\frac{RT}{4F}\ln\frac{P_{O_2}(h)}{P_{O_2}(l)} \quad (11)$$

If the surface process is much faster than the bulk process, i.e. η/E is negligible, a linear relation for $J_{O_2} E^{-1}$ vs 1/L should hold and it should pass through the origin point. However, using the datum in Fig. 18, we cannot obtain a linear relation for $J_{O_2} E^{-1}$ vs 1/L. This means that in case of YBCO the surface barrier cannot be negligible comparing with E. The surface barrier is due to a stable shell of $YBa_2Cu_4O_8$ as discussed by Shi et al[6]. The difference of oxygen flux for samples with different thickness may be caused by the oxygen gradient force since for the same $P_{O_2}(h)$ and $P_{O_2}(l)$ the thicker the membrane is, the weaker the gradient force. A further work to improve the oxygen permeation flux of YBCO by elements doping can be found in Ref. [29].

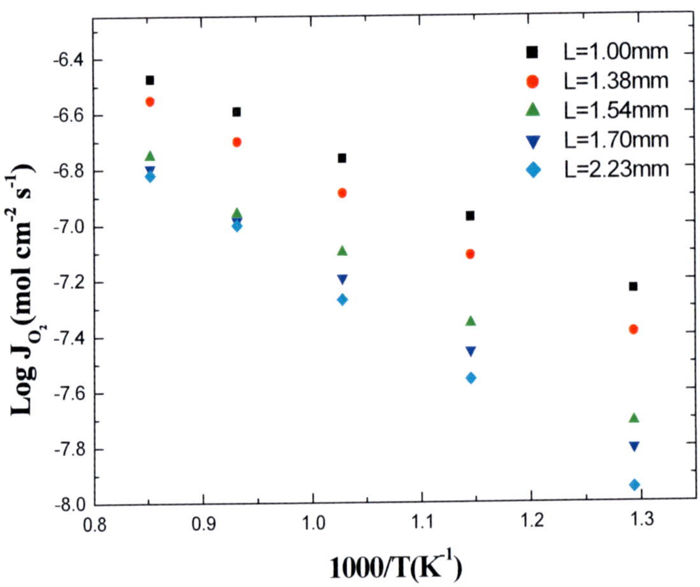

Figure 10. Temperature dependence of the oxygen permeation flux through dense YBCO membranes with different thickness.

Using the oxygen permeation measuring system the performance of YBCO membrane reactor in a partial oxidation of methane (POM) to syngas processes was also studied. Methane conversion X, CO and H_2 selectivity S, and the oxygen permeation flux J_{O_2} were calculated[30]. The results are shown in Fig. 11. As can be seen, at 900°C, CH_4 conversion, CO and H_2 selectivity reach almost 100%, 95%, and 86% respectively. The CH_4 conversion increases monotonously with the temperature rise. While, CO selectivity increases firstly with the increase of temperature and increases only a little from 850 to 875°C. Beyond 875°C the CO selectivity drops with the further increase of temperature. The H_2 selectivity augments slightly with the increase of temperature in the range 800 to 875°C, but decreases at 900°C due to the decrease of CO selectivity. The decrease of CO selectivity can be understood since the CO selectivity is related to not only the temperature but also the CH_4/O_2 radio. A higher temperature is favorable to endothermic reforming reaction and to increasing CO selectivity. However, CO selectivity will decrease when the CH_4/O_2 radio decreases ($CH_4/O_2 \leqslant 2$)[31]. At 900°C the CH_4/O_2 radio will become small than 2 because the oxygen permeation flux will become larger at high temperature. CH_4/O_2 radio plays an important role for CO selectivity. A small CH_4/O_2 radio will lead to a poor CO selectivity. Therefore, CO selectivity decreased at high temperature.

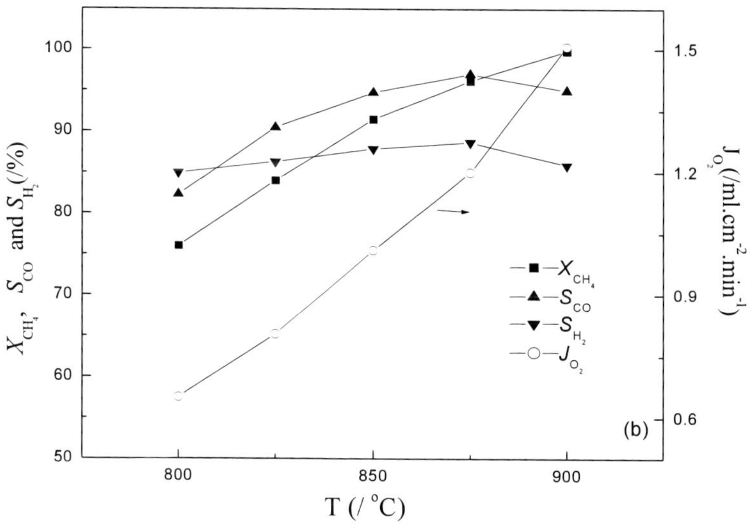

Figure 11. Temperature dependences of CH_4 conversion, CO, H_2 selectivity, and oxygen permeation flux. Reaction condition: feed flow 50ml/min, CH_4 6.0%(v%), SV=8000h^{-1}, Ni/ZrO_2 catalyst.

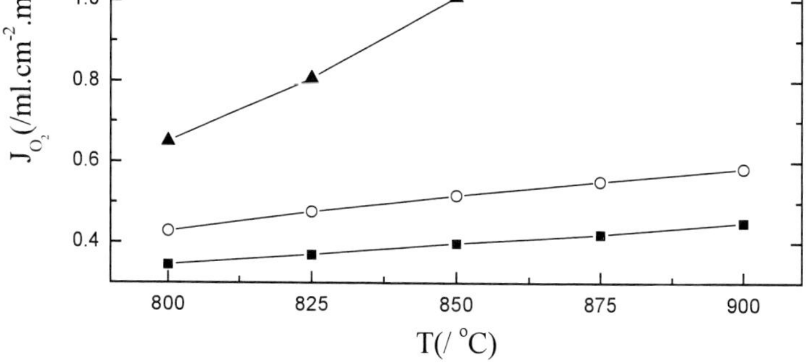

Figure 12. The dependence of oxygen permeation flux on temperature and atmosphere.

It was point out that an oxygen permeation flux higher than $1 ml\cdot min^{-1}\cdot cm^{-2}$ is required for actual applications[32]. Fig. 12 presents the dependencies of oxygen permeation fluxes of YBCO membranes on temperature under different feed atmosphere. It can be seen that the oxygen permeation fluxes under CH_4/He atmosphere can reach $1.5 ml\cdot min^{-1}\cdot cm^{-2}$ at 900°C, increasing about 3~4 times compared to that under pure He atmosphere. The increase of oxygen permeation fluxes under CH_4/He atmosphere can been explained as follows. When methane oxidation occurs the permeated oxygen is consumed fully resulting a very lower oxygen partial pressure in the reactor, which in turn induces a higher oxygen gradient between the two sides of the membrane and increases the oxygen permeation flux (or oxygen diffusion rate) greatly. Similar findings were reported by Wang[33] and Kharton[34] et al.

Although YBCO shows considerable oxygen permeation flux, its stability in reducing atmosphere is questionable because of the reduction of copper from the YBCO membrane[30]. A further study have shown that replacing part of Cu by Co can improve the stability of YBCO membrane in the POM process[35].

5. The Influence of Oxygen Diffusion on Electricity

A YBCO thick film prepared by Sol-gel method was used to measure its resistance dependent on oxygen partial pressure[36]. Fig. 13 shows the dependence of resistivity on oxygen pressure at 650°C. We can see that when the atmosphere was shift from lower to higher oxygen partial pressure atmosphere, the resistivity drops down drastically and reach its balance value very fast. This proves that resistivity of YBCO is very sensitive to increasing oxygen pressure. When the atmosphere was shift from higher to lower oxygen partial pressure atmosphere, a more longer time is needed for the resistivity reaching its equilibrium value. This experiment also proved the result that the rate of oxygen absorption is remarkably faster than the rate of oxygen desorption.

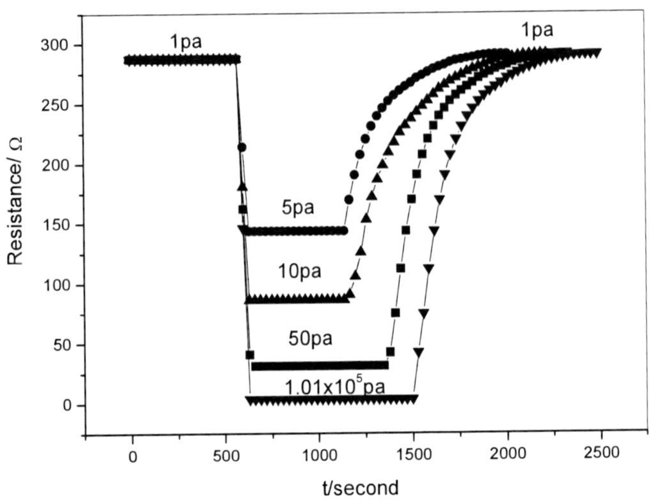

Figure 13. The dependence of resistivity on oxygen pressure at 650°C.

6. Conclusion

The static oxygen adsorption, thermogravimetric (TG), oxygen permeability and resistance measurement experiments were used to study the oxygen diffusion of YBCO. The static oxygen adsorption indicates that oxygen content has a close relationship with temperature and surrounding oxygen pressure. The lower the temperature and the higher the surrounding oxygen pressure, the higher the oxygen content in the YBCO. The isothermal oxygen in-diffusion and out-diffusion experiment and resistance measurement experiments indicates that oxygen in-diffusion is faster than that of oxygen out-diffusion. The oxygen desorption activation energy of YBCO changes with temperature. In the lower temperature range (500~650°C) the activation energy increases obviously with temperature, while in the higher temperature range (650~800°C) the activation energy varies very smoothly. The change from a orthorhombic to a tetragonal phase may be responsible for the variation of activation energy with temperature. The oxygen permeation flux of YBCO membrane increases with the increase of temperature as show in the oxygen permeation experiment. An oxygen flux of 3.36×10^{-7} $mol\cdot cm^{-2}\cdot s^{-1}$ was observed at 900°C for a 1.00mm thick YBCO membrane. The performance of YBCO membrane reactor in a partial oxidation of methane (POM) to syngas processes was also studied. Methane conversion, CO and H_2 selectivity can reach almost 100%, 95%, and 86% respectively at 900°C. However, the stability of YBCO in reducing atmosphere is questionable because of the reduction of copper from the YBCO membrane. Further study to improve the stability of YBCO membrane in the POM process is needed.

References

[1] K. Kishio, K. Suzuki, T. Hasegawa, T. Yamamoto, K. Kitazawa, and K. Fueki, Study on chemical diffusion of oxygen in $Ba_2YCu_3O_{7-\delta}$, *Journal of Solid State Chemistry* 82 (1989), 192-202.

[2] D. Shi, J. Krucpzak, M. Tang, N.Chen, and R. Bhada, Oxygen diffusion and phase transformation in $YBa_2Cu_3O_{7-x}$, *J. Appl. Phys.* 66 (1989), 4325-4328.

[3] J. Zhang, M. Yang, and T. Chen, Oxygen diffusion in tetragonal $Y_1Ba_2Cu_3O_{7-x}$ crystals *Mat. Lett.*, 6 (1988), 379-384.

[4] K.N. Tu, N.C. Yeh, S.I. Park, and C.C. Tsuei, Diffusion of oxygen in superconducting $YBa_2Cu_3O_{7-\delta}$ ceramic oxides, *Phys. Rev. B*, 39 (1989), 304-314.

[5] J. MacManus, D. Fray, and J. Evetts, An electrochemical method for the determination of the chemical diffusion coefficient of oxygen in $Y_1Ba_2Cu_3O_{7-x}$ as a function of vacancy concentration, *Physica C* 169(1990), 193-200.

[6] L.T. Shi, and K.N. Tu, Formation of a stable shell of $YBa_2Cu_4O_8$ upon oxygen outdiffusion from superconducting $YBa_2Cu_3O_{7-\delta}$ oxides, *Appl. Phys. Lett.* 59 (1991), 2040-2042.

[7] T.M. Gür, and R.A. Huggins, High Temperature Oxygen transport and Electrochemical Behavior of $YBa_2Cu_3O_x$, *J. Electrochem. Soc.* 140(1993), 1990-2000.

[8] J.S. Choi, M. Sarikaya, I.A. Aksay, and R. Kikuchi, Theory of oxygen diffusion in the $YBa_2Cu_3O_{7-x}$ superconducting compound, *Phys. Rev. B* 42 (1990), 4244-4255.

[9] K.N. Tu, S.I. Park, and C.C. Tsuei, Diffusion of oxygen in superconducting $YBa_2Cu_3O_{7-\delta}$ oxide upon annealing in helium and oxygen ambients, *Appl. Phys. Lett.,* 51 (1987), 2158-2160.

[10] X.M. Xie, T.G. Chen, and Z.L. Wu, Oxygen diffusion in the superconducting $YBa_2Cu_3O_{7-x}$, *Phys. Rev. B* 40 (1989), 4549-4556.

[11] H.L. Tuller and E. Opila, Defects and oxygen diffusion in high T_c superconductors, *Solid State Ionics* 40-41(1990), 790-794.

[12] J. Maier, P. Murugaraj, and G. Pfundtner, Defect chemistry and anisotropic transport in $YBa_2Cu_3O_{6...7}$, *Solid State Ionics*, 40-41(1990), 802-806.

[13] S. Tsukui, M. Adachi, R. Oshima, H. Nakajima, F. Toujou, K. Tsukamoto, and T. Tabata, Oxygen tracer diffusion in the $YBa_2Cu_3O_y$ superconductor, *Physica C* 351 (2001) 357-362.

[14] K. Conder, Oxygen diffusion in the superconductors of the YBaCuO family: isotope exchange measurements and models, *Materials Sci and Engin*., R32(2001), 41~102.

[15] S.J. Rothman, J.L. Routbort, and J.E. Baker, Tracer diffusion of oxygen in $YBa_2Cu_3O_{7-\delta}$, Phys. Rev. B 40(1989), 8852-8860.

[16] Y. Ikuma and S. Akiyoshi, Diffusion of oxygen in $YBa_2Cu_3O_{7-y}$, *J. Appl. Phys.* 64(1988), 3915-3917.

[17] S. Tsukui, T. Yamamoto, M. Adach, Y. Shono, K. Kawabata, N. Fukuoka, S. Nakanishi, A. Yanase, and Y. Yoshioka, Direct Observation of ^{18}O Tracer Diffusion in a $YBa_2Cu_3O_y$ Single Crystal by Secondary Ion Mass Spectrometry, *Jpn. J. Appl. Phys.* 30(1991), L973-L976

[18] D.L. Yang, H.X. Lu, H.Z. Song, J. Mo, G.X. Li, C.P. Chen, Y.Q. Guo, and X. Hu, Experimental study of oxygen diffusion and permeation through $YBa_2Cu_3O_{7-x}$ membranes, *Journal of Membrane Science* 233 (2004), 45~50.

[19] Z.L. Zhu, D.L. Yang, Y.Q. Guo, Q.Q. Liu, Z.S. Gao, X. Hu, Oxygen desorption activation energy of $YBa_2Cu_3O_{7-x}$ obtained by thermogravimetry with different heating rates, *Physica C* 383(2002), 169-174.

[20] J. Hu, X. Hu, H.S. Hao, L.J. Guo, H.Z. Song, and D.L. Yang, A transient thermogravimetric study on the oxygen permeation at high temperature of the superconducting materiel $YBa_2Cu_3O_{7-x}$, *Solid State Ionics* 176 (2005), 478-194.

[21] J.D. Jorgensen, B.W. Veal, A.P. Paulikas, L.J. Nowicki, G.W. Crabtree, H. Claus and W.K. Kwok, Structural properties of oxygen-deficient $YBa_2Cu_3O_{7-\delta}$, *Phys. Rev. B* 41(1990), 1863-1877.

[22] Z. Zhang, and C.R.A. Catlaw, Molecular dynamics study of oxygen diffusion in $YBa_2Cu_3O_{6.91}$, *Phys. Rev. B* 46 (1992) 457-462.

[23] G. Ottaviani, C. Nobili, F. Nava, M. Affronte, T. Manfredini, F.C. Matacotta, and E. Galli, Oxygen in-diffusion processes in tetragonal $YBa_2Cu_3O_{7-x}$ oxide, *Phys. Rev. B* 39 (1989), 9069-9073.

[24] J. Molenda, A.Stoklosa and T. Bak, Transport properties of $YBa_2Cu_3O_{7-y}$ at high temperatures, *Physica C* 175 (1991) 555-565.

[25] K.N. Tu, C.C. Tauei, S.I. Park, and A. Levi, Oxygen diffusion in superconducting $YBa_2Cu_3O_{7-\delta}$ oxides in ambient helium and oxygen, Phys. Rev. B 38 (1988) 772-775.

[26] X. J. Jin, and L. Li, In situ X-ray studies of oxygen in- and out-diffusion in c-axis-oriented $YBa_2Cu_3O_{7-\delta}$ films, *J. Mater. Sci. Lett.*, 19(2000), 855-857.

[27] Y. Zeng and Y. S. Lin, A transient TGA study on oxygen permeation properties of perovskite-type ceramic membrane, *Solid State Ionics* 110(1998), 209-221

[28] C.S. Chen, Z.P. Zhang, G.S. Jiang, C.G. Fan, W. Liu, and H.J.M. Bouwmeester, Oxygen permeation through $La_{0.4}Sr_{0.6}Co_{0.2}Fe_{0.8}O_{3-\delta}$ membrane, *Chem. Mater.* 13 (2001), 2797-2800.

[29] H.S. Song, M.M. Huang, D.L Yang, X. Hu, Y.X. Li, Oxygen permeability of perovskite-type$Y_{1-x}M_xBa_2Cu_3O_{7-\delta}$ (M = La, Ca) membranes, *Mat. Sci. and Eng. B* 137 (2007), 284–288

[30] J. Hu, T.L. Xing, Q.C. Jia, H.S. Hao, D.L. Yang, Y.G. Guo, and X. Hu, Methane Partial Oxidation to Syngas in $YBa_2Cu_3O_{7-x}$ Membrane Reactor, *Applied Catalysis A-General* 306 (2006), 29-33.

[31] S. Freni, G. Calogero, and S. Cavallaro, Hydrogen production from methane through catalytic partial oxidation reactions *J. Power Sources* 87 (2000), 28-38.

[32] B.C.H. Steele, Oxygen ion conductors and their technological applications, *Mater. Sci.& Eng. B* 13 (1992), 79-87.

[33] H.H. Wang, Y. Cong, and W.S. Yang, Investigation on the partial oxidation of methane to syngas in a tubular $Ba_{0.5}Sr_{0.5}Co_{0.8}Fe_{0.2}O_{3-\delta}$ membrane reactor, *Catal. Today* 82 (2003), 157-166.

[34] V.V. Kharton, A.A. Yaremchenko, and A.A. Valente, Methane oxidation over Fe-, Co-, Ni- and V-containing mixed conductors, *Solid State Ionics* 176(2005), 781-791.

[35] J. Hu, T.L. Xing, Q.C. Jia, H.Z. Yang, H.W. Sun, and X. Hu, Improving the POM performance of $YBa_2Cu_3O_{7-\delta}$ membrane reactor by Co doping, *Catalysis Communications* 8 (2007), 1301-1304

[36] Z.L. Zhu, L.J. Guo, H.Y. Wang, G.X. Tian, H.X. Lu, D.L. Yang, X.J. Li and X. Hu, Potential Application of Superconducting Material $YBa_2Cu_3O_{7-x}$ as Oxygen Resistance Sensors, *International Journal of Modern Physics B* 19 (2005), 3923-3932.

INDEX

A

absorption spectra, 25
absorption spectroscopy, 129
AC, 13, 115, 117, 119, 121, 123, 125, 127, 128, 131, 133, 135, 137, 139, 141, 143, 145, 146, 149, 151, 153, 155, 157, 159, 161
accessibility, 74, 89, 90, 106
accommodation, 201
accuracy, 125
acetic acid, 102
acetone, 102, 105, 107, 176
achievement, 182
acid, vii, 2, 47, 60, 102, 104, 107, 164, 167, 168, 181
acidity, vii, 2, 48, 57, 58, 59, 61, 63
activation, x, 71, 74, 81, 82, 83, 84, 86, 88, 89, 90, 97, 105, 118, 120, 128, 132, 133, 134, 148, 149, 151, 153, 155, 157, 158, 176, 182, 224, 227, 228, 231, 234, 239, 240
activation energy, x, 118, 120, 128, 132, 133, 134, 148, 149, 151, 153, 155, 157, 158, 224, 227, 228, 231, 234, 239, 240
active site, 3, 79
additives, viii, 2, 47, 59, 61, 62, 63, 71, 167, 176, 178, 181
adsorption, x, 77, 89, 227, 228, 229, 230, 239
agent, 75, 78, 169, 199, 200
aggregates, 75, 78
aggregation, 39, 75, 78, 80
aging, 76, 95, 96, 97, 98, 100, 105, 106
Al2O3 particles, 56, 58
alcohol(s), 26, 47, 82, 104
alloys, 73
alternative, 70, 71, 72, 73, 178, 181, 185
aluminium, 76, 81, 86, 176
ammonium, 76, 167, 191
anion, 185
annealing, 3, 15, 61, 167, 240
Argentina, 107
argon, 4, 43, 52, 129, 168, 177
argument, 9, 141
aromatic hydrocarbons, 98
Arrhenius equation, 234
ascorbic acid, 168
assessment, 142
atomic distances, 217, 218
atomic positions, 213, 217, 220
atoms, ix, 42, 45, 74, 79, 84, 86, 89, 169, 171, 176, 185, 189, 194, 196, 197, 209, 210, 211, 212, 213, 214, 216, 217, 219, 220, 221, 222, 223, 224
attacks, 40
attention, 18, 71, 73, 80, 190, 197
attribution, 83
Au nanoparticles, 75
automobiles, 164
availability, 70, 73

B

batteries, ix, 2, 18, 35, 63, 163, 164, 165, 181, 185, 190, 191, 202
behavior, 86, 96, 148, 154, 158, 165, 179, 188, 217, 223, 230, 234
Beijing, 1, 65
beliefs, vii, 2
bending, 26
beneficial effect, 178, 200
benefits, 70, 72, 73
binary oxides, 190, 191
binding energy(ies), 45, 48, 56, 170, 171, 176, 177, 194, 195
bioethanol, 73
biomass growth, 70, 73
bonding, 27, 37, 56, 60
bonds, 74, 98, 102, 195
breathing, 27, 34
buffer, 76, 77
burn(ing), 100, 101

by-products, 73

C

Canada, 67
candidates, 117, 156, 157
capacitance, 127
capillary, 20
carbon, ix, 4, 21, 23, 26, 70, 71, 72, 73, 78, 97, 101, 102, 106, 163, 164, 165, 167, 168, 169, 172, 174, 175, 177, 178, 180, 181, 182, 184, 187, 191, 195, 196
carbon dioxide, 21, 73
carbon monoxide, 21, 70
carbon nanotubes, 165
carbonyl groups, 24
carboxylic groups, 80
carrier, 21, 70, 78, 122
cast, 4
catalyst(s), viii, 58, 69, 71, 72, 73, 74, 75, 76, 77, 78, 80, 81, 82, 86, 87, 89, 90, 91, 92, 93, 94, 95, 96, 97, 98, 99, 100, 101, 102, 104, 105, 106, 107, 237
catalyst deactivation, 72, 73, 94, 100
catalytic activity, 74, 78, 94, 95, 96, 99, 100, 101, 106
cathode materials, vii, viii, 1, 2, 9, 11, 14, 15, 16, 43, 46, 47, 52, 57, 62, 63
cation, 117, 129, 136, 148, 185, 217, 224
C-C, 24, 34, 74
cell, 4, 7, 8, 13, 14, 43, 45, 48, 49, 50, 57, 60, 116, 117, 118, 119, 130, 131, 143, 149, 166, 168, 173, 174, 175, 176, 178, 179, 180, 181, 182, 183, 184, 185, 187, 188, 193, 198, 199, 200, 201, 217, 228, 229, 231
cell cycle, 14, 49, 200
ceramic(s), viii, ix, 61, 115, 116, 118, 119, 120, 121, 123, 124, 125, 128, 129, 130, 132, 133, 143, 151, 153, 154, 156, 157, 158, 185, 191, 239, 241
cerium, 77
chalcogenides, 165
changing environment, 80
chemical composition, 129, 130
chemisorption, 77, 89, 90, 92, 96, 97, 106
China, 1, 4, 47, 227
Chinese, 1
chloride, 171
chromatography, 20, 78
clean energy, 70
cleaning, 85, 90, 97
cleavage, 39
clusters, 84, 117, 118
CO_2, 21, 23, 24, 26, 27, 36, 40, 41, 42, 52, 70, 71, 72, 73, 78, 91, 92, 93, 94, 95, 104, 105, 106
coal, 70
cobalt, 2, 14, 32, 71, 185, 191, 193, 194, 195, 199, 200
coke, viii, 69, 71, 74, 95, 96, 97, 98, 99, 100, 101, 106, 107
coke formation, viii, 69
collaboration, 158
collisions, 81
colloidal particles, 80
combined effect, 10
combustion, 71, 72, 91, 92, 100
compatibility, vii, 1, 18
components, viii, 18, 33, 34, 37, 42, 62, 69, 75, 82, 84, 106, 107, 128, 130, 177, 189, 194, 195, 201, 223
composites, 165, 176, 178
composition, 17, 18, 26, 32, 34, 37, 55, 86, 129, 180, 191, 202, 232, 235
compounds, viii, 2, 26, 30, 41, 61, 74, 78, 103, 164, 166, 185, 186, 191, 194, 210
computers, 164
concentration, 18, 38, 40, 41, 71, 75, 80, 91, 92, 100, 102, 129, 229, 230, 232, 239
condensation, 102, 105
conduction, viii, 61, 63, 115, 116, 117, 118, 119, 120, 121, 125, 127, 128, 132, 133, 134, 138, 139, 141, 143, 146, 147, 148, 149, 152, 153, 154, 155, 156, 157, 158, 175, 189, 223, 224, 226
conductivity, ix, 8, 10, 17, 60, 61, 78, 115, 116, 117, 118, 119, 120, 121, 122, 128, 129, 130, 131, 132, 134, 144, 146, 147, 148, 152, 153, 154, 155, 156, 157, 158, 164, 167, 168, 171, 174, 179, 182, 183, 184, 202, 209, 210, 224, 235
conductor, 179, 209, 213, 223, 232
configuration, 187, 188
consent, 74
consumption, 77, 84, 85, 86, 93
contact time, 71
contaminant(s), 23
contamination, 48, 170
control, 75, 91
conversion, viii, x, 18, 40, 69, 70, 73, 91, 92, 94, 95, 96, 100, 101, 102, 104, 171, 227, 236, 237, 239
cooling, 131, 168
copper, ix, x, 163, 167, 168, 169, 171, 175, 176, 178, 181, 209, 210, 217, 219, 221, 222, 223, 224, 227, 228, 238, 239
correlation(s), ix, 119, 122, 209, 210, 211, 212, 216, 219, 221, 222, 228
corrosion, vii, 3, 4, 39, 43, 46, 54, 58
costs, 70, 71, 72, 73
coupling, 102
covalent bond(ing), 223

coverage, 47, 95, 105
covering, 57, 74, 107
crack, 15
CRR, 71, 92, 100
crystal structure, 116, 118, 130, 154, 210, 211, 217, 218, 221, 225
crystal structures, 118, 130, 210, 211, 225
crystalline, vii, 1, 42, 73, 191, 194, 211, 224
crystallinity, 56, 60, 83, 129, 168, 169, 176, 189, 191, 193, 195, 200
crystallites, 56, 77, 85, 86
crystallization, 73, 76, 82, 96
crystals, 118, 119, 210, 224, 239
cubic system, ix, 185, 209, 216, 225
cultivation, 73
cycles, 5, 6, 7, 8, 10, 11, 12, 13, 14, 15, 34, 48, 49, 50, 51, 57, 60, 62, 76, 164, 165, 173, 174, 175, 176, 178, 181, 183, 184, 185, 188, 190, 199, 200, 201, 231
cycling, vii, ix, 1, 2, 4, 6, 7, 8, 9, 10, 11, 12, 13, 14, 16, 17, 18, 29, 34, 43, 45, 46, 48, 49, 50, 57, 59, 60, 61, 62, 63, 163, 164, 166, 167, 173, 174, 175, 177, 178, 181, 183, 185, 189, 190, 196, 199, 200, 201

D

damping, x, 209, 224
danger, 72
data analysis, 214
data processing, 170
decay, 50
decomposition, vii, 1, 14, 17, 18, 23, 27, 36, 39, 41, 42, 44, 45, 46, 48, 53, 54, 57, 60, 104, 107, 167, 178, 196
deconvolution, 172
deduction, 154
defects, vii, 19, 42, 54, 164, 201
deformation, 34, 130
degradation, 2, 12, 17, 29, 30, 32, 38, 39, 40, 41, 169
Degussa, 47
dehydrate, 19
dehydration, 103, 104, 107
demand, 70, 72, 73
density, 12, 13, 49, 62, 63, 117, 118, 120, 123, 124, 128, 137, 142, 166, 223, 235
deposition, viii, 27, 30, 69, 71, 74, 86, 97, 99, 101, 106, 107, 176, 177
deposition rate, 101
deposits, 28, 98, 99, 100, 105, 106, 107, 180, 181, 197
derivatives, 2, 28, 199
desorption, x, 227, 230, 231, 232, 234, 238, 239, 240
detection, 61
deviation, 132, 138, 211
diamond-like carbon (DLC), 3
diamonds, 147, 152
dielectric, viii, 115, 116, 122, 123, 124, 125, 126, 127, 128, 131, 134, 135, 136, 137, 138, 139, 140, 141, 143, 150, 151, 153, 154, 157, 158
dielectric constant, 122, 123, 124, 127
differential scanning calorimetry, 51
diffraction, ix, 5, 20, 30, 31, 57, 130, 169, 209, 210, 211, 213, 215, 216, 218, 219, 220, 223
diffusion, vii, viii, x, 8, 9, 16, 84, 87, 89, 100, 115, 116, 123, 124, 128, 132, 140, 154, 156, 164, 165, 166, 167, 169, 181, 183, 189, 199, 223, 227, 228, 231, 232, 238, 239, 240
diffusion process, 240
diffusion rates, 232
dimensionality, 201
dipole, 116, 122, 123, 135, 136, 157
dipole moment(s), 116, 122, 123, 135, 136, 157
discharges, 80, 173
disorder, 209, 210, 223
dispersion, 55, 73, 74, 78, 80, 87, 89, 90, 96, 139, 214
displacement, 116, 122, 135, 136, 156, 157, 211
dissociation, 117, 124, 129, 140, 142, 143
distilled water, 3, 4, 19, 43, 47, 186
distribution, viii, 74, 75, 79, 81, 83, 87, 88, 92, 98, 101, 107, 115, 123, 124, 127, 132, 136, 137, 157, 176, 187, 210, 213, 215, 223
distribution function, 127
dopants, 2
doping, 117, 121, 154, 236, 241
DPO, 71, 72, 92, 101
drying, 15, 43, 77
DSC, 4, 15, 16, 51
DVD, 164

E

earth, 82
economic growth, 72
education, 65
effluent, 235
electric conductivity, 119, 120, 121, 134, 147, 149, 156, 175, 232
electric current, 141
electric power, 164
electrochemical reaction, 165, 179, 190, 196, 199
electrochemistry, 186
electrodes, ix, 13, 17, 18, 26, 28, 30, 34, 35, 40, 43, 44, 50, 131, 163, 164, 165, 173, 176, 177, 182, 187, 189, 190, 191, 199, 201

electrolyte, vii, ix, 1, 2, 4, 13, 14, 15, 16, 17, 18, 19, 20, 29, 30, 35, 36, 37, 38, 39, 40, 41, 42, 43, 44, 45, 46, 47, 48, 49, 50, 51, 52, 53, 54, 55, 56, 57, 58, 59, 60, 61, 62, 63, 120, 121, 156, 163, 164, 173, 176, 178, 182, 185, 187, 188, 196, 199, 200, 201
electron(s), ix, 4, 8, 17, 19, 20, 28, 34, 42, 48, 54, 56, 77, 121, 136, 163, 169, 175, 176, 178, 181, 201, 202, 214
electron charge, 175
electron microscopy, 19, 20, 77, 169, 176
electronic structure, vii, 1, 43, 45
emission, 21, 129, 171
encapsulation, viii, 69, 75, 106
encouragement, 158
endothermic, 70, 92, 93, 237
endurance, 18
energy, viii, ix, x, 2, 19, 20, 29, 30, 31, 42, 43, 45, 48, 55, 56, 62, 63, 69, 70, 71, 72, 73, 115, 116, 117, 118, 122, 123, 124, 125, 128, 129, 133, 136, 139, 140, 141, 142, 143, 148, 149, 151, 152, 155, 157, 164, 165, 169, 170, 171, 202, 209, 210, 211, 214, 215, 219, 223, 224, 227, 228, 234, 239
energy consumption, 70
energy density, 2, 48, 73
energy efficiency, 72
energy transfer, 223
environment, 177, 194
environmental impact, 72
equilibrium, 94, 229, 230, 238
equipment, 196
esters, 37
etching, 180
ethanol, viii, 19, 69, 70, 73, 74, 75, 102, 103, 104, 105, 106, 107, 168, 193
ethers, 26
ethylene, 4, 17, 21, 24, 102, 103, 104, 105, 107, 191
ethylene glycol, 191
ethylene oxide, 24
evacuation, 97
evaporation, 179
evidence, 9, 14, 27, 39, 47, 56, 60, 61, 74, 89, 169, 171, 175, 224
evolution, 95, 105, 106, 190
EXAFS, 210, 223, 226
excitation, ix, 170, 209, 224
exothermic, 15, 16, 47, 51, 71, 92, 94, 104, 185
exothermic peaks, 15
experimental condition, 61, 96, 197, 211
exploitation, 74
exposure, 20, 34, 43, 95, 170, 177, 180
extraction, ix, 17, 29, 30, 39, 40, 41, 44, 163, 165, 173, 178, 181, 185
extrapolation, 77

F

failure, 165
family, 2, 240
ferrite, ix, 163, 185, 186, 187, 189, 191, 193, 195, 196, 198, 199, 200
FFT, 77, 88
film(s), 12, 13, 17, 18, 19, 20, 26, 27, 28, 29, 30, 32, 33, 34, 37, 38, 39, 40, 41, 42, 50, 57, 59, 60, 77, 169, 177, 199, 238, 240
film formation, 40
film thickness, 20, 42
filtration, 77, 168
financial support, 107
first generation, 76, 87, 89, 106
flame, 78
flexibility, 75
fossil fuels, viii, 63, 69, 72, 73
Fourier, 20, 77
fractures, 4
France, 163
FT-IR, 20, 23, 24, 27, 30, 34, 37, 38, 53, 54, 60, 98, 195, 196
FT-IR spectroscopy, 27, 34
fuel cell, viii, 69, 71, 72, 73

G

gases, 22, 36, 42
gasification, 71
gasoline, 73
Gaussian, 123, 137
gel, 43, 191, 238
generation, 47, 70, 87
germanium, 213
Germany, 47
glass, 20, 27, 168, 224
glucose, 178
gold, 75, 176, 178, 180, 181
grain boundaries, 123, 125, 133, 154
grains, 118, 123, 133, 143, 154, 156, 232
granules, 78
graphite, 62, 106, 191, 199
greenhouse gas(es), 70, 72, 73
groups, 9, 20, 23, 34, 37, 40, 41, 66, 71, 75, 80, 98, 195
growth, 6, 13, 28, 34, 42, 60, 63, 70, 71, 75, 76, 81, 90, 98, 106, 121, 158, 167, 191, 200, 201

H

hazards, 73
health, 73
heat, 6, 15, 19, 37, 59, 70, 72, 92, 130, 164
heating rate, x, 3, 16, 76, 77, 82, 90, 167, 168, 176, 199, 227, 231, 233, 234, 240
heavy metals, 164
helium, 21, 235, 240
hematite, 186
hexane, 77, 169
higher quality, 56
homeostasis, 230
homogeneity, 78, 86, 129
host, 70
HRTEM, 54, 55, 57, 58, 81, 87, 88, 89, 90, 98, 169, 171, 186, 193, 201, 202
hydrocarbons, 70, 72, 97
hydrogen, viii, 69, 70, 73, 77, 89, 90, 93, 96, 100, 102, 167, 181
hydrophilic groups, 75
hydrophobic groups, 75
hydrothermal synthesis, 192
hydroxide(s), 76, 80, 82, 167, 192
hypothesis, 90, 92
hysteresis loop, 81, 188

I

identification, 23, 87, 95, 171
images, 54, 77, 169, 170, 171, 186, 193, 200, 201
imaging, 5, 15, 55, 77
impedance analysis, viii, 115, 125, 127, 128, 132, 133, 134, 138, 143, 147, 153, 157
impregnation, 74, 77, 99, 106
impurities, 14, 15, 59, 60, 63, 123, 124, 129, 174, 189
in situ, 43, 99, 107, 167, 168
inclusion, viii, 69, 75, 202
India, 203
indication, 86, 94, 104
industrial application, viii, 70
industry, 72
inelastic, x, 209, 210, 223, 224
infinite, 166
infrastructure, 70
inhibition, 100
insertion, ix, 163, 166, 173, 178, 179, 181, 202
instability, 90, 185
instruments, 4, 43
insulation, 8, 167
integrity, 178, 201
intensity, ix, 27, 30, 31, 44, 50, 83, 143, 144, 145, 151, 171, 173, 178, 180, 189, 195, 209, 211, 212, 214, 215, 216, 218, 219, 220, 221, 222, 223
interaction(s), vii, 1, 2, 15, 18, 27, 44, 47, 60, 62, 74, 75, 84, 85, 89, 90, 119, 140
interface, ix, 16, 17, 43, 59, 125, 126, 127, 154, 158, 163, 178, 199, 201
intermetallics, 165
interphase, 13, 54
iodine, 216, 219, 221, 222
ion transport, vii, 17, 59, 61
ionic conduction, viii, 8, 60, 61, 115, 116, 117, 118, 119, 120, 121, 122, 125, 128, 132, 133, 138, 139, 141, 142, 143, 148, 149, 151, 154, 155, 156, 157, 158
ionization, 21
ions, vii, viii, ix, 1, 2, 6, 7, 8, 9, 12, 14, 16, 18, 29, 30, 34, 36, 38, 40, 41, 42, 44, 47, 55, 60, 63, 74, 76, 77, 80, 85, 115, 116, 117, 118, 120, 121, 122, 124, 128, 132, 135, 136, 137, 139, 140, 143, 148, 149, 154, 156, 157, 163, 165, 167, 179, 180, 181, 182, 183, 185, 188, 189, 190, 201, 202, 217, 223, 224, 232
IR spectra, 25, 26, 37, 53, 92, 98, 106
IR spectroscopy, 92
iron, ix, 71, 163, 167, 168, 170, 171, 172, 173, 174, 177, 180, 181, 182, 184, 185, 186, 188, 189, 191, 199
isothermal, x, 86, 94, 97, 227, 228, 230, 231, 232, 239
isotherms, 77, 81, 82
isotope, 211, 240
Israel, 202
Italy, 69, 107
iteration, 137, 153

J

Japan, 47, 110, 115, 209

K

KBr, 20, 25
kinetics, ix, 42, 163, 164, 165, 178, 181, 201, 231

L

lanthanum, ix, 115, 116, 117, 118, 119, 120, 121, 123, 128, 129, 132, 135, 140, 143, 149, 153, 154, 157, 158
lanthanum gallates, ix, 116, 117, 118, 119, 157, 158
lattice parameters, 6, 30, 130, 131, 143

laws, 154
leakage, 60
lifetime, viii, 69, 73, 99
limitation, 121, 173, 223
liquids, 36
literature, 37, 75, 155, 165, 186
lithium, ix, 2, 4, 11, 14, 16, 17, 18, 23, 24, 26, 27, 29, 30, 34, 35, 36, 38, 39, 40, 41, 42, 43, 47, 50, 51, 54, 57, 60, 62, 63, 67, 163, 164, 165, 166, 167, 173, 178, 180, 181, 182, 183, 184, 185, 186, 187, 188, 189, 190, 191, 196, 197, 199, 201, 202
lithium insertion, 178, 182, 201
lithium ion batteries, ix, 2, 16, 18, 23, 47, 51, 57, 62, 63, 67, 163, 164, 165, 167, 197, 199
location, 177
low temperatures, 117, 124
lying, 223, 224

M

magnesium, vii, 1, 3, 143
magnet, 168
magnetic properties, 120
magnetite, 168, 191
manganese, 167, 193, 194, 198, 200
manners, 197
market, 70, 73, 164
masking, 20
mass loss, 231, 233, 234
mass spectrometry, 78
material surface, 26
materials science, vii
matrix, 74, 81, 85, 86, 87, 88, 89, 96, 98, 106, 107, 185
meanings, 126, 210
measurement, ix, x, 21, 52, 127, 128, 131, 197, 209, 210, 211, 213, 214, 215, 217, 219, 223, 225, 227, 232, 239
media, 20
melt, 129
membranes, 227, 235, 236, 238, 240, 241
memory, 140
Merck, 176
mesoporous materials, 81, 191
metal content, 87
metal hydroxides, 76
metal ions, 2, 185, 189
metal nanoparticles, viii, 70, 74, 75, 81, 90, 98, 101, 105, 107, 198
metal oxide(s), 2, 17, 41, 42, 46, 47, 61, 62, 63, 191
metals, 17, 71, 73
methane, viii, x, 21, 69, 70, 71, 72, 74, 92, 96, 97, 100, 102, 105, 227, 236, 238, 239, 241
methanol, 21, 23, 73, 168
methyl group(s), 98
methylene, 195
Mg^{2+}, 6, 8, 9, 10, 11, 14, 16
micelles, 75, 80, 90
microemulsion, 73
microscope, 4, 48, 54, 77, 169
microstructure, 169
migration, viii, 7, 8, 9, 10, 14, 60, 74, 115, 116, 117, 118, 122, 123, 128, 129, 133, 135, 136, 139, 141, 142, 143, 151, 156, 157
milligrams, 168, 173
Ministry of Education, 225
mixing, 80, 130, 165, 176, 181
mobility, viii, 69, 106
model system, 87
modeling, 125
models, 210, 213, 215, 217, 220, 221, 223, 225, 240
modern society, 164
moisture, 4, 14, 20, 22, 32, 34, 46, 60
moisture content, 20
mole, 165, 173, 185, 186, 196, 197, 229
molecular dynamics, 223
molecular structure, 18, 75
molecular weight, 53, 54, 60
molecules, 17, 40, 41, 42, 53, 80
molybdenum, 213
Moon, 109
morphology, vii, 1, 29, 39, 43, 47, 54, 74, 80, 191, 193, 200
movement, 143, 217
multiple factors, 105
multiplicity, 195

N

Na^+, 189
NaCl, 189
nanobelts, ix, 163, 187, 188, 190
nanocomposites, 82, 89, 106, 169, 178, 190
nanocrystals, 190, 224
nanomaterials, ix, 163, 164, 165, 181, 201
nanometer, 186
nanoparticles, viii, 69, 75, 76, 78, 81, 84, 86, 87, 88, 89, 90, 97, 98, 101, 106, 107, 164, 165, 167, 169, 171, 174, 177, 186, 194
nanorods, 186, 188, 190
nation, 73
NATO, 160
natural gas, 70
network, 76, 102, 105, 175
New York, 107, 159, 161, 202, 206
nickel, 2, 9, 71, 167, 191

nitrogen, 230, 231, 232
NMR, 76
noble metals, 71, 73
noise, 23, 31
nucleating agent, 167
nucleation, 167
nuclei, 56
nucleophilicity, 17
numerical analysis, 136

O

observations, 47, 54, 60, 63
occlusion, 74, 106
oil, 75
optimization, 72
organic compounds, 75, 167
organic polymers, 165
organic solvent(s), 20, 51, 164
orientation, 169, 201
oxalate, 167, 181
oxidation, vii, viii, x, 7, 10, 11, 17, 30, 40, 44, 50, 69, 71, 72, 75, 77, 84, 89, 90, 92, 95, 96, 97, 99, 101, 102, 105, 167, 168, 170, 171, 173, 175, 178, 183, 188, 194, 227, 232, 236, 238, 239, 241
oxides, viii, 17, 32, 47, 63, 69, 82, 92, 106, 107, 115, 116, 118, 119, 120, 122, 123, 124, 125, 126, 127, 128, 129, 130, 132, 135, 136, 141, 153, 156, 157, 171, 174, 188, 198, 239, 240
oxygen, vii, viii, x, 1, 17, 20, 21, 22, 34, 36, 40, 41, 42, 45, 72, 86, 87, 92, 95, 100, 101, 102, 115, 116, 117, 118, 119, 120, 121, 123, 124, 125, 127, 128, 129, 131, 133, 135, 137, 139, 140, 141, 142, 143, 145, 147, 149, 151, 153, 155, 156, 157, 159, 161, 168, 171, 180, 185, 186, 191, 227, 228, 229, 230, 231, 232, 234, 235, 236, 237, 238, 239, 240, 241
oxygen absorption, 231, 232, 238

P

parameter, viii, 6, 20, 30, 115, 117, 124, 127, 136, 137, 143, 157, 166, 168, 193
Paris, 163
particles, 2, 3, 4, 14, 19, 20, 28, 29, 30, 32, 33, 34, 39, 41, 45, 47, 55, 56, 62, 63, 71, 74, 75, 77, 78, 79, 80, 86, 87, 88, 89, 90, 97, 98, 100, 101, 106, 164, 166, 167, 169, 171, 175, 176, 178, 179, 182, 183, 184, 191, 193, 199, 200, 201
passivation, 32, 101
pathways, 102
performance, vii, viii, 1, 2, 3, 6, 7, 12, 13, 15, 16, 17, 18, 19, 29, 35, 41, 43, 46, 47, 48, 49, 50, 54, 59, 60, 61, 62, 63, 71, 94, 101, 165, 166, 167, 174, 175, 176, 178, 181, 190, 191, 199, 200, 201, 236, 239, 241
periodicity, 143
permeability, x, 227, 234, 239, 241
permeation, 227, 228, 234, 235, 236, 237, 238, 239, 240, 241
perovskite, 74, 116, 117, 118, 119, 120, 130, 131, 143, 148, 149, 228, 241
perovskite oxide, 116
PET, 47
pH, 58, 75, 76, 80, 81, 106, 192
phase transformation, 164, 239
phase transitions, 118, 132, 140, 210
phosphates, 167, 175, 176, 179
phosphorus, 170, 177, 180
photoelectron spectroscopy, vii, 1
physics, vii, 46, 128
plants, 72, 73
plasma, 4
point defects, 140
poison, 97
Poland, 129
polarization, 7, 8, 9, 10, 48, 50, 178, 199, 212
pollutants, 73
polymer(s), 4, 26, 60, 61, 62, 191
polymerization, 71, 101
polynomials, 210, 216
polypropylene, 4, 47
poor, 56, 175, 181, 183, 237
population, 70
porosity, 86, 201
power, vii, 1, 18, 35, 41, 42, 44, 45, 58, 62, 63, 73, 78, 201
precipitation, 3, 42, 74, 75, 76, 80, 81, 192
pressure, x, 77, 89, 173, 227, 228, 229, 230, 231, 232, 238, 239
prevention, 11
primary products, 72, 92
probability, 9, 212, 213, 221, 222
probe, 58, 128, 131, 132, 133, 134, 144, 147
problem-solving, 65
production, viii, 69, 70, 73, 81, 93, 94, 100, 102, 105, 107, 241
program, 130
promote, 102, 104, 191
promoter, 100, 101, 107
propylene, 17
protective coating, 185
prototype, 118
PTFE, 20, 47, 52, 57
pulses, 77
pyrolysis, 26, 92, 176, 181

Q

quartz, 78

R

radiation, 4, 20, 44, 77, 130, 170
radio, 237
radius, 7, 117, 127, 141, 166
Raman, 20, 23, 24, 27, 28, 30, 31, 32, 33, 34, 38, 39, 52, 53, 60, 65
Raman spectra, 20, 23, 27, 28, 31, 32, 33, 38, 39
Raman spectroscopy, 20, 23, 27, 30
range, ix, x, 21, 78, 84, 93, 98, 104, 119, 130, 131, 135, 137, 150, 151, 166, 167, 168, 173, 177, 187, 193, 195, 199, 209, 210, 211, 212, 214, 223, 224, 227, 231, 234, 236, 239
reactants, 74, 91, 106
reaction rate, 232
reaction temperature, 15, 16, 51, 104
reactivity, vii, 17, 32, 42, 71, 78, 92, 94, 100, 106, 164, 171, 180, 186, 197, 200, 201
reagents, 74, 92
reality, 70
recall, 86
reconstruction, 199
recovery, 95, 96
reduction, x, 7, 17, 40, 41, 42, 50, 58, 70, 74, 78, 81, 82, 83, 84, 85, 86, 89, 90, 92, 96, 97, 101, 105, 107, 143, 148, 149, 167, 168, 170, 171, 173, 175, 178, 183, 199, 227, 232, 238, 239
reflection, 214
reflexes, 83
regeneration, viii, 69, 86, 95, 99
relationship(s), 14, 212, 229, 231, 239
relaxation process(es), viii, 115, 116, 119, 122, 123, 124, 125, 126, 127, 128, 132, 134, 135, 136, 137, 138, 139, 140, 141, 142, 143, 150, 151, 152, 153, 154, 156, 157, 158
relaxation times, viii, 115, 123, 124, 126, 127, 132, 136, 151, 157
relevance, viii, 69
reliability, 220
renewable energy, viii, 69, 73
residues, 106
resistance, viii, x, 8, 10, 14, 60, 69, 74, 128, 131, 133, 134, 136, 142, 143, 146, 147, 148, 155, 157, 158, 227, 238, 239
resolution, 20, 24, 43, 54, 77, 118, 130, 157, 194, 223, 224
resources, 70
retention, 11, 13, 21, 22, 48, 49, 50, 60, 176, 189, 200, 201
rhodium, viii, 69, 71, 86
room temperature, vii, 1, 4, 18, 20, 25, 26, 43, 59, 61, 76, 77, 97, 121, 130, 143, 144, 168, 182, 183, 210, 213, 214, 215, 217, 224
Royal Society, 111
rubber, 20, 47, 52

S

safety, 15, 42, 43, 45, 46, 47, 51, 60, 62, 63, 73, 164, 185
salt(s), 17, 18, 36, 40, 42, 47, 54, 76, 78, 80, 86, 164, 167, 168, 189, 191
sample, x, 4, 15, 20, 21, 22, 27, 43, 77, 81, 82, 83, 85, 86, 87, 95, 96, 97, 99, 104, 105, 131, 146, 168, 169, 170, 176, 177, 178, 179, 180, 186, 196, 212, 214, 224, 227, 229, 230, 231, 232, 233, 234
satellite, 143, 195
savings, 72
scaling, 72
scanning calorimetry, 4
scatter(ing), ix, 83, 171, 209, 210, 211, 212, 213, 214, 215, 216, 217, 218, 219, 221, 222, 223, 224, 225, 228
search, ix, 163
second generation, 76, 81, 82, 87, 106
security, 73
segregation, 86
selected area electron diffraction, 55
selecting, 125
selectivity, viii, x, 69, 73, 74, 92, 94, 104, 227, 236, 237, 239
semicircle, 13, 14, 127, 128, 140, 153
separation, vii, 2, 47, 78, 194
series, 21, 34, 42, 78, 102, 125, 126, 154
severity, 19
shape, 14, 56, 80, 89, 171, 175, 177, 186, 193, 196, 197, 201, 210
sharing, 3
shear, 141
signals, 23, 28, 131, 177, 195
signs, 17
silica, 20, 47, 75
silicon, 165
silver, 222
similarity, 26, 196, 224
single crystals, 118, 119, 158
sintering, vii, viii, 69, 71, 72, 73, 74, 75, 86, 89, 96, 97, 98, 101, 106, 121, 130, 143, 144, 145, 146, 168, 200
SiO2, 57, 60, 62, 63

sites, 9, 74, 79, 97, 101, 102, 104, 105, 106, 107, 116, 124, 128, 180, 185, 189, 211, 212, 213, 216, 217, 219, 220, 223, 234
skeleton, 34, 36
sodium, 191, 192
sodium hydroxide, 192
software, 168, 193
sol-gel, 74, 75
solid oxide fuel cells, 117
solid phase, 73
solid solutions, 60, 100, 118
solid state, 128, 164, 213, 225
solid waste, 73
solvent(s), vii, 1, 17, 18, 19, 20, 21, 22, 23, 24, 25, 26, 27, 28, 29, 30, 31, 32, 33, 34, 35, 36, 37, 38, 39, 40, 41, 42, 47, 52, 54, 164, 173, 185, 187, 199, 200
solvent molecules, 17, 23, 34, 40, 41, 42
Spain, 163, 202
species, 6, 11, 14, 17, 18, 19, 20, 21, 22, 23, 26, 27, 30, 31, 32, 33, 34, 36, 37, 38, 43, 44, 52, 53, 54, 57, 60, 71, 72, 74, 83, 84, 85, 86, 87, 92, 99, 100, 101, 106, 170, 177, 187, 194, 195
specific heat, 72
specific surface, 18, 19, 27, 35, 38, 230
spectroscopy, 20, 53, 55, 125, 128, 129, 132, 134, 135, 136, 140, 141, 147, 148, 155, 157, 158
spectrum, 21, 22, 23, 25, 28, 32, 34, 37, 44, 56, 96, 106, 171, 172, 177, 181, 187, 194, 195, 223, 224
speculation, ix, 116, 141, 143, 154, 157
speed, 230
spin, 7, 44, 211
sputtering, 44, 177, 180, 181
stability, vii, viii, x, 1, 2, 3, 7, 10, 11, 14, 15, 16, 18, 29, 42, 46, 48, 59, 60, 61, 62, 69, 71, 73, 80, 82, 94, 95, 96, 99, 101, 105, 106, 165, 171, 227, 238, 239
stabilization, 101, 142
stages, 12
standards, 70
statistical analysis, 87
steel, 20, 173, 228
stoichiometry, x, 77, 91, 171, 194, 227, 228, 232
storage, 15, 17, 34, 37, 41, 60, 63, 73
strategies, viii, 69, 70, 97, 99, 185
strength, 58, 75, 80, 81, 90, 106
stress, 6
stretching, 26, 34, 98, 195
strong interaction, 86, 97, 140
strontium, 143
structural changes, 17
structural defects, 20
substitution, 2, 60, 72, 120
substrates, 37
sucrose, 176, 177, 181, 182
sulfuric acid, 129
Sun, 66, 110, 113, 203, 205, 206, 241
superacids, viii, 2, 58, 59, 60, 61, 62, 63
superconductor(s), x, 227, 240
suppression, 7, 10, 45, 58, 146, 164
surface area, 3, 74, 77, 78, 82, 84, 86, 87, 89, 90, 92, 96, 105, 147, 164, 190, 200, 201
surface chemistry, 2, 15, 43
surface layer, 8, 14, 17, 18, 27, 28, 29, 34, 79
surface modification, vii, 1, 3, 7, 15, 18, 29, 37, 41, 42, 47, 62
surface properties, 2
surface reactions, 18, 19, 36, 232
surfactant(s), 75, 76, 79, 80, 90, 191
surprise, 57
suspensions, 77
sustainability, 73
switching, 99
symbols, 83, 152
symmetry, 3, 118, 120, 154, 185, 217
synthesis, viii, 69, 74, 75, 76, 79, 80, 81, 87, 90, 99, 100, 101, 120, 122, 191, 193
systems, ix, 71, 73, 74, 89, 118, 163, 164, 165, 176, 188

T

technical assistance, 107
technology, 70, 72, 164
temperature, vii, viii, ix, x, 1, 15, 21, 23, 51, 59, 61, 69, 72, 73, 75, 76, 77, 78, 82, 84, 85, 86, 89, 91, 92, 93, 94, 96, 98, 100, 104, 105, 106, 107, 117, 120, 121, 123, 124, 125, 127, 130, 131, 132, 133, 134, 135, 137, 143, 144, 145, 146, 147, 150, 151, 152, 153, 155, 158, 167, 168, 174, 181, 183, 184, 186, 191, 193, 200, 209, 210, 211, 213, 214, 215, 216, 217, 219, 220, 221, 222, 224, 227, 228, 229, 230, 231, 232, 233, 234, 235, 236, 237, 238, 239, 240
temperature dependence, x, 123, 133, 209, 224
temperature gradient, 91, 92
TGA, 241
theory, vii, 122, 123, 135
thermal decomposition, 182
thermal properties, 58
thermal stability, viii, 15, 69, 73, 74, 88, 90, 106
thermal treatment, 82, 95, 167, 193
thermodynamic equilibrium, 104, 105
thermogravimetric, 230, 232, 239, 240
thermogravimetry, x, 227, 232, 240
thin films, 40, 197

time, 8, 20, 21, 22, 32, 43, 48, 54, 60, 63, 72, 82, 90, 92, 93, 94, 96, 100, 101, 102, 105, 107, 122, 123, 142, 156, 170, 177, 180, 181, 183, 224, 228, 231, 238
tin, 165
toxicity, 168
transference, 21, 60
transformation, viii, 7, 69, 73, 82, 84, 165, 185
transition(s), 2, 6, 7, 9, 10, 11, 16, 43, 44, 57, 70, 71, 83, 132, 135, 136, 137, 140, 141, 152, 164, 165, 185, 190, 191, 234
transition metal ions, 2, 57, 71, 141, 164, 165, 185, 190, 191
transition temperature, 152
transmission, 48, 54
Transmission Electron Microscopy (TEM), 56, 87, 106, 169, 170, 186, 193, 200
transport, vii, viii, ix, 7, 17, 34, 115, 119, 120, 125, 128, 129, 134, 157, 158, 163, 165, 201, 232, 235, 239, 240
transportation, viii, 9, 69, 168
trend, 80, 90, 94, 190
tunneling, 28, 34, 42
tunneling effect, 28, 34
twinning, 118

U

Ukraine, 115
uniform, 14, 101, 181, 193, 232
urban areas, 164

V

vacancies, 16, 60, 116, 117, 118, 119, 120, 123, 124, 140, 141, 142, 143, 156
vacuum, 4, 20, 25, 43, 47, 77, 228
valence, 228
validity, 75
values, 5, 6, 8, 9, 30, 43, 73, 78, 79, 86, 94, 116, 117, 124, 125, 126, 128, 129, 132, 133, 134, 136, 137, 138, 139, 140, 141, 142, 143, 144, 146, 147, 148, 150, 151, 152, 153, 154, 155, 157, 158, 166, 167, 168, 170, 171, 174, 177, 178, 183, 190, 193, 194, 196, 200, 212, 214, 216, 219, 221, 224
van der Waals forces, 78
vanadium, 165
variable(s), 183, 197, 228
variation, vii, x, 1, 13, 20, 27, 30, 45, 55, 57, 58, 152, 178, 180, 190, 199, 200, 227, 231, 232, 234, 239
vector, 131
vehicles, 72, 164
velocity, viii, 69
vibration, 24, 27, 37, 209, 211, 217, 223, 224
vinylidene fluoride, 4

W

Warsaw, 129
wavelengths, 213
weight gain, 233
weight loss, 233
weight ratio, 4, 78, 164, 168, 176, 186
workers, 75
working conditions, viii, 69, 73

X

XPS, 43, 44, 48, 56, 58, 170, 171, 172, 176, 177, 180, 181, 191, 194, 195
X-ray diffraction (XRD), vii, ix, 1, 4, 5, 9, 19, 20, 30, 31, 33, 34, 43, 55, 57, 58, 77, 83, 84, 96, 118, 120, 130, 143, 144, 145, 157, 168, 169, 171, 176, 177, 178, 186, 189, 190, 192, 193, 209, 210, 212, 217, 218, 219

Y

yield, viii, 69, 92, 134, 136, 137, 149, 153, 168, 185, 235
YPO_4, vii, 1, 3, 4, 5, 11, 12, 13, 15, 16, 58, 61, 63
yttria-stabilized zirconia, 116, 120
yttrium phosphate, vii, 1, 3

Z

zinc, 216
zirconia, 82, 84, 87, 89, 90, 99, 105, 107
ZnO, 3, 46, 61